Fundamentals of Electrical Engineering I

By:

Don Johnson

Fundamentals of Electrical Engineering I

By:
Don Johnson

Online:
<http://cnx.org/content/col10040/1.9/ >

C O N N E X I O N S

Rice University, Houston, Texas

Table of Contents

Chapter 1

Introduction

1.1 Themes[1]

From its beginnings in the late nineteenth century, electrical engineering has blossomed from focusing on electrical circuits for power, telegraphy and telephony to focusing on a much broader range of disciplines. However, the underlying themes are relevant today: **Power** creation and transmission and **information** have been the underlying themes of electrical engineering for a century and a half. This course concentrates on the latter theme: the *representation, manipulation, transmission, and reception of information by electrical means.* This course describes what information is, how engineers quantify information, and how electrical **signals** represent information.

Information can take a variety of forms. When you speak to a friend, your thoughts are translated by your brain into motor commands that cause various vocal tract components–the jaw, the tongue, the lips–to move in a coordinated fashion. Information arises in your thoughts and is represented by speech, which must have a well defined, broadly known structure so that someone else can understand what you say. Utterances convey information in sound pressure waves, which propagate to your friend's ear. There, sound energy is converted back to neural activity, and, if what you say makes sense, she understands what you say. Your words could have been recorded on a compact disc (CD), mailed to your friend and listened to by her on her stereo. Information can take the form of a text file you type into your word processor. You might send the file via e-mail to a friend, who reads it and understands it. From an information theoretic viewpoint, all of these scenarios are equivalent, although the forms of the information representation–sound waves, plastic and computer files–are very different.

Engineers, who don't care about information *content*, categorize information into two different forms: **analog** and **digital**. Analog information is continuous valued; examples are audio and video. Digital information is discrete valued; examples are text (like what you are reading now) and DNA sequences.

The conversion of information-bearing signals from one energy form into another is known as *energy conversion* or *transduction*. All conversion systems are inefficient since some input energy is lost as heat, but this loss does not necessarily mean that the conveyed information is lost. Conceptually we could use any form of energy to represent information, but electric signals are uniquely well-suited for information representation, transmission (signals can be broadcast from antennas or sent through wires), and manipulation (circuits can be built to reduce noise and computers can be used to modify information). Thus, we will be concerned with how to

- *represent* all forms of information with electrical signals,
- *encode* information as voltages, currents, and electromagnetic waves,
- *manipulate* information-bearing electric signals with circuits and computers, and
- *receive* electric signals and convert the information expressed by electric signals back into a useful form.

[1]This content is available online at <http://cnx.org/content/m0000/2.18/>.

Telegraphy represents the earliest electrical information system, and it dates from 1837. At that time, electrical science was largely empirical, and only those with experience and intuition could develop telegraph systems. Electrical science came of age when James Clerk Maxwell[2] proclaimed in 1864 a set of equations that he claimed governed all electrical phenomena. These equations predicted that light was an electromagnetic wave, and that energy could propagate. Because of the complexity of Maxwell's presentation, the development of the telephone in 1876 was due largely to empirical work. Once Heinrich Hertz confirmed Maxwell's prediction of what we now call radio waves in about 1882, Maxwell's equations were simplified by Oliver Heaviside and others, and were widely read. This understanding of fundamentals led to a quick succession of inventions–the wireless telegraph (1899), the vacuum tube (1905), and radio broadcasting–that marked the true emergence of the communications age.

During the first part of the twentieth century, circuit theory and electromagnetic theory were all an electrical engineer needed to know to be qualified and produce first-rate designs. Consequently, circuit theory served as the foundation and the framework of all of electrical engineering education. At mid-century, three "inventions" changed the ground rules. These were the first public demonstration of the first electronic computer (1946), the invention of the transistor (1947), and the publication of *A Mathematical Theory of Communication* by Claude Shannon[3] (1948). Although conceived separately, these creations gave birth to the information age, in which digital and analog communication systems interact and compete for design preferences. About twenty years later, the laser was invented, which opened even more design possibilities. Thus, the primary focus shifted from *how* to build communication systems (the circuit theory era) to *what* communications systems were intended to accomplish. Only once the intended system is specified can an implementation be selected. Today's electrical engineer must be mindful of the system's ultimate goal, and understand the tradeoffs between digital and analog alternatives, and between hardware and software configurations in designing information systems.

> VISION IMPAIRED ACCESS: Thanks to the translation efforts of Rice University's Disability Support Services[4] , this collection is now available in a Braille-printable version. Please click here[5] to download a .zip file containing all the necessary .dxb and image files.

1.2 Signals Represent Information[6]

Whether analog or digital, information is represented by the fundamental quantity in electrical engineering: the **signal**. Stated in mathematical terms, *a signal is merely a function.* Analog signals are continuous-valued; digital signals are discrete-valued. The independent variable of the signal could be time (speech, for example), space (images), or the integers (denoting the sequencing of letters and numbers in the football score).

1.2.1 Analog Signals

Analog signals are usually signals defined over continuous independent variable(s). Speech (Section 4.10) is produced by your vocal cords exciting acoustic resonances in your vocal tract. The result is pressure waves propagating in the air, and the speech signal thus corresponds to a function having independent variables of space and time and a value corresponding to air pressure: $s(\mathbf{x}, t)$ (Here we use vector notation $\mathbf{x}$ to denote spatial coordinates). When you record someone talking, you are evaluating the speech signal at a particular spatial location, $\mathbf{x}_0$ say. An example of the resulting waveform $s(\mathbf{x}_0, t)$ is shown in this figure (Figure 1.1: Speech Example).

[2] http://www-groups.dcs.st-andrews.ac.uk/~history/Mathematicians/Maxwell.html
[3] http://www.lucent.com/minds/infotheory/
[4] http://www.dss.rice.edu/
[5] http://cnx.org/content/m0000/latest/FundElecEngBraille.zip
[6] This content is available online at <http://cnx.org/content/m0001/2.26/>.

Speech Example

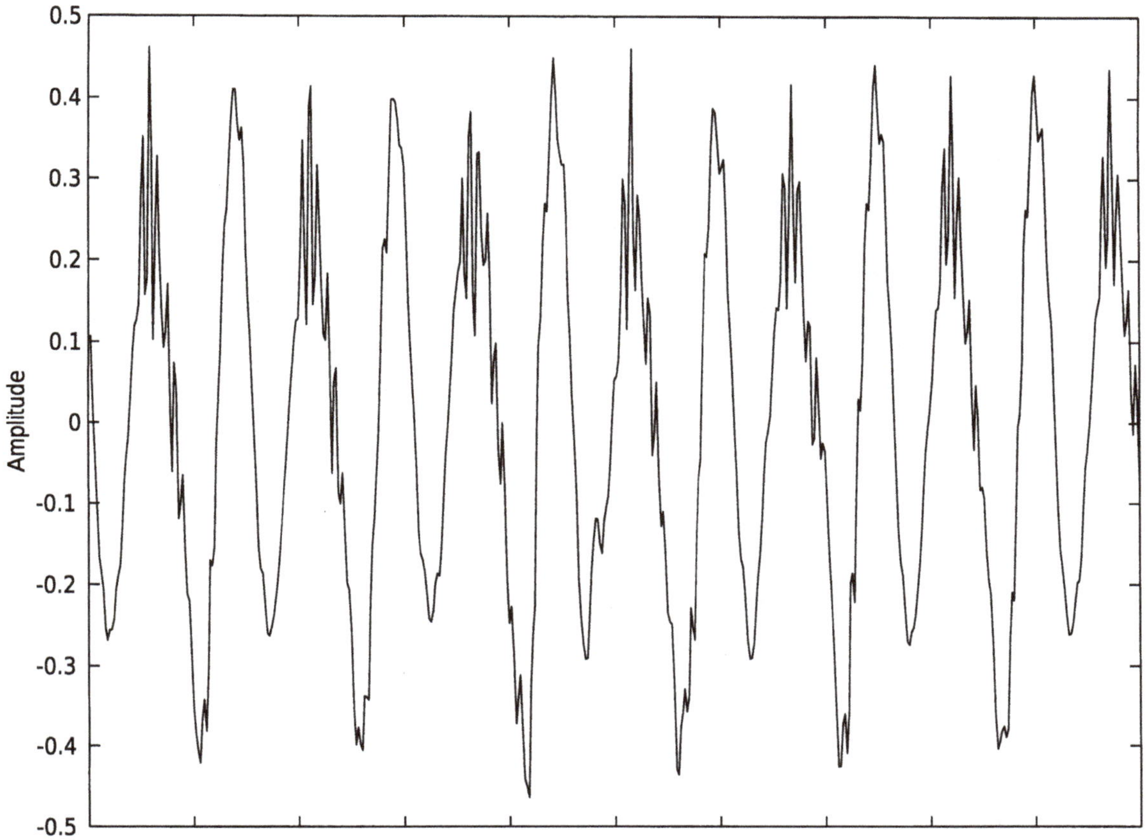

Figure 1.1: A speech signal's amplitude relates to tiny air pressure variations. Shown is a recording of the vowel "e" (as in "speech").

Photographs are static, and are continuous-valued signals defined over space. Black-and-white images have only one value at each point in space, which amounts to its optical reflection properties. In Figure 1.2 (Lena), an image is shown, demonstrating that it (and all other images as well) are functions of two independent spatial variables.

Lena

(a)

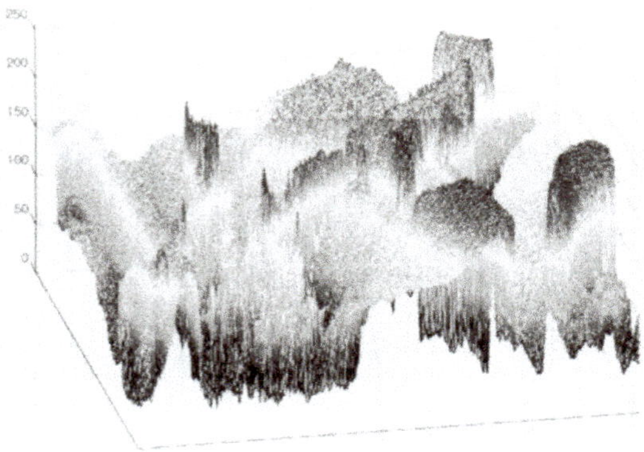

(b)

Figure 1.2: On the left is the classic *Lena* image, which is used ubiquitously as a test image. It contains straight and curved lines, complicated texture, and a face. On the right is a perspective display of the Lena image as a signal: a function of two spatial variables. The colors merely help show what signal values are about the same size. In this image, signal values range between 0 and 255; why is that?

Color images have values that express how reflectivity depends on the optical spectrum. Painters long ago found that mixing together combinations of the so-called primary colors–red, yellow and blue–can produce very realistic color images. Thus, images today are usually thought of as having three values at every point in space, but a different set of colors is used: How much of red, *green* and blue is present. Mathematically, color pictures are multivalued–vector-valued–signals: $s(\mathbf{x}) = (r(\mathbf{x}), g(\mathbf{x}), b(\mathbf{x}))^T$.

Interesting cases abound where the analog signal depends not on a continuous variable, such as time, but on a discrete variable. For example, temperature readings taken every hour have continuous–analog–values, but the signal's independent variable is (essentially) the integers.

1.2.2 Digital Signals

The word "digital" means discrete-valued and implies the signal has an integer-valued independent variable. Digital information includes numbers and symbols (characters typed on the keyboard, for example). Computers rely on the digital representation of information to manipulate and transform information. Symbols do not have a numeric value, and each is represented by a unique number. The ASCII character code has the upper- and lowercase characters, the numbers, punctuation marks, and various other symbols represented by a seven-bit integer. For example, the ASCII code represents the letter a as the number 97 and the letter A as 65. Figure 1.3 shows the international convention on associating characters with integers.

ASCII Table

00	nul	01	soh	02	stx	03	etx	04	eot	05	enq	06	ack	07	bel
08	bs	09	ht	0A	nl	0B	vt	0C	np	0D	cr	0E	so	0F	si
10	dle	11	dc1	12	dc2	13	dc3	14	dc4	15	nak	16	syn	17	etb
18	car	19	em	1A	sub	1B	esc	1C	fs	1D	gs	1E	rs	1F	us
20	sp	21	!	22	"	23	#	24	$	25	%	26	&	27	’
28	(	29	)	2A	*	2B	+	2C	,	2D	-	2E	.	2F	/
30	0	31	1	32	2	33	3	34	4	35	5	36	6	37	7
38	8	39	9	3A	:	3B	;	3C	<	3D	=	3E	>	3F	?
40	@	41	A	42	B	43	C	44	D	45	E	46	F	47	G
48	H	49	I	4A	J	4B	K	4C	L	4D	M	4E	N	4F	0
50	P	51	Q	52	R	53	S	54	T	55	U	56	V	57	W
58	X	59	Y	5A	Z	5B	[	5C	\	5D	]	5E	^	5F	_
60	’	61	a	62	b	63	c	64	d	65	e	66	f	67	g
68	h	69	i	6A	j	6B	k	6C	l	6D	m	6E	n	6F	o
70	p	71	q	72	r	73	s	74	t	75	u	76	v	77	w
78	x	79	y	7A	z	7B	{	7C	|	7D	}	7E	~	7F	del

Figure 1.3: The ASCII translation table shows how standard keyboard characters are represented by integers. In pairs of columns, this table displays first the so-called 7-bit code (how many characters in a seven-bit code?), then the character the number represents. The numeric codes are represented in hexadecimal (base-16) notation. Mnemonic characters correspond to control characters, some of which may be familiar (like *cr* for carriage return) and some not (*bel* means a "bell").

1.3 Structure of Communication Systems[7]

Fundamental model of communication

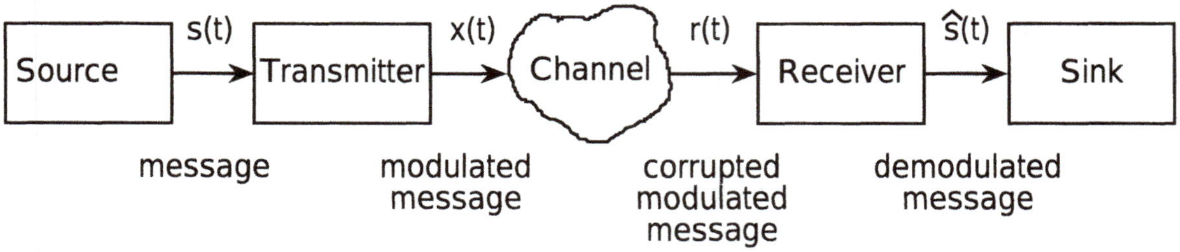

Figure 1.4: The Fundamental Model of Communication.

Definition of a system

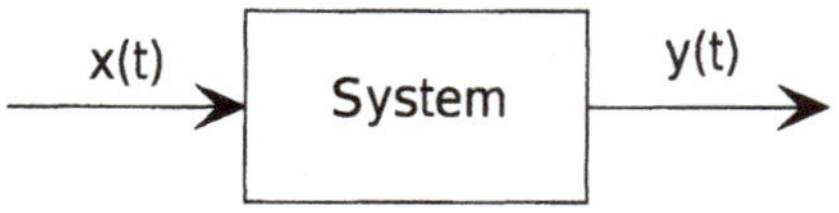

Figure 1.5: A system operates on its input signal $x(t)$ to produce an output $y(t)$.

The fundamental model of communications is portrayed in Figure 1.4 (Fundamental model of communication). In this fundamental model, each message-bearing signal, exemplified by $s(t)$, is analog and is a function of time. A **system** operates on zero, one, or several signals to produce more signals or to simply absorb them (Figure 1.5 (Definition of a system)). In electrical engineering, we represent a system as a box, receiving input signals (usually coming from the left) and producing from them new output signals. This graphical representation is known as a **block diagram**. We denote input signals by lines having arrows pointing into the box, output signals by arrows pointing away. As typified by the communications model, how information flows, how it is corrupted and manipulated, and how it is ultimately received is summarized by interconnecting block diagrams: The outputs of one or more systems serve as the inputs to others.

In the communications model, the **source** produces a signal that will be absorbed by the **sink**. Examples of time-domain signals produced by a source are music, speech, and characters typed on a keyboard. Signals can also be functions of two variables—an image is a signal that depends on two spatial variables—or more—television pictures (video signals) are functions of two spatial variables and time. Thus, information sources produce signals. *In physical systems, each signal corresponds to an electrical voltage or current.* To be able to design systems, we must understand electrical science and technology. However, we first need to understand the big picture to appreciate the context in which the electrical engineer works.

In communication systems, messages—signals produced by sources—must be recast for transmission. The block diagram has the message $s(t)$ passing through a block labeled **transmitter** that produces the signal $x(t)$. In the case of a radio transmitter, it accepts an input audio signal and produces a signal that

[7] This content is available online at <http://cnx.org/content/m0002/2.16/>.

physically is an electromagnetic wave radiated by an antenna and propagating as Maxwell's equations predict. In the case of a computer network, typed characters are encapsulated in packets, attached with a destination address, and launched into the Internet. From the communication systems "big picture" perspective, the *same* block diagram applies although the systems can be very different. In any case, the transmitter should *not* operate in such a way that the message $s(t)$ cannot be recovered from $x(t)$. In the mathematical sense, the inverse system must exist, else the communication system cannot be considered reliable. (It is ridiculous to transmit a signal in such a way that *no one* can recover the original. However, clever systems exist that transmit signals so that only the "in crowd" can recover them. Such crytographic systems underlie secret communications.)

Transmitted signals next pass through the next stage, the evil **channel**. Nothing good happens to a signal in a channel: It can become corrupted by noise, distorted, and attenuated among many possibilities. The channel cannot be escaped (the real world is cruel), and transmitter design *and* receiver design focus on how best to jointly fend off the channel's effects on signals. The channel is another system in our block diagram, and produces $r(t)$, the signal *received* by the receiver. If the channel were benign (good luck finding such a channel in the real world), the receiver would serve as the inverse system to the transmitter, and yield the message with no distortion. However, because of the channel, the receiver must do its best to produce a received message $\hat{s}(t)$ that resembles $s(t)$ as much as possible. Shannon[8] showed in his 1948 paper that reliable—for the moment, take this word to mean error-free—digital communication was possible over arbitrarily noisy channels. It is this result that modern communications systems exploit, and why many communications systems are going "digital." The module on Information Communication (Section 6.1) details Shannon's theory of information, and there we learn of Shannon's result and how to use it.

Finally, the received message is passed to the information **sink** that somehow makes use of the message. In the communications model, the source is a system having no input but producing an output; a sink has an input and no output.

Understanding signal generation and how systems work amounts to understanding signals, the nature of the information they represent, how information is transformed between analog and digital forms, and how information can be processed by systems operating on information-bearing signals. This understanding demands two different fields of knowledge. One is electrical science: How are signals represented and manipulated electrically? The second is signal science: What is the structure of signals, no matter what their source, what is their information content, and what capabilities does this structure force upon communication systems?

1.4 The Fundamental Signal[9]

1.4.1 The Sinusoid

The most ubiquitous and important signal in electrical engineering is the **sinusoid**.

Sine Definition

$$s(t) = A\cos(2\pi ft + \phi) \text{ or } A\cos(\omega t + \phi) \quad (1.1)$$

A is known as the sinusoid's **amplitude**, and determines the sinusoid's size. The amplitude conveys the sinusoid's physical units (volts, lumens, etc). The **frequency** f has units of Hz (Hertz) or s^{-1}, and determines how rapidly the sinusoid oscillates per unit time. The temporal variable t always has units of seconds, and thus the frequency determines how many oscillations/second the sinusoid has. AM radio stations have carrier frequencies of about 1 MHz (one mega-hertz or 10^6 Hz), while FM stations have carrier frequencies of about 100 MHz. Frequency can also be expressed by the symbol ω, which has units of radians/second. Clearly, $\omega = 2\pi f$. In communications, we most often express frequency in Hertz. Finally, ϕ is the **phase**, and determines the sine wave's behavior at the origin ($t = 0$). It has units of radians, but we can express it in

[8] http://www.lucent.com/minds/infotheory/

[9] This content is available online at <http://cnx.org/content/m0003/2.15/>.

degrees, realizing that in computations we must convert from degrees to radians. Note that if $\phi = -\left(\frac{\pi}{2}\right)$, the sinusoid corresponds to a sine function, having a zero value at the origin.

$$A\sin(2\pi ft + \phi) = A\cos\left(2\pi ft + \phi - \frac{\pi}{2}\right) \tag{1.2}$$

Thus, the only difference between a sine and cosine signal is the phase; we term either a sinusoid.

We can also define a discrete-time variant of the sinusoid: $A\cos(2\pi fn + \phi)$. Here, the independent variable is n and represents the integers. Frequency now has no dimensions, and takes on values between 0 and 1.

Exercise 1.1 ***(Solution on p. 11.)***

Show that $\cos(2\pi fn) = \cos(2\pi(f+1)n)$, which means that a sinusoid having a frequency larger than one corresponds to a sinusoid having a frequency less than one.

ANALOG OR DIGITAL?: Notice that we shall call either sinusoid an analog signal. Only when the discrete-time signal takes on a finite set of values can it be considered a digital signal.

Exercise 1.2 ***(Solution on p. 11.)***

Can you think of a simple signal that has a finite number of values but is defined in continuous time? Such a signal is also an analog signal.

1.4.2 Communicating Information with Signals

The basic idea of communication engineering is to use a signal's parameters to represent either real numbers or other signals. The technical term is to **modulate** the **carrier** signal's parameters to transmit information from one place to another. To explore the notion of modulation, we can send a real number (today's temperature, for example) by changing a sinusoid's amplitude accordingly. If we wanted to send the daily temperature, we would keep the frequency constant (so the receiver would know what to expect) and change the amplitude at midnight. We could relate temperature to amplitude by the formula $A = A_0(1 + kT)$, where A_0 and k are constants that the transmitter and receiver must *both* know.

If we had two numbers we wanted to send at the same time, we could modulate the sinusoid's frequency as well as its amplitude. This modulation scheme assumes we can estimate the sinusoid's amplitude and frequency; we shall learn that this is indeed possible.

Now suppose we have a sequence of parameters to send. We have exploited all of the sinusoid's two parameters. What we can do is modulate them for a limited time (say T seconds), and send two parameters every T. This simple notion corresponds to how a modem works. Here, typed characters are encoded into eight bits, and the individual bits are encoded into a sinusoid's amplitude and frequency. We'll learn how this is done in subsequent modules, and more importantly, we'll learn what the limits are on such digital communication schemes.

1.5 Introduction Problems[10]

Problem 1.1: RMS Values

The *rms* (root-mean-square) value of a periodic signal is defined to be

$$s = \sqrt{\frac{1}{T}\int_0^T s^2(t)\,dt}$$

where T is defined to be the signal's *period*: the smallest positive number such that $s(t) = s(t+T)$.

[10] This content is available online at <http://cnx.org/content/m10353/2.16/>.

a) What is the period of $s(t) = A\sin(2\pi f_0 t + \phi)$?
b) What is the rms value of this signal? How is it related to the peak value?
c) What is the period and rms value of the depicted (Figure 1.6) *square wave*, generically denoted by $\mathrm{sq}(t)$?
d) By inspecting any device you plug into a wall socket, you'll see that it is labeled "110 volts AC". What is the expression for the voltage provided by a wall socket? What is its rms value?

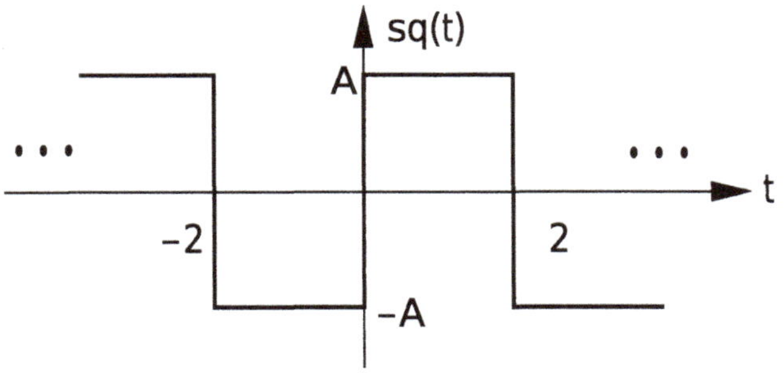

Figure 1.6

Problem 1.2: Modems
The word "modem" is short for "modulator-demodulator." Modems are used not only for connecting computers to telephone lines, but also for connecting digital (discrete-valued) sources to generic channels. In this problem, we explore a simple kind of modem, in which binary information is represented by the presence or absence of a sinusoid (presence representing a "1" and absence a "0"). Consequently, the modem's transmitted signal that represents a single bit has the form

$$x(t) = A\sin(2\pi f_0 t), 0 \le t \le T$$

Within each bit interval T, the amplitude is either A or zero.

a) What is the smallest transmission interval that makes sense with the frequency f_0?
b) Assuming that ten cycles of the sinusoid comprise a single bit's transmission interval, what is the datarate of this transmission scheme?
c) Now suppose instead of using "on-off" signaling, we allow one of several *different* values for the amplitude during any transmission interval. If N amplitude values are used, what is the resulting datarate?
d) The classic communications block diagram applies to the modem. Discuss how the transmitter must interface with the message source since the source is producing letters of the alphabet, not bits.

Problem 1.3: Advanced Modems
To transmit symbols, such as letters of the alphabet, RU computer modems use two frequencies (1600 and 1800 Hz) and several amplitude levels. A transmission is sent for a period of time T (known as the transmission or baud interval) and equals the sum of two amplitude-weighted carriers.

$$x(t) = A_1\sin(2\pi f_1 t) + A_2\sin(2\pi f_2 t), 0 \le t \le T$$

We send successive symbols by choosing an appropriate frequency and amplitude combination, and sending them one after another.

a) What is the smallest transmission interval that makes sense to use with the frequencies given above? In other words, what should T be so that an integer number of cycles of the carrier occurs?

b) Sketch (using Matlab) the signal that modem produces over several transmission intervals. Make sure you axes are labeled.

c) Using your signal transmission interval, how many amplitude levels are needed to transmit ASCII characters at a datarate of 3,200 bits/s? Assume use of the extended (8-bit) ASCII code.

NOTE: We use a discrete set of values for A_1 and A_2. If we have N_1 values for amplitude A_1, and N_2 values for A_2, we have $N_1 N_2$ possible symbols that can be sent during each T second interval. To convert this number into bits (the fundamental unit of information engineers use to qualify things), compute $\log_2 (N_1 N_2)$.

Solutions to Exercises in Chapter 1

Solution to Exercise 1.1 (p. 8)

As $\cos(\alpha+\beta) = \cos(\alpha)\cos(\beta) - \sin(\alpha)\sin(\beta)$, $\cos(2\pi(f+1)n) = \cos(2\pi fn)\cos(2\pi n) - \sin(2\pi fn)\sin(2\pi n) = \cos(2\pi fn)$.

Solution to Exercise 1.2 (p. 8)

A square wave takes on the values 1 and -1 alternately. See the plot in the module Elemental Signals (Section 2.2.6: Square Wave).

Chapter 2

Signals and Systems

2.1 Complex Numbers[1]

While the fundamental signal usd in electrical engineering is the sinusoid, it can be expressed mathematically in terms of an even more fundamental signal: the **complex exponential**. Representing sinusoids in terms of complex exponentials is *not* a mathematical oddity. Fluency with complex numbers and rational functions of complex variables is a critical skill all engineers master. Understanding information and power system designs and developing new systems all hinge on using complex numbers. In short, they are critical to modern electrical engineering, a realization made over a century ago.

2.1.1 Definitions

The notion of the square root of -1 originated with the quadratic formula: the solution of certain quadratic equations mathematically exists only if the so-called imaginary quantity $\sqrt{-1}$ could be defined. Euler[2] first used i for the imaginary unit but that notation did not take hold until roughly Ampère's time. Ampère[3] used the symbol i to denote current (intensité de current). It wasn't until the twentieth century that the importance of complex numbers to circuit theory became evident. By then, using i for current was entrenched and electrical engineers chose j for writing complex numbers.

An **imaginary number** has the form $jb = \sqrt{-(b^2)}$. A **complex number**, z, consists of the ordered pair (a,b), a is the real component and b is the imaginary component (the j is suppressed because the imaginary component of the pair is always in the second position). The imaginary number jb equals (0,b). Note that a and b are real-valued numbers.

Figure 2.1 (The Complex Plane) shows that we can locate a complex number in what we call the **complex plane**. Here, a, the real part, is the x-coordinate and b, the imaginary part, is the y-coordinate. From analytic geometry, we know that locations in the plane can be expressed as the sum of vectors, with the vectors corresponding to the x and y directions. Consequently, a complex number z can be expressed as the (vector) sum $z = a + jb$ where j indicates the y-coordinate. This representation is known as the **Cartesian form of** z. An imaginary number can't be numerically added to a real number; rather, this notation for a complex number represents vector addition, but it provides a convenient notation when we perform arithmetic manipulations.

Some obvious terminology. The **real part** of the complex number $z = a + jb$, written as $\text{Re}(z)$, equals a. We consider the real part as a function that works by selecting that component of a complex number *not* multiplied by j. The **imaginary part** of z, $\text{Im}(z)$, equals b: that part of a complex number that is multiplied by j. Again, both the real and imaginary parts of a complex number are real-valued.

[1] This content is available online at <http://cnx.org/content/m0081/2.27/>.
[2] http://www-groups.dcs.st-and.ac.uk/~history/Mathematicians/Euler.html
[3] http://www-groups.dcs.st-and.ac.uk/~history/Mathematicians/Ampere.html

The Complex Plane

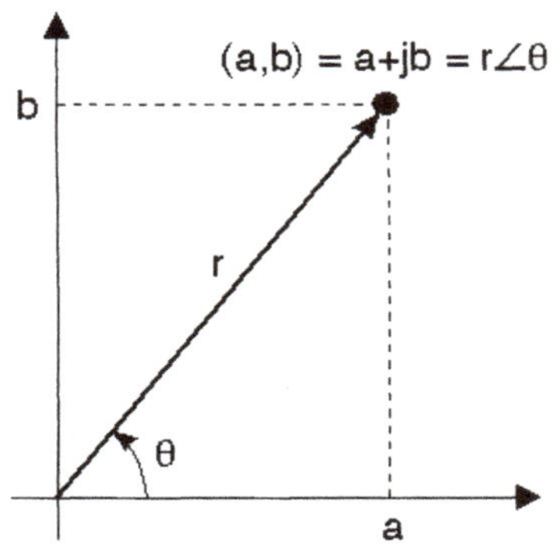

Figure 2.1: A complex number is an ordered pair (a,b) that can be regarded as coordinates in the plane. Complex numbers can also be expressed in polar coordinates as $r\angle\theta$.

The **complex conjugate** of z, written as z^*, has the same real part as z but an imaginary part of the opposite sign.

$$\begin{aligned} z &= \mathrm{Re}\,(z) + j\mathrm{Im}\,(z) \\ z^* &= \mathrm{Re}\,(z) - j\mathrm{Im}\,(z) \end{aligned} \tag{2.1}$$

Using Cartesian notation, the following properties easily follow.

- If we add two complex numbers, the real part of the result equals the sum of the real parts and the imaginary part equals the sum of the imaginary parts. This property follows from the laws of vector addition.

$$a_1 + jb_1 + a_2 + jb_2 = a_1 + a_2 + j\,(b_1 + b_2)$$

  In this way, the real and imaginary parts remain separate.
- The product of j and a real number is an imaginary number: ja. The product of j and an imaginary number is a real number: $j\,(jb) = -b$ because $j^2 = -1$. Consequently, multiplying a complex number by j *rotates* the number's position by 90 degrees.

Exercise 2.1 *(Solution on p. 34.)*

Use the definition of addition to show that the real and imaginary parts can be expressed as a sum/difference of a complex number and its conjugate. $\mathrm{Re}\,(z) = \frac{z+z^*}{2}$ and $\mathrm{Im}\,(z) = \frac{z-z^*}{2j}$.

Complex numbers can also be expressed in an alternate form, **polar form**, which we will find quite useful. Polar form arises arises from the geometric interpretation of complex numbers. The Cartesian form of a complex number can be re-written as

$$a + jb = \sqrt{a^2+b^2}\left(\frac{a}{\sqrt{a^2+b^2}} + j\frac{b}{\sqrt{a^2+b^2}}\right)$$

By forming a right triangle having sides a and b, we see that the real and imaginary parts correspond to the cosine and sine of the triangle's base angle. We thus obtain the **polar form** for complex numbers.

$$\begin{aligned} z &= a + jb = r\angle\theta \\ r &= |z| = \sqrt{a^2+b^2} \\ a &= r\cos(\theta) \\ b &= r\sin(\theta) \\ \theta &= \arctan\left(\tfrac{b}{a}\right) \end{aligned}$$

The quantity r is known as the **magnitude** of the complex number z, and is frequently written as $|z|$. The quantity θ is the complex number's **angle.** In using the arc-tangent formula to find the angle, we must take into account the quadrant in which the complex number lies.

Exercise 2.2 *(Solution on p. 34.)*

Convert $3 - 2j$ to polar form.

2.1.2 Euler's Formula

Surprisingly, the polar form of a complex number z can be expressed mathematically as

$$z = re^{j\theta} \tag{2.2}$$

To show this result, we use **Euler's relations** that express exponentials with imaginary arguments in terms of trigonometric functions.

$$e^{j\theta} = \cos(\theta) + j\sin(\theta) \tag{2.3}$$

$$\cos(\theta) = \frac{e^{j\theta} + e^{-(j\theta)}}{2} \tag{2.4}$$

$$\sin(\theta) = \frac{e^{j\theta} - e^{-(j\theta)}}{2j}$$

The first of these is easily derived from the Taylor's series for the exponential.

$$e^x = 1 + \frac{x}{1!} + \frac{x^2}{2!} + \frac{x^3}{3!} + \dots$$

Substituting $j\theta$ for x, we find that

$$e^{j\theta} = 1 + j\frac{\theta}{1!} - \frac{\theta^2}{2!} - j\frac{\theta^3}{3!} + \dots$$

because $j^2 = -1$, $j^3 = -j$, and $j^4 = 1$. Grouping separately the real-valued terms and the imaginary-valued ones,

$$e^{j\theta} = 1 - \frac{\theta^2}{2!} + \dots + j\left(\frac{\theta}{1!} - \frac{\theta^3}{3!} + \dots\right)$$

The real-valued terms correspond to the Taylor's series for $\cos(\theta)$, the imaginary ones to $\sin(\theta)$, and Euler's first relation results. The remaining relations are easily derived from the first. Because of , we see that multiplying the exponential in (2.3) by a real constant corresponds to setting the radius of the complex number by the constant.

2.1.3 Calculating with Complex Numbers

Adding and subtracting complex numbers expressed in Cartesian form is quite easy: You add (subtract) the real parts and imaginary parts separately.

$$(z_1 \pm z_2) = (a_1 \pm a_2) + j(b_1 \pm b_2) \tag{2.5}$$

To multiply two complex numbers in Cartesian form is not quite as easy, but follows directly from following the usual rules of arithmetic.

$$\begin{aligned} z_1 z_2 &= (a_1 + jb_1)(a_2 + jb_2) \\ &= a_1 a_2 - b_1 b_2 + j(a_1 b_2 + a_2 b_1) \end{aligned} \tag{2.6}$$

Note that we are, in a sense, multiplying two vectors to obtain another vector. Complex arithmetic provides a unique way of defining vector multiplication.

Exercise 2.3 *(Solution on p. 34.)*

What is the product of a complex number and its conjugate?

Division requires mathematical manipulation. We convert the division problem into a multiplication problem by multiplying both the numerator and denominator by the conjugate of the denominator.

$$\begin{aligned}\frac{z_1}{z_2} &= \frac{a_1+jb_1}{a_2+jb_2}\\ &= \frac{a_1+jb_1}{a_2+jb_2}\frac{a_2-jb_2}{a_2-jb_2}\\ &= \frac{(a_1+jb_1)(a_2-jb_2)}{{a_2}^2+{b_2}^2}\\ &= \frac{a_1a_2+b_1b_2+j(a_2b_1-a_1b_2)}{{a_2}^2+{b_2}^2}\end{aligned} \quad (2.7)$$

Because the final result is so complicated, it's best to remember *how* to perform division—multiplying numerator and denominator by the complex conjugate of the denominator—than trying to remember the final result.

The properties of the exponential make calculating the product and ratio of two complex numbers much simpler when the numbers are expressed in polar form.

$$\begin{aligned}z_1z_2 &= r_1e^{j\theta_1}r_2e^{j\theta_2}\\ &= r_1r_2e^{j(\theta_1+\theta_2)}\end{aligned} \quad (2.8)$$

$$\frac{z_1}{z_2} = \frac{r_1e^{j\theta_1}}{r_2e^{j\theta_2}} = \frac{r_1}{r_2}e^{j(\theta_1-\theta_2)}$$

To multiply, the radius equals the product of the radii and the angle the sum of the angles. To divide, the radius equals the ratio of the radii and the angle the difference of the angles. When the original complex numbers are in Cartesian form, it's usually worth translating into polar form, then performing the multiplication or division (especially in the case of the latter). Addition and subtraction of polar forms amounts to converting to Cartesian form, performing the arithmetic operation, and converting back to polar form.

Example 2.1

When we solve circuit problems, the crucial quantity, known as a transfer function, will always be expressed as the ratio of polynomials in the variable $s = j2\pi f$. What we'll need to understand the circuit's effect is the transfer function in polar form. For instance, suppose the transfer function equals

$$\frac{s+2}{s^2+s+1} \quad (2.9)$$

$$s = j2\pi f \quad (2.10)$$

Performing the required division is most easily accomplished by first expressing the numerator and denominator each in polar form, then calculating the ratio. Thus,

$$\frac{s+2}{s^2+s+1} = \frac{j2\pi f+2}{-4\pi^2 f^2 + j2\pi f + 1} \tag{2.11}$$

$$= \frac{\sqrt{4+4\pi^2 f^2} e^{j\arctan(\pi f)}}{\sqrt{(1-4\pi^2 f^2)^2 + 4\pi^2 f^2} e^{j\arctan\left(\frac{2\pi f}{1-4\pi^2 f^2}\right)}} \tag{2.12}$$

$$= \sqrt{\frac{4+4\pi^2 f^2}{1-4\pi^2 f^2 + 16\pi^4 f^4}} e^{j\left(\arctan(\pi f) - \arctan\left(\frac{2\pi f}{1-4\pi^2 f^2}\right)\right)} \tag{2.13}$$

2.2 Elemental Signals[4]

Elemental signals are the building blocks with which we build complicated signals. By definition, elemental signals have a simple structure. Exactly what we mean by the "structure of a signal" will unfold in this section of the course. Signals are nothing more than functions defined with respect to some independent variable, which we take to be time for the most part. Very interesting signals are not functions solely of time; one great example of which is an image. For it, the independent variables are x and y (two-dimensional space). Video signals are functions of three variables: two spatial dimensions and time. Fortunately, most of the ideas underlying modern signal theory can be exemplified with one-dimensional signals.

2.2.1 Sinusoids

Perhaps the most common real-valued signal is the sinusoid.

$$s(t) = A\cos(2\pi f_0 t + \phi) \tag{2.14}$$

For this signal, A is its amplitude, f_0 its frequency, and ϕ its phase.

2.2.2 Complex Exponentials

The most important signal is complex-valued, the complex exponential.

$$\begin{aligned} s(t) &= Ae^{j(2\pi f_0 t + \phi)} \\ &= Ae^{j\phi} e^{j2\pi f_0 t} \end{aligned} \tag{2.15}$$

Here, j denotes $\sqrt{-1}$. $Ae^{j\phi}$ is known as the signal's **complex amplitude**. Considering the complex amplitude as a complex number in polar form, its magnitude is the amplitude A and its angle the signal phase. The complex amplitude is also known as a **phasor**. The complex exponential cannot be further decomposed into more elemental signals, and is the *most important signal in electrical engineering*! Mathematical manipulations at first appear to be more difficult because complex-valued numbers are introduced. In fact, early in the twentieth century, mathematicians thought engineers would not be sufficiently sophisticated to handle complex exponentials even though they greatly simplified solving circuit problems. Steinmetz [5] introduced complex exponentials to electrical engineering, and demonstrated that "mere" engineers could use them to good effect and even obtain right answers! See Complex Numbers (Section 2.1) for a review of complex numbers and complex arithmetic.

The complex exponential defines the notion of frequency: it is the *only* signal that contains only one frequency component. The sinusoid consists of two frequency components: one at the frequency $+f_0$ and the other at $-f_0$.

[4]This content is available online at <http://cnx.org/content/m0004/2.26/>.

[5]http://www.invent.org/hall_of_fame/139.html

EULER RELATION: This decomposition of the sinusoid can be traced to Euler's relation.

$$\cos(2\pi ft) = \frac{e^{j2\pi ft} + e^{-(j2\pi ft)}}{2} \tag{2.16}$$

$$\sin(2\pi ft) = \frac{e^{j2\pi ft} - e^{-(j2\pi ft)}}{2j} \tag{2.17}$$

$$e^{j2\pi ft} = \cos(2\pi ft) + j\sin(2\pi ft) \tag{2.18}$$

DECOMPOSITION: The complex exponential signal can thus be written in terms of its real and imaginary parts using Euler's relation. Thus, sinusoidal signals can be expressed as either the real or the imaginary part of a complex exponential signal, the choice depending on whether cosine or sine phase is needed, or as the sum of two complex exponentials. These two decompositions are mathematically equivalent to each other.

$$A\cos(2\pi ft + \phi) = \mathrm{Re}\left(Ae^{j\phi}e^{j2\pi ft}\right) \tag{2.19}$$

$$A\sin(2\pi ft + \phi) = \mathrm{Im}\left(Ae^{j\phi}e^{j2\pi ft}\right) \tag{2.20}$$

Using the complex plane, we can envision the complex exponential's temporal variations as seen in the above figure (Figure 2.2). The magnitude of the complex exponential is A, and the initial value of the complex exponential at $t = 0$ has an angle of ϕ. As time increases, the locus of points traced by the complex exponential is a circle (it has constant magnitude of A). The number of times per second we go around the circle equals the frequency f. The time taken for the complex exponential to go around the circle once is known as its **period** T, and equals $\frac{1}{f}$. The projections onto the real and imaginary axes of the rotating vector representing the complex exponential signal are the cosine and sine signal of Euler's relation ((2.16)).

2.2.3 Real Exponentials

As opposed to complex exponentials which oscillate, real exponentials (Figure 2.3) decay.

$$s(t) = e^{-\left(\frac{t}{\tau}\right)} \tag{2.21}$$

The quantity τ is known as the exponential's **time constant**, and corresponds to the time required for the exponential to decrease by a factor of $\frac{1}{e}$, which approximately equals 0.368. *A decaying complex exponential is the product of a real and a complex exponential.*

$$\begin{aligned} s(t) &= Ae^{j\phi}e^{-\left(\frac{t}{\tau}\right)}e^{j2\pi ft} \\ &= Ae^{j\phi}e^{\left(-\left(\frac{1}{\tau}\right)+j2\pi f\right)t} \end{aligned} \tag{2.22}$$

In the complex plane, this signal corresponds to an exponential spiral. For such signals, we can define **complex frequency** as the quantity multiplying t.

2.2.4 Unit Step

The unit step function (Figure 2.4) is denoted by $u(t)$, and is defined to be

$$u(t) = \begin{cases} 0 \text{ if } t < 0 \\ 1 \text{ if } t > 0 \end{cases} \tag{2.23}$$

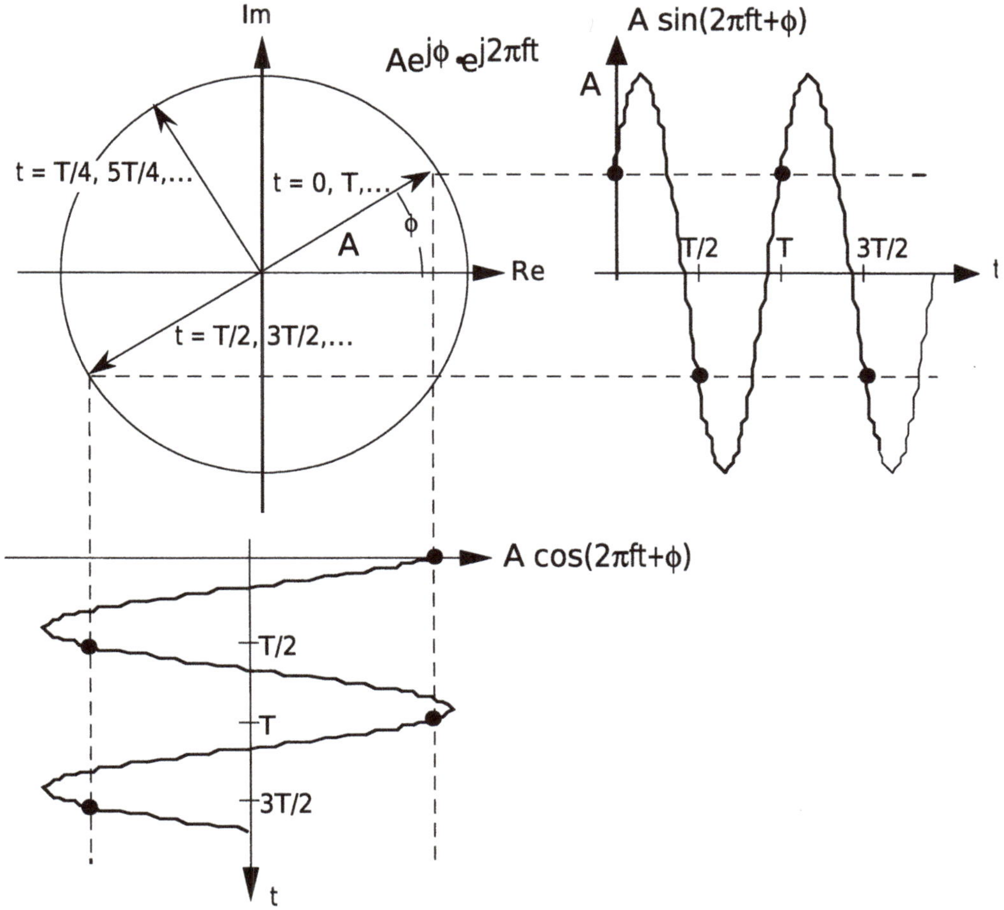

Figure 2.2: Graphically, the complex exponential scribes a circle in the complex plane as time evolves. Its real and imaginary parts are sinusoids. The rate at which the signal goes around the circle is the frequency f and the time taken to go around is the *period* T. A fundamental relationship is $T = \frac{1}{f}$.

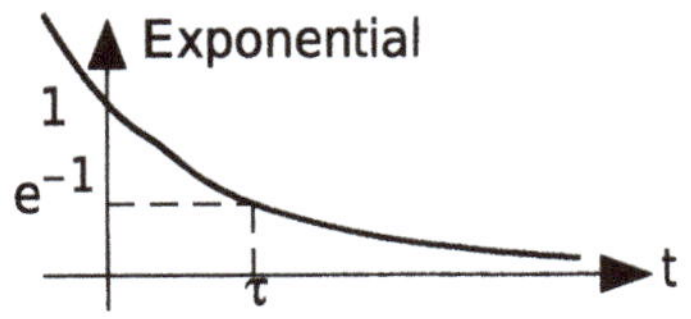

Figure 2.3: The real exponential.

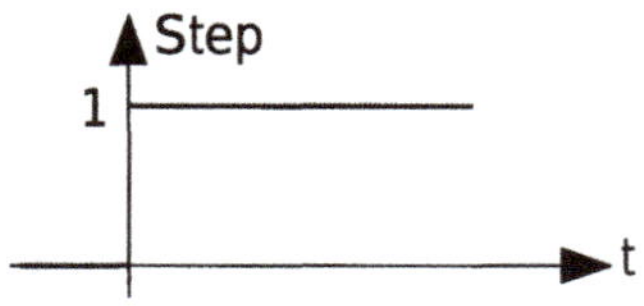

Figure 2.4: The unit step.

ORIGIN WARNING: This signal is discontinuous at the origin. Its value at the origin need not be defined, and doesn't matter in signal theory.

This kind of signal is used to describe signals that "turn on" suddenly. For example, to mathematically represent turning on an oscillator, we can write it as the product of a sinusoid and a step: $s(t) = A\sin(2\pi ft)\,u(t)$.

2.2.5 Pulse

The unit pulse (Figure 2.5) describes turning a unit-amplitude signal on for a duration of Δ seconds, then turning it off.

$$p_\Delta(t) = \begin{cases} 0 \text{ if } t < 0 \\ 1 \text{ if } 0 < t < \Delta \\ 0 \text{ if } t > \Delta \end{cases} \tag{2.24}$$

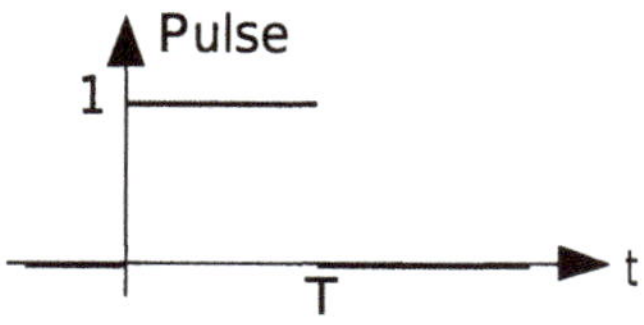

Figure 2.5: The pulse.

We will find that this is the second most important signal in communications.

2.2.6 Square Wave

The square wave (Figure 2.6) $\mathrm{sq}(t)$ is a periodic signal like the sinusoid. It too has an amplitude and a period, which must be specified to characterize the signal. We find subsequently that the sine wave is a simpler signal than the square wave.

2.3 Signal Decomposition[6]

A signal's complexity is not related to how wiggly it is. Rather, a signal expert looks for ways of decomposing a given signal into a *sum of simpler signals*, which we term the **signal decomposition**. Though we will

[6]This content is available online at <http://cnx.org/content/m0008/2.12/>.

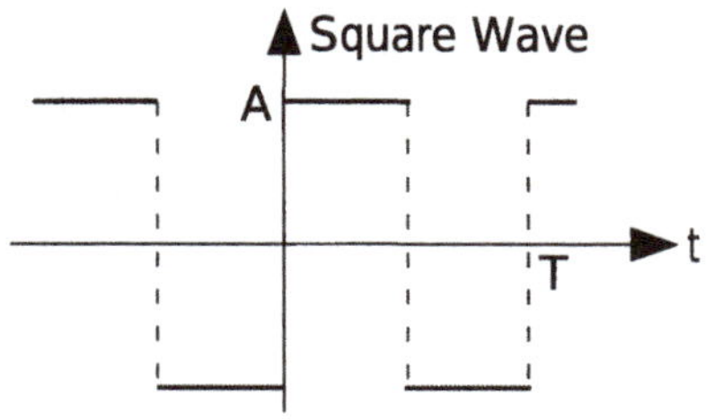

Figure 2.6: The square wave.

never compute a signal's complexity, it essentially equals the number of terms in its decomposition. In writing a signal as a sum of component signals, we can change the component signal's gain by multiplying it by a constant and by delaying it. More complicated decompositions could contain derivatives or integrals of simple signals.

Example 2.2

As an example of signal complexity, we can express the pulse $p_{\Delta}(t)$ as a sum of delayed unit steps.

$$p_{\Delta}(t) = u(t) - u(t - \Delta) \tag{2.25}$$

Thus, the pulse is a more complex signal than the step. Be that as it may, the pulse is very useful to us.

Exercise 2.4 ***(Solution on p. 34.)***

Express a square wave having period T and amplitude A as a superposition of delayed and amplitude-scaled pulses.

Because the sinusoid is a superposition of two complex exponentials, the sinusoid is more complex. We could not prevent ourselves from the pun in this statement. Clearly, the word "complex" is used in two different ways here. The complex exponential can also be written (using Euler's relation (2.16)) as a sum of a sine and a cosine. We will discover that virtually every signal can be decomposed into a sum of complex exponentials, and that this decomposition is *very* useful. Thus, the complex exponential is more fundamental, and Euler's relation does not adequately reveal its complexity.

2.4 Discrete-Time Signals[7]

So far, we have treated what are known as **analog** signals and systems. Mathematically, analog signals are functions having continuous quantities as their independent variables, such as space and time. Discrete-time signals (Section 5.5) are functions defined on the integers; they are sequences. One of the fundamental results of signal theory (Section 5.3) will detail conditions under which an analog signal can be converted into a discrete-time one and retrieved *without error.* This result is important because discrete-time signals can be manipulated by systems instantiated as computer programs. Subsequent modules describe how virtually all analog signal processing can be performed with software.

As important as such results are, discrete-time signals are more general, encompassing signals derived from analog ones *and* signals that aren't. For example, the characters forming a text file form a sequence, which is also a discrete-time signal. We must deal with such symbolic valued (p. 167) signals and systems as well.

[7]This content is available online at <http://cnx.org/content/m0009/2.23/>.

As with analog signals, we seek ways of decomposing real-valued discrete-time signals into simpler components. With this approach leading to a better understanding of signal structure, we can exploit that structure to represent information (create ways of representing information with signals) and to extract information (retrieve the information thus represented). For symbolic-valued signals, the approach is different: We develop a common representation of all symbolic-valued signals so that we can embody the information they contain in a unified way. From an information representation perspective, the most important issue becomes, for both real-valued and symbolic-valued signals, efficiency; What is the most parsimonious and compact way to represent information so that it can be extracted later.

2.4.1 Real- and Complex-valued Signals

A discrete-time signal is represented symbolically as $s(n)$, where $n = \{\dots, -1, 0, 1, \dots\}$. We usually draw discrete-time signals as stem plots to emphasize the fact they are functions defined only on the integers. We can delay a discrete-time signal by an integer just as with analog ones. A delayed unit sample has the expression $\delta(n-m)$, and equals one when $n = m$.

Discrete-Time Cosine Signal

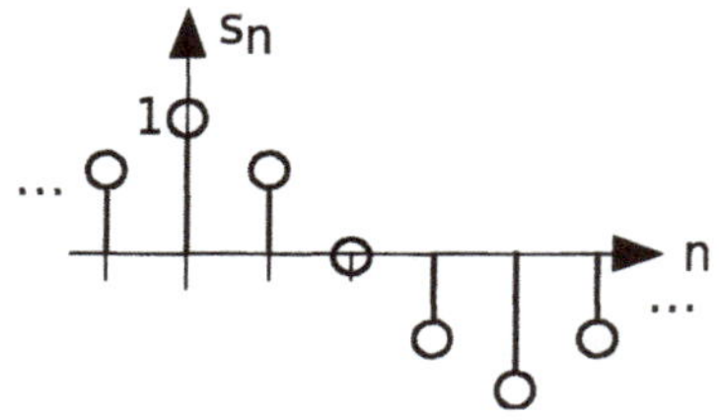

Figure 2.7: The discrete-time cosine signal is plotted as a stem plot. Can you find the formula for this signal?

2.4.2 Complex Exponentials

The most important signal is, of course, the **complex exponential sequence**.

$$s(n) = e^{j2\pi fn} \tag{2.26}$$

2.4.3 Sinusoids

Discrete-time sinusoids have the obvious form $s(n) = A\cos(2\pi fn + \phi)$. As opposed to analog complex exponentials and sinusoids that can have their frequencies be any real value, frequencies of their discrete-time counterparts yield unique waveforms *only* when f lies in the interval $\left(-\left(\frac{1}{2}\right), \frac{1}{2}\right]$. This property can be easily understood by noting that adding an integer to the frequency of the discrete-time complex exponential has no effect on the signal's value.

$$\begin{aligned} e^{j2\pi(f+m)n} &= e^{j2\pi fn}e^{j2\pi mn} \\ &= e^{j2\pi fn} \end{aligned} \tag{2.27}$$

This derivation follows because the complex exponential evaluated at an integer multiple of 2π equals one.

2.4.4 Unit Sample

The second-most important discrete-time signal is the **unit sample**, which is defined to be

$$\delta(n) = \begin{cases} 1 \text{ if } n = 0 \\ 0 \text{ otherwise} \end{cases} \tag{2.28}$$

Unit Sample

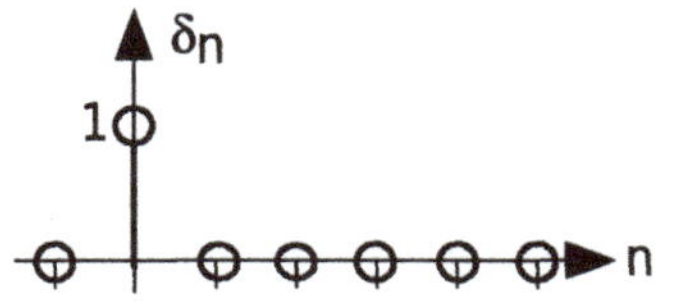

Figure 2.8: The unit sample.

Examination of a discrete-time signal's plot, like that of the cosine signal shown in Figure 2.7 (Discrete-Time Cosine Signal), reveals that all signals consist of a sequence of delayed and scaled unit samples. Because the value of a sequence at each integer m is denoted by $s(m)$ and the unit sample delayed to occur at m is written $\delta(n-m)$, we can decompose *any* signal as a sum of unit samples delayed to the appropriate location and scaled by the signal value.

$$s(n) = \sum_{m=-\infty}^{\infty} (s(m)\,\delta(n-m)) \tag{2.29}$$

This kind of decomposition is unique to discrete-time signals, and will prove useful subsequently.

Discrete-time systems can act on discrete-time signals in ways similar to those found in analog signals and systems. Because of the role of software in discrete-time systems, many more different systems can be envisioned and "constructed" with programs than can be with analog signals. In fact, a special class of analog signals can be converted into discrete-time signals, processed with software, and converted back into an analog signal, all without the incursion of error. For such signals, systems can be easily produced in software, with equivalent analog realizations difficult, if not impossible, to design.

2.4.5 Symbolic-valued Signals

Another interesting aspect of discrete-time signals is that their values do not need to be real numbers. We do have real-valued discrete-time signals like the sinusoid, but we also have signals that denote the sequence of characters typed on the keyboard. Such characters certainly aren't real numbers, and as a collection of possible signal values, they have little mathematical structure other than that they are members of a set. More formally, each element of the symbolic-valued signal $s(n)$ takes on one of the values $\{a_1, \ldots, a_K\}$ which comprise the **alphabet** A. This technical terminology does not mean we restrict symbols to being members of the English or Greek alphabet. They could represent keyboard characters, bytes (8-bit quantities), integers that convey daily temperature. Whether controlled by software or not, discrete-time systems are ultimately constructed from digital circuits, which consist *entirely* of analog circuit elements. Furthermore, the transmission and reception of discrete-time signals, like e-mail, is accomplished with analog signals and systems. Understanding how discrete-time and analog signals and systems intertwine is perhaps the main goal of this course.

2.5 Introduction to Systems[8]

Signals are manipulated by systems. Mathematically, we represent what a system does by the notation $y(t) = S(x(t))$, with x representing the input signal and y the output signal.

Definition of a system

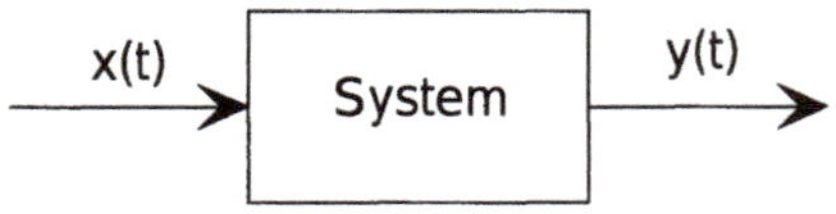

Figure 2.9: The system depicted has input $x(t)$ and output $y(t)$. Mathematically, systems operate on function(s) to produce other function(s). In many ways, systems are like functions, rules that yield a value for the dependent variable (our output signal) for each value of its independent variable (its input signal). The notation $y(t) = S(x(t))$ corresponds to this block diagram. We term $S(\cdot)$ the input-output relation for the system.

This notation mimics the mathematical symbology of a function: A system's input is analogous to an independent variable and its output the dependent variable. For the mathematically inclined, a system is a **functional**: a function of a function (signals are functions).

Simple systems can be connected together–one system's output becomes another's input–to accomplish some overall design. Interconnection topologies can be quite complicated, but usually consist of weaves of three basic interconnection forms.

2.5.1 Cascade Interconnection

cascade

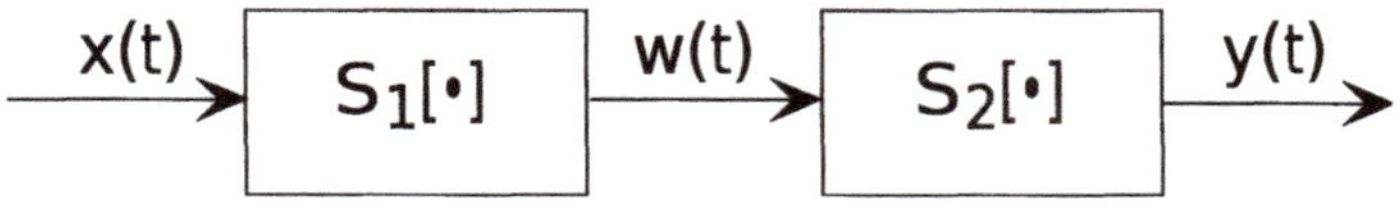

Figure 2.10: The most rudimentary ways of interconnecting systems are shown in the figures in this section. This is the cascade configuration.

The simplest form is when one system's output is connected only to another's input. Mathematically, $w(t) = S_1(x(t))$, and $y(t) = S_2(w(t))$, with the information contained in $x(t)$ processed by the first, then the second system. In some cases, the ordering of the systems matter, in others it does not. For example, in the fundamental model of communication (Figure 1.4: Fundamental model of communication) the ordering most certainly matters.

[8] This content is available online at <http://cnx.org/content/m0005/2.18/>.

2.5.2 Parallel Interconnection

parallel

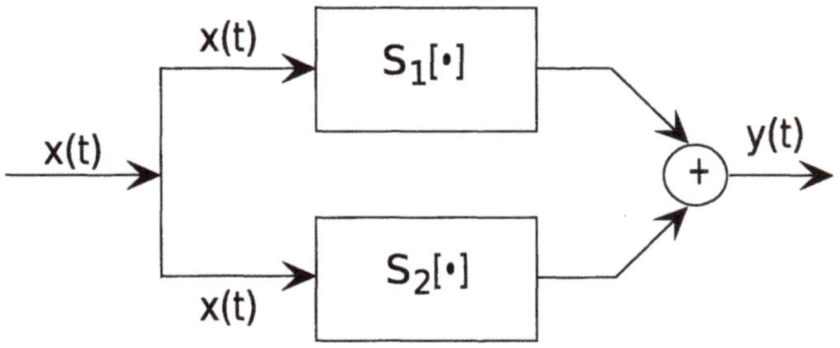

Figure 2.11: The parallel configuration.

A signal $x(t)$ is routed to two (or more) systems, with this signal appearing as the input to all systems simultaneously and with equal strength. Block diagrams have the convention that signals going to more than one system are not split into pieces along the way. Two or more systems operate on $x(t)$ and their outputs are added together to create the output $y(t)$. Thus, $y(t) = S_1(x(t)) + S_2(x(t))$, and the information in $x(t)$ is processed separately by both systems.

2.5.3 Feedback Interconnection

feedback

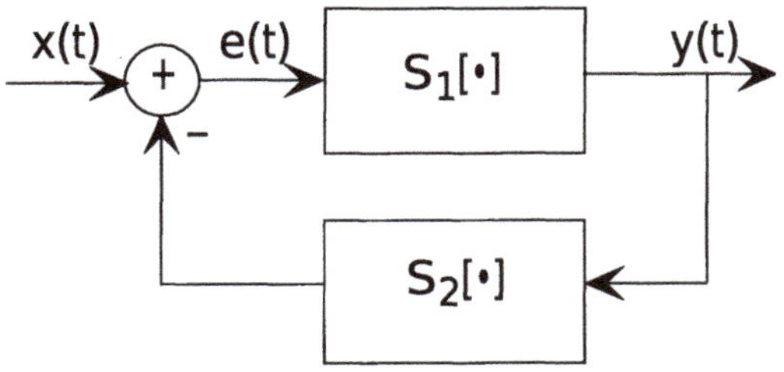

Figure 2.12: The feedback configuration.

The subtlest interconnection configuration has a system's output also contributing to its input. Engineers would say the output is "fed back" to the input through system 2, hence the terminology. The mathematical statement of the feedback interconnection (Figure 2.12: feedback) is that the feed-forward system produces the output: $y(t) = S_1(e(t))$. The input $e(t)$ equals the input signal minus the output of some other system's output to $y(t)$: $e(t) = x(t) - S_2(y(t))$. Feedback systems are omnipresent in control problems, with the error signal used to adjust the output to achieve some condition defined by the input (controlling) signal. For example, in a car's cruise control system, $x(t)$ is a constant representing what speed you want, and $y(t)$

is the car's speed as measured by a speedometer. In this application, system 2 is the identity system (output equals input).

2.6 Simple Systems[9]

Systems manipulate signals, creating output signals derived from their inputs. Why the following are categorized as "simple" will only become evident towards the end of the course.

2.6.1 Sources

Sources produce signals without having input. We like to think of these as having controllable parameters, like amplitude and frequency. Examples would be oscillators that produce periodic signals like sinusoids and square waves and noise generators that yield signals with erratic waveforms (more about noise subsequently). Simply writing an expression for the signals they produce specifies sources. A sine wave generator might be specified by $y(t) = A\sin(2\pi f_0 t)\, u(t)$, which says that the source was turned on at $t = 0$ to produce a sinusoid of amplitude A and frequency f_0.

2.6.2 Amplifiers

An amplifier (Figure 2.13: amplifier) multiplies its input by a constant known as the amplifier **gain**.

$$y(t) = Gx(t) \tag{2.30}$$

amplifier

Figure 2.13: An amplifier.

The gain can be positive or negative (if negative, we would say that the amplifier *inverts* its input) and can be greater than one or less than one. If less than one, the amplifier actually *attenuates*. A real-world example of an amplifier is your home stereo. You control the gain by turning the volume control.

2.6.3 Delay

A system serves as a time delay (Figure 2.14: delay) when the output signal equals the input signal at an earlier time.

$$y(t) = x(t - \tau) \tag{2.31}$$

[9] This content is available online at <http://cnx.org/content/m0006/2.23/>.

delay

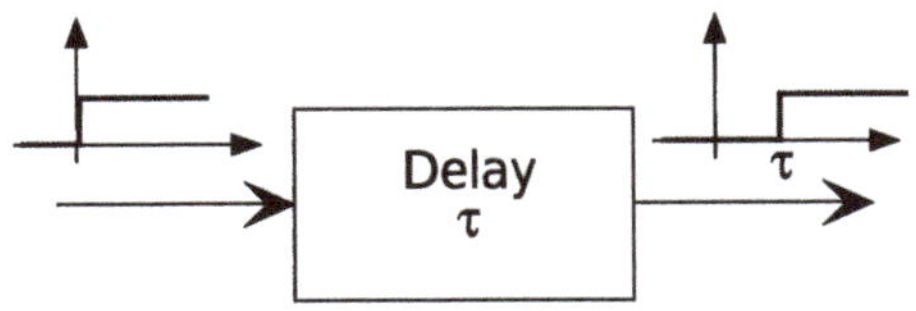

Figure 2.14: A delay.

Here, τ is the delay. The way to understand this system is to focus on the time origin: The output at time $t = \tau$ equals the input at time $t = 0$. Thus, if the delay is positive, the output emerges later than the input, and plotting the output amounts to shifting the input plot to the right. The delay can be negative, in which case we say the system *advances* its input. Such systems are difficult to build (they would have to produce signal values derived from what the input *will be*), but we will have occasion to advance signals in time.

2.6.4 Time Reversal

Here, the output signal equals the input signal flipped about the time origin.

$$y(t) = x(-t) \tag{2.32}$$

time reversal

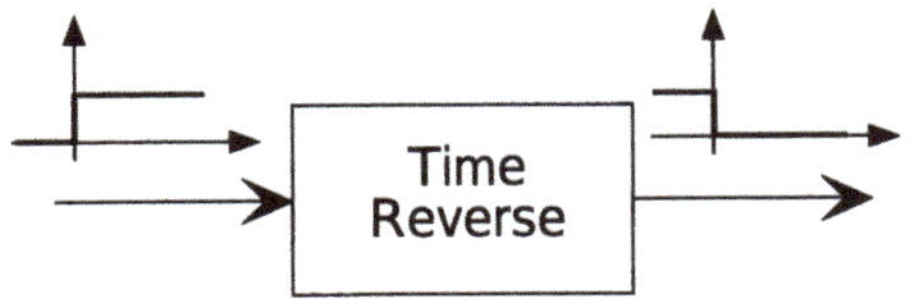

Figure 2.15: A time reversal system.

Again, such systems are difficult to build, but the notion of time reversal occurs frequently in communications systems.

Exercise 2.5 ***(Solution on p. 34.)***

Mentioned earlier was the issue of whether the ordering of systems mattered. In other words, if we have two systems in cascade, does the output depend on which comes first? Determine if the ordering matters for the cascade of an amplifier and a delay and for the cascade of a time-reversal system and a delay.

2.6.5 Derivative Systems and Integrators

Systems that perform calculus-like operations on their inputs can produce waveforms significantly different than present in the input. Derivative systems operate in a straightforward way: A first-derivative system would have the input-output relationship $y(t) = \frac{d}{dt}x(t)$. Integral systems have the complication that the integral's limits must be defined. It is a signal theory convention that the elementary integral operation have a lower limit of $-\infty$, and that the value of *all* signals at $t = -\infty$ equals zero. A simple integrator would have input-output relation

$$y(t) = \int_{-\infty}^{t} x(\alpha)\, d\alpha \tag{2.33}$$

2.6.6 Linear Systems

Linear systems are a *class* of systems rather than having a specific input-output relation. Linear systems form the foundation of system theory, and are the most important class of systems in communications. They have the property that when the input is expressed as a weighted sum of component signals, the output equals the same weighted sum of the outputs produced by each component. When $S(\cdot)$ is linear,

$$S(G_1 x_1(t) + G_2 x_2(t)) = G_1 S(x_1(t)) + G_2 S(x_2(t)) \tag{2.34}$$

for all choices of signals and gains.

This general input-output relation property can be manipulated to indicate specific properties shared by all linear systems.

- $S(Gx(t)) = GS(x(t))$ The colloquialism summarizing this property is "Double the input, you double the output." Note that this property is consistent with alternate ways of expressing gain changes: Since $2x(t)$ also equals $x(t) + x(t)$, the linear system definition provides the same output no matter which of these is used to express a given signal.
- $S(0) = 0$ If the input is *identically zero for all time*, the output of a linear system must be zero. This property follows from the simple derivation $S(0) = S(x(t) - x(t)) = S(x(t)) - S(x(t)) = 0$.

Just why linear systems are so important is related not only to their properties, which are divulged throughout this course, but also because they lend themselves to relatively simple mathematical analysis. Said another way, "They're the only systems we thoroughly understand!"

We can find the output of any linear system to a complicated input by decomposing the input into simple signals. The equation above (2.34) says that when a system is linear, its output to a decomposed input is the sum of outputs to each input. For example, if

$$x(t) = e^{-t} + \sin(2\pi f_0 t)$$

the output $S(x(t))$ of any linear system equals

$$y(t) = S\left(e^{-t}\right) + S(\sin(2\pi f_0 t))$$

2.6.7 Time-Invariant Systems

Systems that don't change their input-output relation with time are said to be time-invariant. The mathematical way of stating this property is to use the signal delay concept described in Simple Systems (Section 2.6.3: Delay).

$$y(t) = S(x(t)) \Rightarrow y(t - \tau) = S(x(t - \tau)) \tag{2.35}$$

If you delay (or advance) the input, the output is similarly delayed (advanced). Thus, a time-invariant system responds to an input you may supply tomorrow the same way it responds to the same input applied today; today's output is merely delayed to occur tomorrow.

The collection of linear, time-invariant systems are *the* most thoroughly understood systems. Much of the signal processing and system theory discussed here concentrates on such systems. For example, electric circuits are, for the most part, linear and time-invariant. Nonlinear ones abound, but characterizing them so that you can predict their behavior for any input remains an unsolved problem.

Linear, Time-Invariant Table

Input-Output Relation	Linear	Time-Invariant
$y(t) = \frac{d}{dt}(x)$	yes	yes
$y(t) = \frac{d^2}{dt^2}(x)$	yes	yes
$y(t) = \left(\frac{d}{dt}(x)\right)^2$	no	yes
$y(t) = \frac{d}{dt}(x) + x$	yes	yes
$y(t) = x_1 + x_2$	yes	yes
$y(t) = x(t - \tau)$	yes	yes
$y(t) = \cos(2\pi f t)\, x(t)$	yes	no
$y(t) = x(-t)$	yes	no
$y(t) = x^2(t)$	no	yes
$y(t) = \lvert x(t) \rvert$	no	yes
$y(t) = mx(t) + b$	no	yes

Figure 2.16

2.7 Signals and Systems Problems[10]

Problem 2.1: Complex Number Arithmetic
Find the real part, imaginary part, the magnitude and angle of the complex numbers given by the following expressions.

a) -1
b) $\frac{1+\sqrt{3}j}{2}$
c) $1 + j + e^{j\frac{\pi}{2}}$
d) $e^{j\frac{\pi}{3}} + e^{j\pi} + e^{-\left(j\frac{\pi}{3}\right)}$

Problem 2.2: Discovering Roots
Complex numbers expose all the roots of real (and complex) numbers. For example, there should be two square-roots, three cube-roots, etc. of any number. Find the following roots.

a) What are the cube-roots of 27? In other words, what is $27^{\frac{1}{3}}$?
b) What are the fifth roots of 3 ($3^{\frac{1}{5}}$)?
c) What are the fourth roots of one?

[10] This content is available online at <http://cnx.org/content/m10348/2.24/>.

Problem 2.3: Cool Exponentials
Simplify the following (cool) expressions.

a) j^j
b) j^{2j}
c) j^{j^j}

Problem 2.4: Complex-valued Signals
Complex numbers and phasors play a very important role in electrical engineering. Solving systems for complex exponentials is much easier than for sinusoids, and linear systems analysis is particularly easy.

a) Find the phasor representation for each, and re-express each as the real and imaginary parts of a complex exponential. What is the frequency (in Hz) of each? In general, are your answers unique? If so, prove it; if not, find an alternative answer for the complex exponential representation.
   i) $3\sin(24t)$
   ii) $\sqrt{2}\cos\left(2\pi 60t + \frac{\pi}{4}\right)$
   iii) $2\cos\left(t + \frac{\pi}{6}\right) + 4\sin\left(t - \frac{\pi}{3}\right)$
b) Show that for linear systems having real-valued outputs for real inputs, that when the input is the real part of a complex exponential, the output is the real part of the system's output to the complex exponential (see Figure 2.17).

$$S\left(\mathrm{Re}\left(Ae^{j2\pi ft}\right)\right) = \mathrm{Re}\left(S\left(Ae^{j2\pi ft}\right)\right)$$

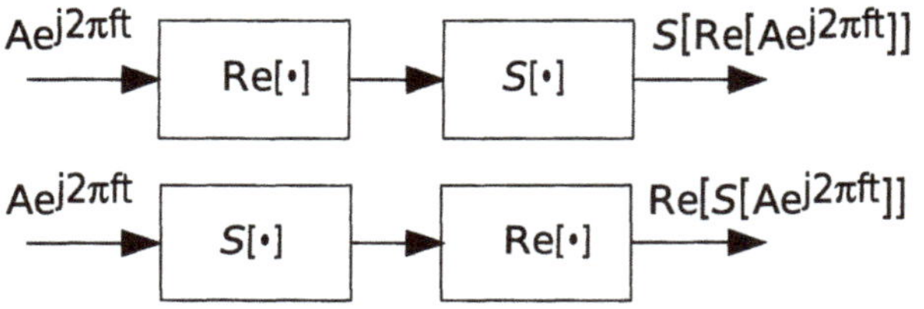

Figure 2.17

Problem 2.5:
For each of the indicated voltages, write it as the real part of a complex exponential ($v(t) = \mathrm{Re}(Ve^{st})$). Explicitly indicate the value of the complex amplitude V and the complex frequency s. Represent each complex amplitude as a vector in the V-plane, and indicate the location of the frequencies in the complex s-plane.

a) $v(t) = \cos(5t)$
b) $v(t) = \sin\left(8t + \frac{\pi}{4}\right)$
c) $v(t) = e^{-t}$
d) $v(t) = e^{-(3t)}\sin\left(4t + \frac{3\pi}{4}\right)$
e) $v(t) = 5e^{(2t)}\sin(8t + 2\pi)$
f) $v(t) = -2$
g) $v(t) = 4\sin(2t) + 3\cos(2t)$
h) $v(t) = 2\cos\left(100\pi t + \frac{\pi}{6}\right) - \sqrt{3}\sin\left(100\pi t + \frac{\pi}{2}\right)$

Problem 2.6:
Express each of the following signals (Figure 2.18) as a linear combination of delayed and weighted step functions and ramps (the integral of a step).

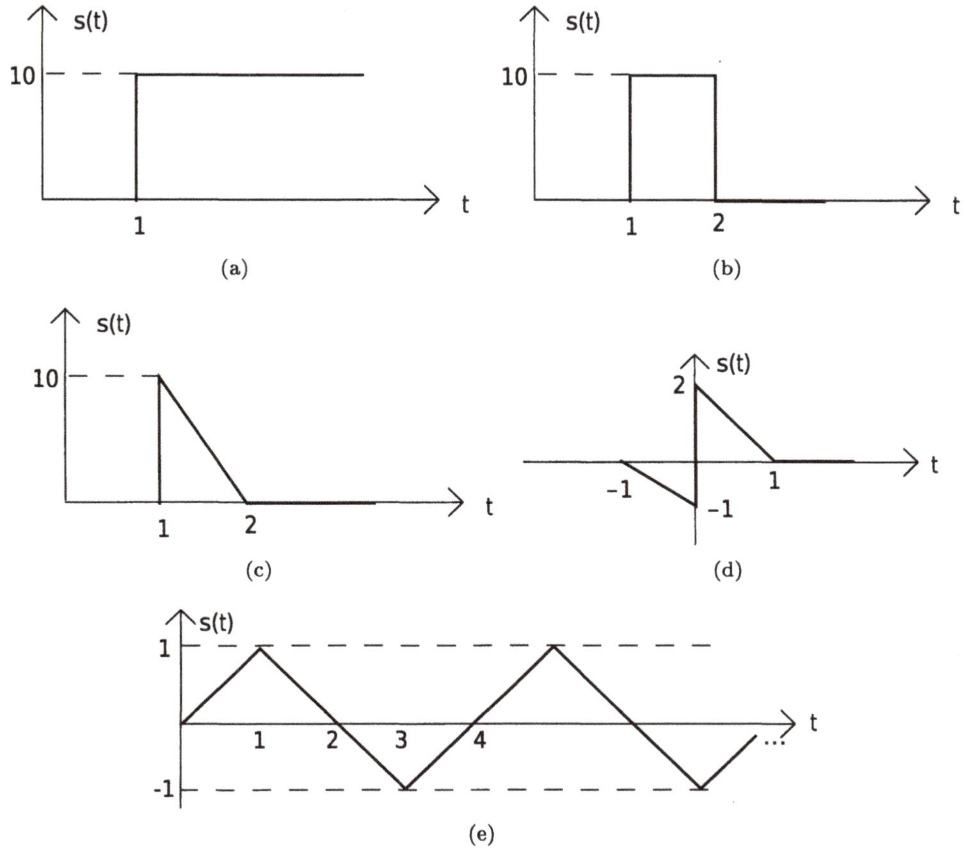

Figure 2.18

Problem 2.7: Linear, Time-Invariant Systems
When the input to a linear, time-invariant system is the signal $x(t)$, the output is the signal $y(t)$ (Figure 2.19).

a) Find and sketch this system's output when the input is the depicted signal (Figure 2.20).
b) Find and sketch this system's output when the input is a unit step.

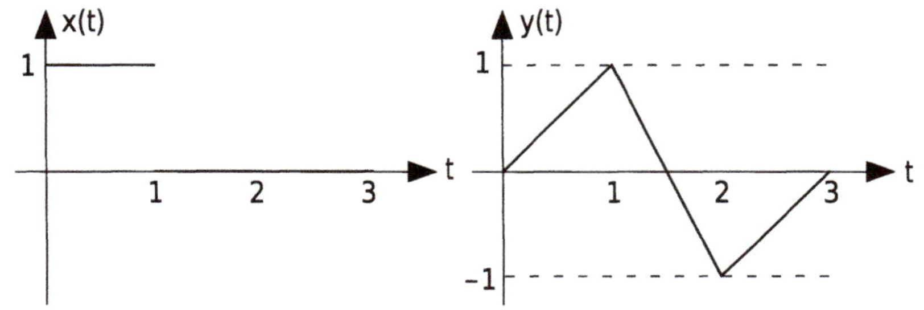

Figure 2.19

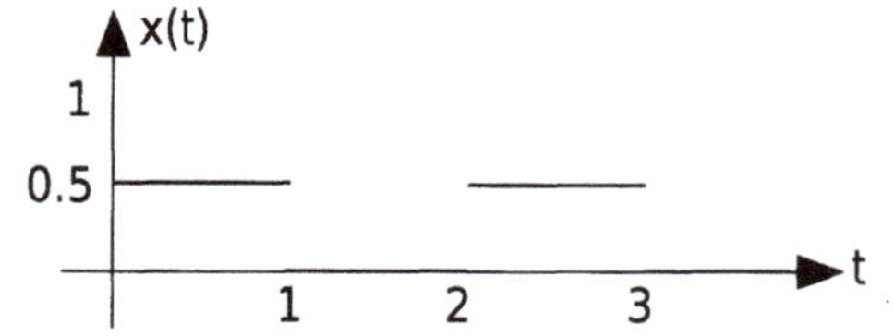

Figure 2.20

Problem 2.8: Linear Systems
The depicted input (Figure 2.21) $x(t)$ to a linear, time-invariant system yields the output $y(t)$.

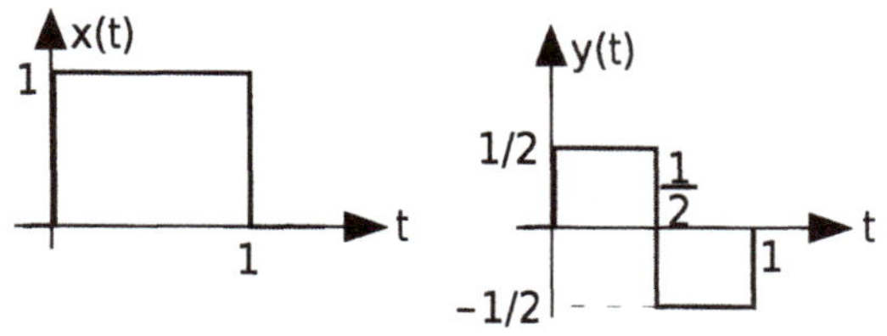

Figure 2.21

a) What is the system's output to a unit step input $u(t)$?
b) What will the output be when the input is the depicted square wave (Figure 2.22)?

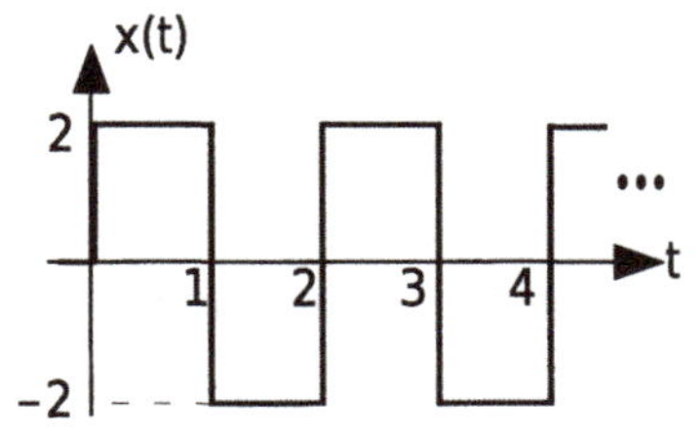

Figure 2.22

Problem 2.9: Communication Channel
A particularly interesting communication channel can be modeled as a linear, time-invariant system. When the transmitted signal $x(t)$ is a pulse, the received signal $r(t)$ is as shown (Figure 2.23).

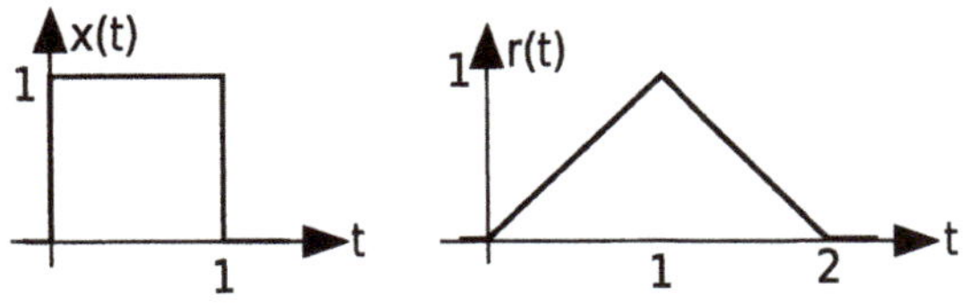

Figure 2.23

a) What will be the received signal when the transmitter sends the pulse sequence (Figure 2.24) $x_1(t)$?
b) What will be the received signal when the transmitter sends the pulse signal (Figure 2.24) $x_2(t)$ that has half the duration as the original?

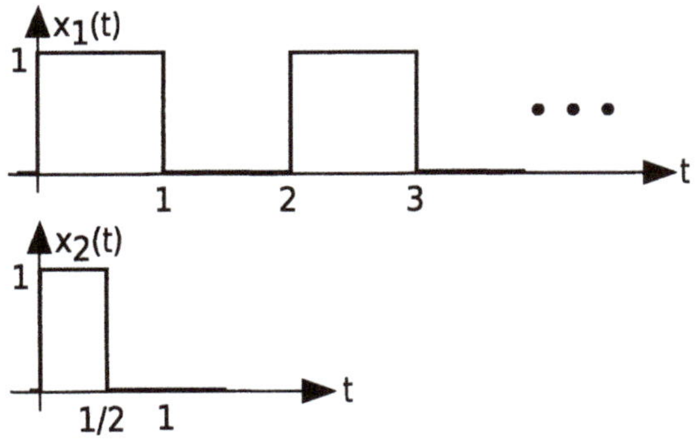

Figure 2.24

Solutions to Exercises in Chapter 2

Solution to Exercise 2.1 (p. 14)
$z + z^* = a + jb + a - jb = 2a = 2\mathrm{Re}(z)$. Similarly, $z - z^* = a + jb - (a - jb) = 2jb = 2j\mathrm{Im}(z)$

Solution to Exercise 2.2 (p. 15)
To convert $3 - 2j$ to polar form, we first locate the number in the complex plane in the fourth quadrant. The distance from the origin to the complex number is the magnitude r, which equals $\sqrt{13} = \sqrt{3^2 + (-2)^2}$. The angle equals $-\left(\arctan\left(\frac{2}{3}\right)\right)$ or -0.588 radians (-33.7 degrees). The final answer is $\sqrt{13}\angle(-33.7)$ degrees.

Solution to Exercise 2.3 (p. 16)
$zz^* = (a + jb)(a - jb) = a^2 + b^2$. Thus, $zz^* = r^2 = (|z|)^2$.

Solution to Exercise 2.4 (p. 21)
$\mathrm{sq}(t) = \sum_{n=-\infty}^{\infty} \left((-1)^n A p_{T/2}\left(t - n\frac{T}{2}\right)\right)$

Solution to Exercise 2.5 (p. 27)
In the first case, order does not matter; in the second it does. "Delay" means $t \rightarrow t - \tau$. "Time-reverse" means $t \rightarrow -t$

Case 1 $y(t) = Gx(t - \tau)$, and the way we apply the gain and delay the signal gives the same result.

Case 2 Time-reverse then delay: $y(t) = x(-(t - \tau)) = x(-t + \tau)$. Delay then time-reverse: $y(t) = x(-t - \tau)$.

Chapter 3

Analog Signal Processing

3.1 Voltage, Current, and Generic Circuit Elements[1]

We know that information can be represented by signals; now we need to understand how signals are physically realized. Over the years, electric signals have been found to be the easiest to use. Voltage and currents comprise the electric instantiations of signals. Thus, we need to delve into the world of electricity and electromagnetism. The systems used to manipulate electric signals directly are called **circuits**, and they refine the information representation or extract information from the voltage or current. In many cases, they make nice examples of linear systems.

A generic circuit element places a constraint between the classic variables of a circuit: voltage and current. **Voltage** is electric potential and represents the "push" that drives electric charge from one place to another. What causes charge to move is a physical separation between positive and negative charge. A battery generates, through electrochemical means, excess positive charge at one terminal and negative charge at the other, creating an electric field. Voltage is defined *across* a circuit element, with the positive sign denoting a positive voltage drop across the element. When a conductor connects the positive and negative potentials, **current** flows, with positive current indicating that positive charge flows from the positive terminal to the negative. Electrons comprise current flow in many cases. Because electrons have a negative charge, electrons move in the opposite direction of positive current flow: Negative charge flowing to the right is equivalent to positive charge moving to the left.

It is important to understand the physics of current flow in conductors to appreciate the innovation of new electronic devices. Electric charge can arise from many sources, the simplest being the electron. When we say that "electrons flow through a conductor," what we mean is that the conductor's constituent atoms freely give up electrons from their outer shells. "Flow" thus means that electrons hop from atom to atom driven along by the applied electric potential. A missing electron, however, is a virtual positive charge. Electrical engineers call these **holes**, and in some materials, particularly certain semiconductors, current flow is actually due to holes. Current flow also occurs in nerve cells found in your brain. Here, neurons "communicate" using propagating voltage pulses that rely on the flow of positive ions (potassium and sodium primarily, and to some degree calcium) across the neuron's outer wall. Thus, current can come from many sources, and circuit theory can be used to understand how current flows in reaction to electric fields.

[1]This content is available online at <http://cnx.org/content/m0011/2.12/>.

Generic Circuit Element

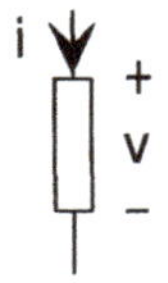

Figure 3.1: The generic circuit element.

Current flows through circuit elements, such as that depicted in Figure 3.1 (Generic Circuit Element), and through conductors, which we indicate by lines in circuit diagrams. For every circuit element we define a voltage and a current. The element has a *v-i* relation defined by the element's physical properties. In defining the *v-i* relation, we have the convention that positive current flows from positive to negative voltage drop. Voltage has units of volts, and both the unit and the quantity are named for Volta[2] . Current has units of amperes, and is named for the French physicist Ampère[3] .

Voltages and currents also carry **power**. Again using the convention shown in Figure 3.1 (Generic Circuit Element) for circuit elements, the **instantaneous power** at each moment of time consumed by the element is given by the product of the voltage and current.

$$p(t) = v(t)\, i(t)$$

A positive value for power indicates that at time t the circuit element is *consuming* power; a negative value means it is *producing* power. With voltage expressed in volts and current in amperes, power defined this way has units of **watts**. Just as in all areas of physics and chemistry, power is the rate at which **energy** is consumed or produced. Consequently, energy is the integral of power.

$$E(t) = \int_{-\infty}^{t} p(\alpha)\, d\alpha$$

Again, positive energy corresponds to consumed energy and negative energy corresponds to energy production. Note that a circuit element having a power profile that is both positive and negative over some time interval could consume or produce energy according to the sign of the integral of power. The units of energy are **joules** since a watt equals joules/second.

Exercise 3.1 *(Solution on p. 106.)*

Residential energy bills typically state a home's energy usage in kilowatt-hours. Is this really a unit of energy? If so, how many joules equals one kilowatt-hour?

3.2 Ideal Circuit Elements[4]

The elementary circuit elements—the resistor, capacitor, and inductor— impose **linear** relationships between voltage and current.

[2]http://www.bioanalytical.com/info/calendar/97/volta.htm
[3]http://www-groups.dcs.st-and.ac.uk/~history/Mathematicians/Ampere.html
[4]This content is available online at <http://cnx.org/content/m0012/2.19/>.

3.2.1 Resistor

Resistor

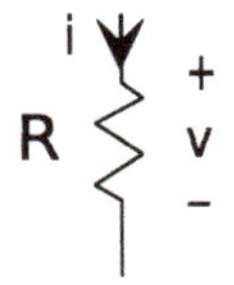

Figure 3.2: Resistor. $v = Ri$

The resistor is far and away the simplest circuit element. In a resistor, the voltage is proportional to the current, with the constant of proportionality R, known as the **resistance**.

$$v(t) = Ri(t)$$

Resistance has units of ohms, denoted by Ω, named for the German electrical scientist Georg Ohm[5] . Sometimes, the *v-i* relation for the resistor is written $i = Gv$, with G, the **conductance**, equal to $\frac{1}{R}$. Conductance has units of Siemens (S), and is named for the German electronics industrialist Werner von Siemens[6] .

When resistance is positive, as it is in most cases, a resistor consumes power. A resistor's instantaneous power consumption can be written one of two ways.

$$p(t) = Ri^2(t) = \frac{1}{R}v^2(t)$$

As the resistance approaches infinity, we have what is known as an **open circuit**: No current flows but a non-zero voltage can appear across the open circuit. As the resistance becomes zero, the voltage goes to zero for a non-zero current flow. This situation corresponds to a **short circuit**. A superconductor physically realizes a short circuit.

3.2.2 Capacitor

Capacitor

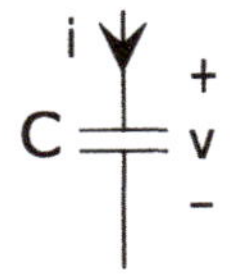

Figure 3.3: Capacitor. $i = C\frac{d}{dt}v(t)$

[5] http://www-groups.dcs.st-and.ac.uk/~history/Mathematicians/Ohm.html
[6] http://w4.siemens.de/archiv/en/persoenlichkeiten/werner_von_siemens.html

The capacitor stores charge and the relationship between the charge stored and the resultant voltage is $q = Cv$. The constant of proportionality, the capacitance, has units of farads (F), and is named for the English experimental physicist Michael Faraday[7] . As current is the rate of change of charge, the *v-i* relation can be expressed in differential or integral form.

$$i(t) = C\frac{d}{dt}v(t) \, \mathrm{or} \, v(t) = \frac{1}{C}\int_{-\infty}^{t} i(\alpha)\, d\alpha \tag{3.1}$$

If the voltage across a capacitor is constant, then the current flowing into it equals zero. In this situation, the capacitor is equivalent to an open circuit. The power consumed/produced by a voltage applied to a capacitor depends on the product of the voltage and its derivative.

$$p(t) = Cv(t)\frac{d}{dt}v(t)$$

This result means that a capacitor's total energy expenditure up to time t is concisely given by

$$E(t) = \frac{1}{2}Cv^2(t)$$

This expression presumes the **fundamental assumption** of circuit theory: *all voltages and currents in any circuit were zero in the far distant past* ($t = -\infty$).

3.2.3 Inductor

Inductor

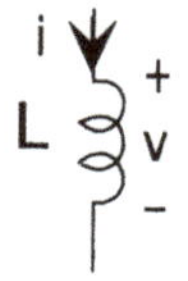

Figure 3.4: Inductor. $v = L\frac{d}{dt}i(t)$

The inductor stores magnetic flux, with larger valued inductors capable of storing more flux. Inductance has units of henries (H), and is named for the American physicist Joseph Henry[8] . The differential and integral forms of the inductor's *v-i* relation are

$$v(t) = L\frac{d}{dt}i(t) \quad \mathrm{or} \quad i(t) = \frac{1}{L}\int_{-\infty}^{t} v(\alpha)\, d\alpha \tag{3.2}$$

The power consumed/produced by an inductor depends on the product of the inductor current and its derivative

$$p(t) = Li(t)\frac{d}{dt}i(t)$$

and its total energy expenditure up to time t is given by

$$E(t) = \frac{1}{2}Li^2(t)$$

[7] http://www.iee.org.uk/publish/faraday/faraday1.html
[8] http://www.si.edu/archives//ihd/jhp/

3.2.4 Sources

Sources

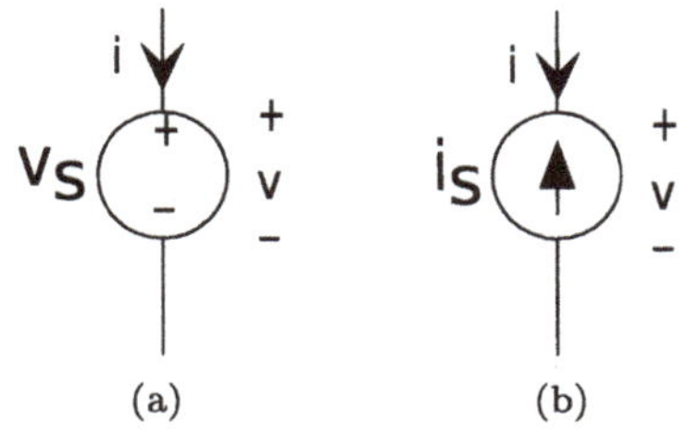

Figure 3.5: The voltage source on the left and current source on the right are like all circuit elements in that they have a particular relationship between the voltage and current defined for them. For the voltage source, $v = v_s$ for any current i; for the current source, $i = -i_s$ for any voltage v.

Sources of voltage and current are also circuit elements, but they are not linear in the strict sense of linear systems. For example, the voltage source's *v-i* relation is $v = v_s$ regardless of what the current might be. As for the current source, $i = -i_s$ regardless of the voltage. Another name for a constant-valued voltage source is a battery, and can be purchased in any supermarket. Current sources, on the other hand, are much harder to acquire; we'll learn why later.

3.3 Ideal and Real-World Circuit Elements[9]

Source and linear circuit elements are *ideal* circuit elements. One central notion of circuit theory is combining the ideal elements to describe how physical elements operate in the real world. For example, the 1 kΩ resistor you can hold in your hand is not exactly an ideal 1 kΩ resistor. First of all, physical devices are manufactured to close tolerances (the tighter the tolerance, the more money you pay), but never have exactly their advertised values. The fourth band on resistors specifies their tolerance; 10% is common. More pertinent to the current discussion is another deviation from the ideal: If a sinusoidal voltage is placed across a physical resistor, the current will not be exactly proportional to it as frequency becomes high, say above 1 MHz. At very high frequencies, the way the resistor is constructed introduces inductance and capacitance effects. Thus, the smart engineer must be aware of the frequency ranges over which his ideal models match reality well.

On the other hand, physical circuit elements can be readily found that well approximate the ideal, but they will always deviate from the ideal in some way. For example, a flashlight battery, like a C-cell, roughly corresponds to a 1.5 V voltage source. However, it ceases to be modeled by a voltage source capable of supplying *any* current (that's what ideal ones can do!) when the resistance of the light bulb is too small.

3.4 Electric Circuits and Interconnection Laws[10]

A **circuit** connects circuit elements together in a specific configuration designed to transform the source signal (originating from a voltage or current source) into another signal—the output—that corresponds to the current or voltage defined for a particular circuit element. A simple resistive circuit is shown in Figure 3.6.

[9]This content is available online at <http://cnx.org/content/m0013/2.9/>.
[10]This content is available online at <http://cnx.org/content/m0014/2.26/>.

This circuit is the electrical embodiment of a system having its input provided by a source system producing $v_{\mathrm{in}}(t)$.

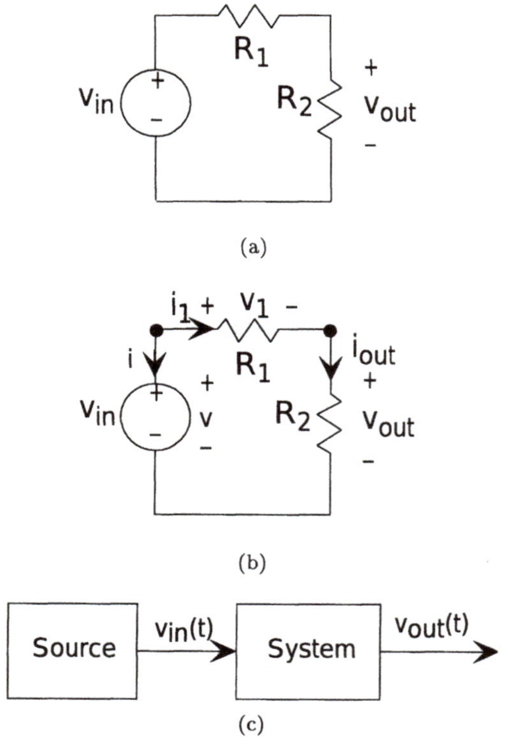

Figure 3.6: The circuit shown in the top two figures is perhaps the simplest circuit that performs a signal processing function. On the bottom is the block diagram that corresponds to the circuit. The input is provided by the voltage source v_{in} and the output is the voltage v_{out} across the resistor label R_2. As shown in the middle, we *analyze* the circuit—understand what it accomplishes—by defining currents and voltages for all circuit elements, and then solving the circuit and element equations.

To understand what this circuit accomplishes, we want to determine the voltage across the resistor labeled by its value R_2. Recasting this problem mathematically, we need to solve some set of equations so that we relate the output voltage v_{out} to the source voltage. It would be simple—a little too simple at this point—if we could instantly write down the one equation that relates these two voltages. Until we have more knowledge about how circuits work, we must write a set of equations that allow us to find *all* the voltages and currents that can be defined for every circuit element. Because we have a three-element circuit, we have a total of six voltages and currents that must be either specified or determined. You can define the directions for current flow and positive voltage drop *any way you like*. When two people solve a circuit their own ways, the signs of their variables may not agree, but current flow and voltage drop values for each element will agree. Do recall in defining your voltage and current variables (Section 3.2) that the *v-i* relations for the elements presume that positive current flow is in the same direction as positive voltage drop. Once you define voltages and currents, we need six nonredundant equations to solve for the six unknown voltages and currents. By specifying the source, we have one; this amounts to providing the source's *v-i* relation. The *v-i* relations for the resistors give us two more. We are only halfway there; where do we get the other three equations we need?

What we need to solve every circuit problem are mathematical statements that express how the circuit elements are interconnected. Said another way, we need the laws that govern the electrical connection of circuit elements. First of all, the places where circuit elements attach to each other are called **nodes**. Two nodes are explicitly indicated in Figure 3.6; a third is at the bottom where the voltage source and resistor R_2 are connected. Electrical engineers tend to draw circuit diagrams—schematics— in a rectilinear fashion. Thus the long line connecting the bottom of the voltage source with the bottom of the resistor is intended to make the diagram look pretty. This line simply means that the two elements are connected together. **Kirchoff's Laws**, one for voltage (Section 3.4.2: Kirchoff's Voltage Law (KVL)) and one for current (Section 3.4.1: Kirchoff's Current Law), determine what a connection among circuit elements means. These laws can help us analyze this circuit.

3.4.1 Kirchoff's Current Law

At every node, the sum of all currents entering a node must equal zero. What this law means physically is that charge cannot accumulate in a node; what goes in must come out. In the example, Figure 3.6, below we have a three-node circuit and thus have three KCL equations.

$$-i - i_1 = 0$$
$$i_1 - i_2 = 0$$
$$i + i_2 = 0$$

Note that the current entering a node is the negative of the current leaving the node.

Given any two of these KCL equations, we can find the other by adding or subtracting them. Thus, one of them is redundant and, in mathematical terms, we can discard any one of them. The convention is to discard the equation for the (unlabeled) node at the bottom of the circuit.

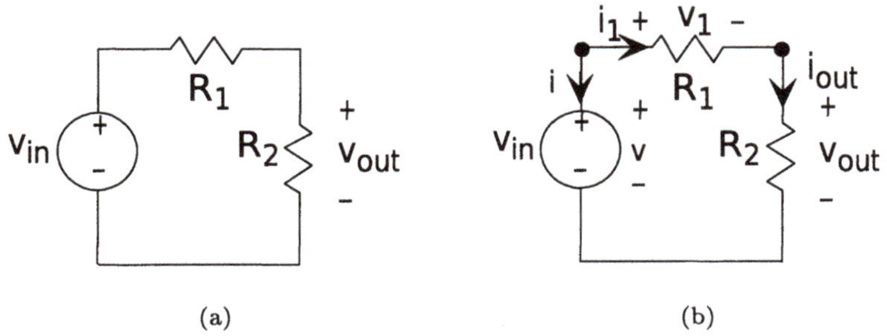

Figure 3.7: The circuit shown is perhaps the simplest circuit that performs a signal processing function. The input is provided by the voltage source v_{in} and the output is the voltage v_{out} across the resistor labelled R_2.

Exercise 3.2 *(Solution on p. 106.)*

In writing KCL equations, you will find that in an n-node circuit, exactly one of them is always redundant. Can you sketch a proof of why this might be true? Hint: It has to do with the fact that charge won't accumulate in one place on its own.

3.4.2 Kirchoff's Voltage Law (KVL)

The voltage law says that the sum of voltages around every closed loop in the circuit must equal zero. A closed loop has the obvious definition: Starting at a node, trace a path through the circuit that returns you to the origin node. KVL expresses the fact that electric fields are conservative: The total work performed in moving a test charge around a closed path is zero. The KVL equation for our circuit is

$$v_1 + v_2 - v = 0$$

In writing KVL equations, we follow the convention that an element's voltage enters with a plus sign when traversing the closed path, we go from the positive to the negative of the voltage's definition.

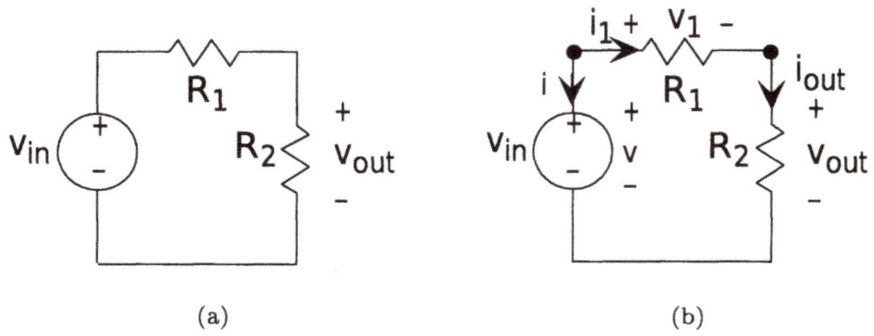

Figure 3.8: The circuit shown is perhaps the simplest circuit that performs a signal processing function. The input is provided by the voltage source v_{in} and the output is the voltage v_{out} across the resistor labelled R_2.

For the example circuit (Figure 3.8), we have three *v-i* relations, two KCL equations, and one KVL equation for solving for the circuit's six voltages and currents.

$$\begin{aligned}
\text{v-i:} \quad & v = v_{in} \\
& v_1 = R_1 i_1 \\
& v_{out} = R_2 i_{out} \\
\text{KCL:} \quad & -i - i_1 = 0 \\
& i_1 - i_{out} = 0 \\
\text{KVL:} \quad & -v + v_1 + v_{out} = 0
\end{aligned}$$

We have exactly the right number of equations! Eventually, we will discover shortcuts for solving circuit problems; for now, we want to eliminate all the variables but v_{out} and determine how it depends on v_{in} and on resistor values. The KVL equation can be rewritten as $v_{in} = v_1 + v_{out}$. Substituting into it the resistor's *v-i* relation, we have $v_{in} = R_1 i_1 + R_2 i_{out}$. Yes, we temporarily eliminate the quantity we seek. Though not obvious, it is the simplest way to solve the equations. One of the KCL equations says $i_1 = i_{out}$, which means that $v_{in} = R_1 i_{out} + R_2 i_{out} = (R_1 + R_2)\, i_{out}$. Solving for the current in the output resistor, we have $i_{out} = \frac{v_{in}}{R_1 + R_2}$. *We have now solved the circuit*: We have expressed one voltage or current in terms of sources and circuit-element values. To find any other circuit quantities, we can back substitute this answer into our original equations or ones we developed along the way. Using the *v-i* relation for the output resistor, we obtain the quantity we seek.

$$v_{out} = \frac{R_2}{R_1 + R_2} v_{in}$$

Exercise 3.3 *(Solution on p. 106.)*
Referring back to Figure 3.6, a circuit should serve some useful purpose. What kind of system does our circuit realize and, in terms of element values, what are the system's parameter(s)?

3.5 Power Dissipation in Resistor Circuits[11]

We can find voltages and currents in simple circuits containing resistors and voltage or current sources. We should examine whether these circuits variables obey the Conservation of Power principle: since a circuit is a closed system, it should not dissipate or create energy. For the moment, our approach is to investigate first a resistor circuit's **power** consumption/creation. Later, we will *prove* that because of **KVL** and **KCL** *all* circuits conserve power.

As defined on p. 36, the instantaneous power consumed/created by every circuit element equals the product of its voltage and current. The total power consumed/created by a circuit equals the sum of each element's power.

$$P = \sum_{k} (v_k i_k)$$

Recall that each element's current and voltage must obey the convention that positive current is defined to enter the positive-voltage terminal. With this convention, a positive value of $v_k i_k$ corresponds to consumed power, a negative value to created power. Because the total power in a circuit must be zero ($P = 0$), some circuit elements must create power while others consume it.

Consider the simple series circuit should in Section 3.4. In performing our calculations, we defined the current i_{out} to flow through the positive-voltage terminals of both resistors and found it to equal $i_{\text{out}} = \frac{v_{\text{in}}}{R_1+R_2}$. The voltage across the resistor R_2 is the output voltage and we found it to equal $v_{\text{out}} = \frac{R_2}{R_1+R_2} v_{\text{in}}$. Consequently, calculating the power for this resistor yields

$$P_2 = \frac{R_2}{(R_1 + R_2)^2} {v_{\text{in}}}^2$$

Consequently, this resistor dissipates power because P_2 is positive. This result should not be surprising since we showed (p. 37) that the power consumed by *any* resistor equals either of the following.

$$\frac{v^2}{R} \text{ or } i^2 R \tag{3.3}$$

Since resistors are positive-valued, *resistors always dissipate power.* But where does a resistor's power go? By Conversation of Power, the dissipated power must be absorbed somewhere. The answer is not directly predicted by circuit theory, but is by physics. Current flowing through a resistor makes it hot; its power is dissipated by heat.

> NOTE: A physical wire has a resistance and hence dissipates power (it gets warm just like a resistor in a circuit). In fact, the resistance of a wire of length L and cross-sectional area A is given by
>
> $$R = \frac{\rho L}{A}$$
>
> The quantity ρ is known as the **resistivity** and presents the resistance of a unit-length of material constituting the wire. Most materials have a positive value for ρ, which means the longer the wire, the greater the resistance and thus the power dissipated. The thicker the wire, the smaller the resistance. Superconductors have no resistance and hence do not dissipate power. If a room-temperature superconductor could be found, electric power could be sent through power lines without loss!

[11]This content is available online at <http://cnx.org/content/m17305/1.5/>.

Exercise 3.4 *(Solution on p. 106.)*

Calculate the power consumed/created by the resistor R_1 in our simple circuit example.

We conclude that both resistors in our example circuit consume power, which points to the voltage source as the producer of power. The current flowing *into* the source's positive terminal is $-i_{\text{out}}$. Consequently, the power calculation for the source yields

$$-(v_{\text{in}} i_{\text{out}}) = -\left(\frac{1}{R_1 + R_2} {v_{\text{in}}}^2\right)$$

We conclude that the source provides the power consumed by the resistors, no more, no less.

Exercise 3.5 *(Solution on p. 106.)*

Confirm that the source produces *exactly* the total power consumed by both resistors.

This result is quite general: sources produce power and the circuit elements, especially resistors, consume it. But where do sources get their power? Again, circuit theory does not model how sources are constructed, but the theory decrees that *all* sources must be provided energy to work.

3.6 Series and Parallel Circuits[12]

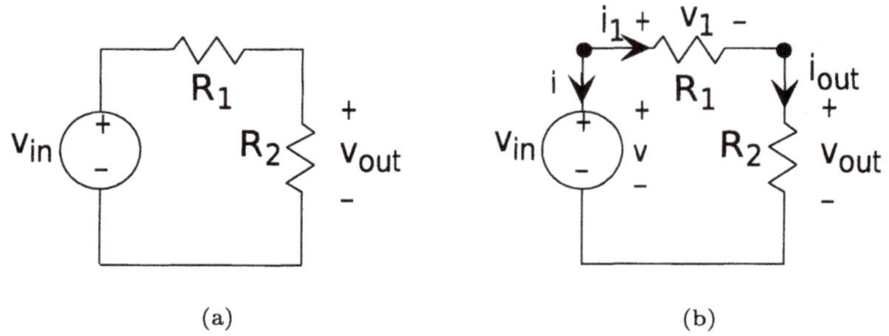

Figure 3.9: The circuit shown is perhaps the simplest circuit that performs a signal processing function. The input is provided by the voltage source v_{in} and the output is the voltage v_{out} across the resistor labelled R_2.

The results shown in other modules (circuit elements (Section 3.4), KVL and KCL (Section 3.4), interconnection laws (Section 3.4)) with regard to this circuit (Figure 3.9), and the values of other currents and voltages in this circuit as well, have profound implications.

Resistors connected in such a way that current from one must flow *only* into another—currents in all resistors connected this way have the same magnitude—are said to be connected in **series**. For the two series-connected resistors in the example, *the voltage across one resistor equals the ratio of that resistor's value and the sum of resistances times the voltage across the series combination*. This concept is so pervasive it has a name: **voltage divider**.

The **input-output relationship** for this system, found in this particular case by voltage divider, takes the form of a ratio of the output voltage to the input voltage.

$$\frac{v_{out}}{v_{in}} = \frac{R_2}{R_1 + R_2}$$

[12]This content is available online at <http://cnx.org/content/m10674/2.7/>.

In this way, we express how the components used to build the system affect the input-output relationship. Because this analysis was made with ideal circuit elements, we might expect this relation to break down if the input amplitude is too high (Will the circuit survive if the input changes from 1 volt to one million volts?) or if the source's frequency becomes too high. In any case, this important way of expressing input-output relationships—as a ratio of output to input—pervades circuit and system theory.

The current i_1 is the current flowing out of the voltage source. Because it equals i_2, we have that $\frac{v_{in}}{i_1} = R_1 + R_2$:

> RESISTORS IN SERIES: The series combination of two resistors acts, as far as the voltage source is concerned, as a single resistor having a value equal to the sum of the two resistances.

This result is the first of several equivalent circuit ideas: In many cases, a complicated circuit when viewed from its terminals (the two places to which you might attach a source) appears to be a single circuit element (at best) or a simple combination of elements at worst. Thus, the equivalent circuit for a series combination of resistors is a single resistor having a resistance equal to the sum of its component resistances.

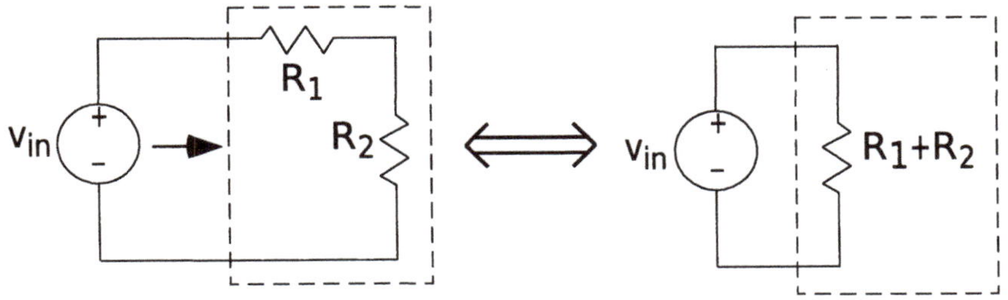

Figure 3.10: The resistor (on the right) is equivalent to the two resistors (on the left) and has a resistance equal to the sum of the resistances of the other two resistors.

Thus, the circuit the voltage source "feels" (through the current drawn from it) is a single resistor having resistance $R_1 + R_2$. Note that in making this equivalent circuit, the output voltage can no longer be defined: The output resistor labeled R_2 no longer appears. Thus, this equivalence is made strictly from the voltage source's viewpoint.

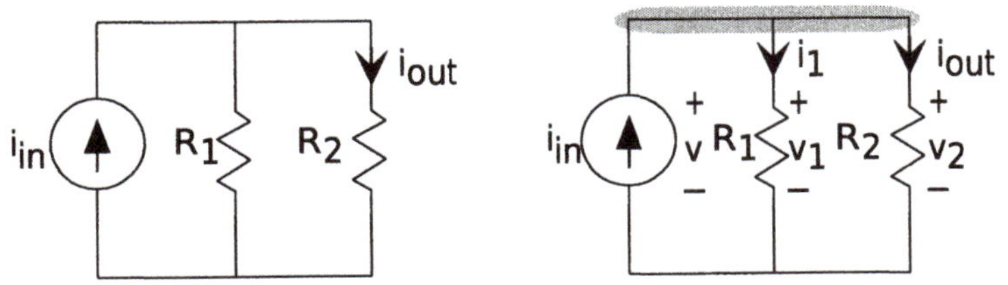

Figure 3.11: A simple parallel circuit.

One interesting simple circuit (Figure 3.11) has two resistors connected side-by-side, what we will term a **parallel** connection, rather than in series. Here, applying KVL reveals that all the voltages are identical:

$v_1 = v$ and $v_2 = v$. This result typifies parallel connections. To write the KCL equation, note that the top node consists of the entire upper interconnection section. The KCL equation is $i_{in} - i_1 - i_2 = 0$. Using the *v-i* relations, we find that

$$i_{out} = \frac{R_1}{R_1 + R_2} i_{in}$$

Exercise 3.6 ***(Solution on p. 106.)***

Suppose that you replaced the current source in Figure 3.11 by a voltage source. How would i_{out} be related to the source voltage? Based on this result, what purpose does this revised circuit have?

This circuit highlights some important properties of parallel circuits. You can easily show that the parallel combination of R_1 and R_2 has the *v-i* relation of a resistor having resistance $\left(\frac{1}{R_1} + \frac{1}{R_2}\right)^{-1} = \frac{R_1 R_2}{R_1 + R_2}$. A shorthand notation for this quantity is $(R_1 \parallel R_2)$. As the reciprocal of resistance is conductance (Section 3.2.1: Resistor), we can say that *for a parallel combination of resistors, the equivalent conductance is the sum of the conductances* .

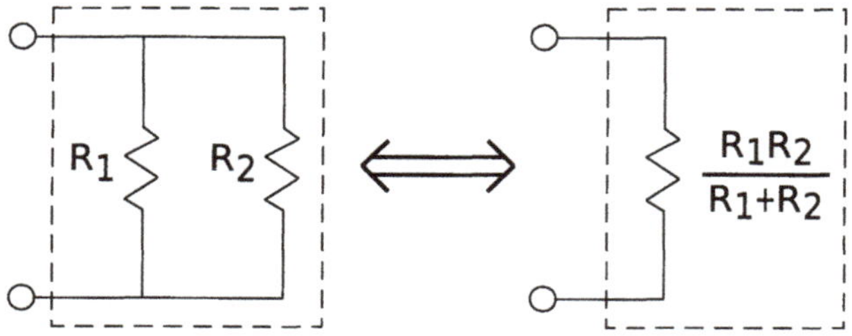

Figure 3.12

Similar to voltage divider (p. 44) for series resistances, we have **current divider** for parallel resistances. The current through a resistor in parallel with another is the ratio of the conductance of the first to the sum of the conductances. Thus, for the depicted circuit, $i_2 = \frac{G_2}{G_1 + G_2} i$. Expressed in terms of resistances, current divider takes the form of the resistance of the *other* resistor divided by the sum of resistances: $i_2 = \frac{R_1}{R_1 + R_2} i$.

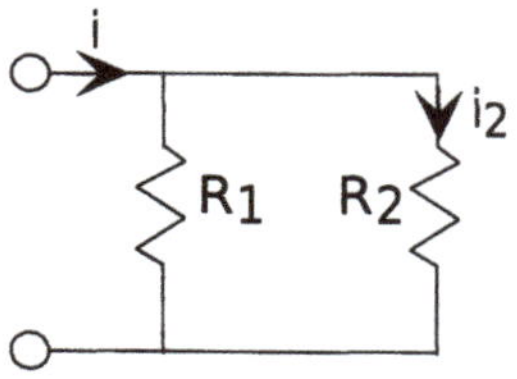

Figure 3.13

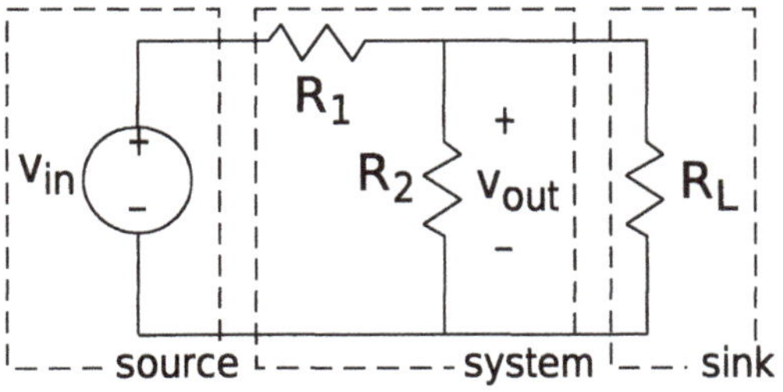

Figure 3.14: The simple attenuator circuit (Figure 3.9) is attached to an oscilloscope's input. The input-output relation for the above circuit without a load is: $v_{out} = \frac{R_2}{R_1+R_2} v_{in}$.

Suppose we want to pass the output signal into a voltage measurement device, such as an oscilloscope or a voltmeter. In system-theory terms, we want to pass our circuit's output to a sink. For most applications, we can represent these measurement devices as a resistor, with the current passing through it driving the measurement device through some type of display. In circuits, a sink is called a **load**; thus, we describe a system-theoretic sink as a load resistance R_L. Thus, we have a complete system built from a cascade of three systems: a source, a signal processing system (simple as it is), and a sink.

We must analyze afresh how this revised circuit, shown in Figure 3.14, works. Rather than defining eight variables and solving for the current in the load resistor, let's take a hint from other analysis (series rules (p. 44), parallel rules (p. 46)). Resistors R_2 and R_L are in a **parallel** configuration: The voltages across each resistor are the same while the currents are not. Because the voltages are the same, we can find the current through each from their *v-i* relations: $i_2 = \frac{v_{out}}{R_2}$ and $i_L = \frac{v_{out}}{R_L}$. Considering the node where all three resistors join, KCL says that the sum of the three currents must equal zero. Said another way, the current entering the node through R_1 must equal the sum of the other two currents leaving the node. Therefore, $i_1 = i_2 + i_L$, which means that $i_1 = v_{out}\left(\frac{1}{R_2} + \frac{1}{R_1}\right)$.

Let R_{eq} denote the equivalent resistance of the parallel combination of R_2 and R_L. Using R_1's *v-i* relation, the voltage across it is $v_1 = \frac{R_1 v_{out}}{R_{eq}}$. The KVL equation written around the leftmost loop has $v_{in} = v_1 + v_{out}$; substituting for v_1, we find

$$v_{in} = v_{out}\left(\frac{R_1}{R_{eq}} + 1\right)$$

or

$$\frac{v_{out}}{v_{in}} = \frac{R_{eq}}{R_1 + R_{eq}}$$

Thus, we have the input-output relationship for our entire system having the form of voltage divider, but it does *not* equal the input-output relation of the circuit without the voltage measurement device. We can not measure voltages reliably unless the measurement device has little effect on what we are trying to measure. We should look more carefully to determine if any values for the load resistance would lessen its impact on the circuit. Comparing the input-output relations before and after, what we need is $R_{eq} \approx R_2$. As $R_{eq} = \left(\frac{1}{R_2} + \frac{1}{R_L}\right)^{-1}$, the approximation would apply if $\left(\frac{1}{R_2} \gg \frac{1}{R_L}\right)$ or $(R_2 \ll R_L)$. This is the condition we seek:

VOLTAGE MEASUREMENT: Voltage measurement devices must have large resistances compared with that of the resistor across which the voltage is to be measured.

Exercise 3.7 ***(Solution on p. 106.)***

Let's be more precise: How much larger would a load resistance need to be to affect the input-output relation by less than 10%? by less than 1%?

Example 3.1

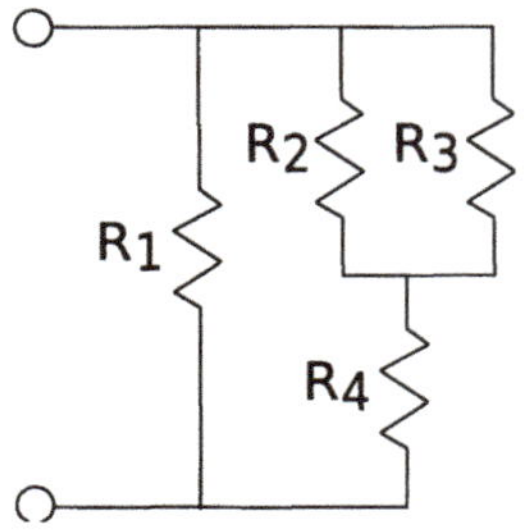

Figure 3.15

We want to find the total resistance of the example circuit. To apply the series and parallel combination rules, it is best to first determine the circuit's structure: What is in series with what and what is in parallel with what at both small- and large-scale views. We have R_2 in parallel with R_3; this combination is in series with R_4. This series combination is in parallel with R_1. Note that in determining this structure, we started *away* from the terminals, and worked toward them. In most cases, this approach works well; try it first. The total resistance expression mimics the structure:

$$R_T = R_1 \parallel (R_2 \parallel R_3 + R_4)$$

$$R_T = \frac{R_1R_2R_3 + R_1R_2R_4 + R_1R_3R_4}{R_1R_2 + R_1R_3 + R_2R_3 + R_2R_4 + R_3R_4}$$

Such complicated expressions typify circuit "simplifications." A simple check for accuracy is the units: Each component of the numerator should have the same units (here Ω^3) as well as in the denominator (Ω^2). The entire expression is to have units of resistance; thus, the ratio of the numerator's and denominator's units should be ohms. Checking units does not guarantee accuracy, but can catch many errors.

Another valuable lesson emerges from this example concerning the difference between cascading systems and cascading circuits. In system theory, systems can be cascaded without changing the input-output relation of intermediate systems. In cascading circuits, this ideal is rarely true *unless* the circuits are so *designed.* Design is in the hands of the engineer; he or she must recognize what have come to be known as loading effects. In our simple circuit, you might think that making the resistance R_L large enough would do the trick. Because the resistors R_1 and R_2 can have virtually any value, you can never make the resistance of your voltage measurement device big enough. Said another way, *a circuit cannot be designed in isolation that will work in cascade with all other circuits.* Electrical engineers deal with this situation through the notion of *specifications*: Under what conditions will the circuit perform as designed? Thus, you will find that oscilloscopes and voltmeters have their internal resistances clearly stated, enabling you to determine whether the voltage you measure closely equals what was present before they were attached to your circuit. Furthermore, since our resistor circuit functions as an attenuator, with the attenuation (a fancy word for gains less than one) depending only on the ratio of the two resistor values $\frac{R_2}{R_1+R_2} = \left(1 + \frac{R_1}{R_2}\right)^{-1}$, we can select *any* values for the two resistances we want to achieve the desired attenuation. The designer of this circuit must thus specify not only what the attenuation is, but also the resistance values employed so that integrators—people who put systems together from component systems—can combine systems together and have a chance of the combination working.

Figure 3.16 (series and parallel combination rules) summarizes the series and parallel combination results. These results are easy to remember and very useful. Keep in mind that for series combinations, voltage and resistance are the key quantities, while for parallel combinations current and conductance are more important. In series combinations, the currents through each element are the same; in parallel ones, the voltages are the same.

---

series and parallel combination rules

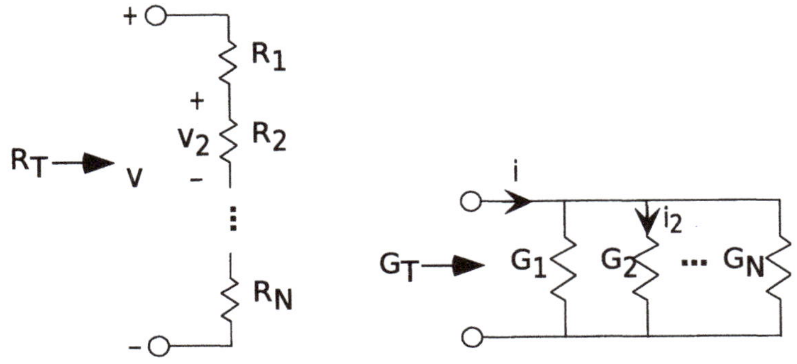

(a) series combination rule (b) parallel combination rule

Figure 3.16: Series and parallel combination rules. (a) $R_T = \sum_{n=1}^{N} (R_n)$ $v_n = \frac{R_n}{R_T} v$ (b) $G_T = \sum_{n=1}^{N} G_n$ $i_n = \frac{G_n}{G_T} i$

Exercise 3.8 *(Solution on p. 106.)*

Contrast a series combination of resistors with a parallel one. Which variable (voltage or current) is the same for each and which differs? What are the equivalent resistances? When resistors are placed in series, is the equivalent resistance bigger, in between, or smaller than the component resistances? What is this relationship for a parallel combination?

3.7 Equivalent Circuits: Resistors and Sources[13]

We have found that the way to think about circuits is to locate and group parallel and series resistor combinations. Those resistors not involved with variables of interest can be collapsed into a single resistance. This result is known as an **equivalent circuit**: from the viewpoint of a pair of terminals, a group of resistors functions as a single resistor, the resistance of which can usually be found by applying the parallel and series rules.

This result generalizes to include sources in a very interesting and useful way. Let's consider our simple attenuator circuit (shown in the figure (Figure 3.17)) from the viewpoint of the output terminals. We want to find the *v-i* relation for the output terminal pair, and then find the equivalent circuit for the boxed circuit. To perform this calculation, use the circuit laws and element relations, but do not attach anything to the output terminals. We seek the relation between v and i that describes the kind of element that lurks within the dashed box. The result is

$$v = (R_1 \parallel R_2)\, i + \frac{R_2}{R_1 + R_2} v_{in} \tag{3.4}$$

[13]This content is available online at <http://cnx.org/content/m0020/2.22/>.

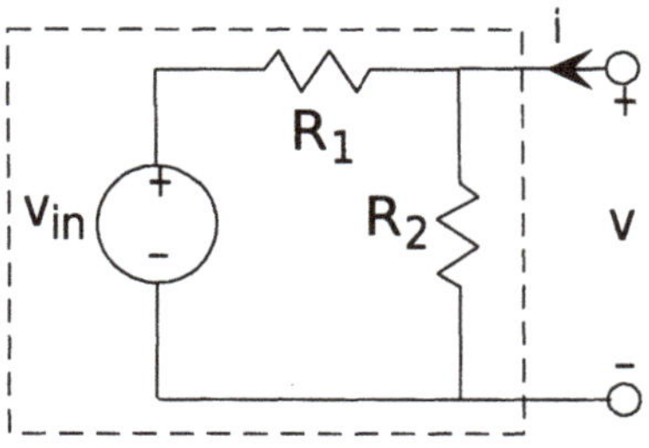

Figure 3.17

If the source were zero, it could be replaced by a short circuit, which would confirm that the circuit does indeed function as a parallel combination of resistors. However, the source's presence means that the circuit is *not* well modeled as a resistor.

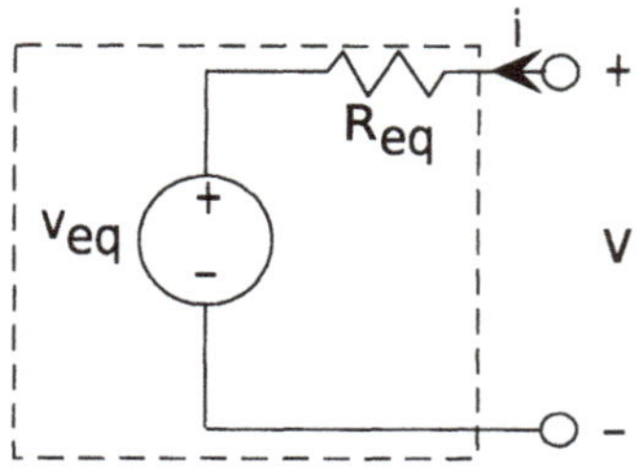

Figure 3.18: The Thévenin equivalent circuit.

If we consider the simple circuit of Figure 3.18, we find it has the *v*-*i* relation at its terminals of

$$v = R_{eq}i + v_{eq} \tag{3.5}$$

Comparing the two *v*-*i* relations, we find that they have the same form. In this case the *Thévenin equivalent resistance* is $R_{eq} = (R_1 \parallel R_2)$ and the *Thévenin equivalent source* has voltage $v_{eq} = \frac{R_2}{R_1+R_2}v_{in}$. Thus, from viewpoint of the terminals, you cannot distinguish the two circuits. Because the equivalent circuit has fewer elements, it is easier to analyze and understand than any other alternative.

For *any* circuit containing resistors and sources, the *v*-*i* relation will be of the form

$$v = R_{eq}i + v_{eq} \tag{3.6}$$

and the **Thévenin equivalent circuit** for any such circuit is that of Figure 3.18. This equivalence applies no matter how many sources or resistors may be present in the circuit. In the example (Example 3.2) below, we know the circuit's construction and element values, and derive the equivalent source and resistance. Because Thévenin's theorem applies in general, we should be able to make measurements or calculations *only from the terminals* to determine the equivalent circuit.

To be more specific, consider the equivalent circuit of this figure (Figure 3.18). Let the terminals be open-circuited, which has the effect of setting the current i to zero. Because no current flows through the resistor, the voltage across it is zero (remember, Ohm's Law says that $v = Ri$). Consequently, by applying KVL we have that the so-called open-circuit voltage v_{oc} equals the Thévenin equivalent voltage. Now consider the situation when we set the terminal voltage to zero (short-circuit it) and measure the resulting current. Referring to the equivalent circuit, the source voltage now appears entirely across the resistor, leaving the short-circuit current to be $i_{sc} = -\left(\frac{v_{eq}}{R_{eq}}\right)$. From this property, we can determine the equivalent resistance.

$$v_{eq} = v_{oc} \tag{3.7}$$

$$R_{eq} = -\left(\frac{v_{oc}}{i_{sc}}\right) \tag{3.8}$$

Exercise 3.9 *(Solution on p. 106.)*

Use the open/short-circuit approach to derive the Thévenin equivalent of the circuit shown in Figure 3.19.

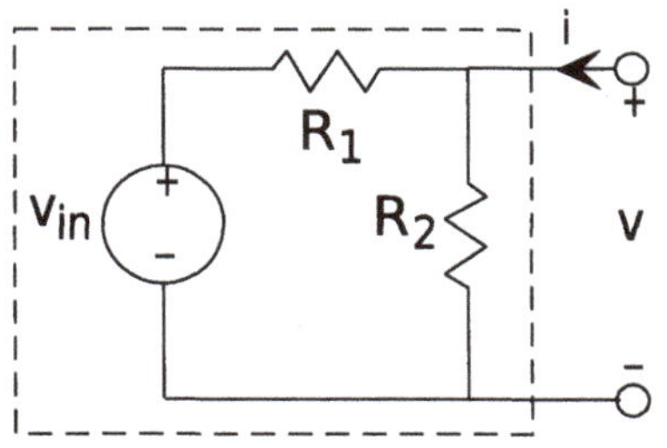

Figure 3.19

Example 3.2

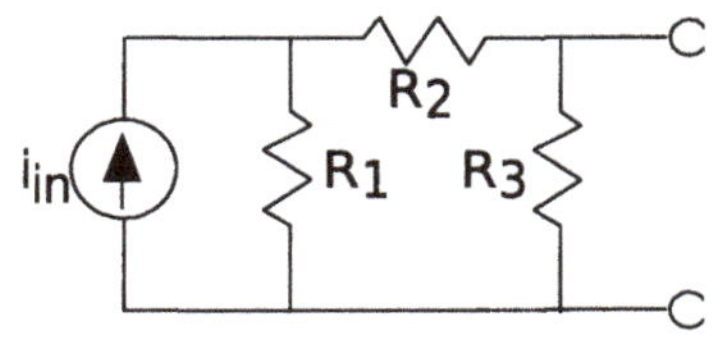

Figure 3.20

For the circuit depicted in Figure 3.20, let's derive its Thévenin equivalent two different ways. Starting with the open/short-circuit approach, let's first find the open-circuit voltage v_{oc}. We have a current divider relationship as R_1 is in parallel with the series combination of R_2 and R_3. Thus, $v_{oc} = \frac{i_{in} R_3 R_1}{R_1 + R_2 + R_3}$. When we short-circuit the terminals, no voltage appears across R_3, and thus no current flows through it. In short, R_3 does not affect the short-circuit current, and can be eliminated. We again have a current divider relationship: $i_{sc} = -\left(\frac{i_{in} R_1}{R_1 + R_2}\right)$. Thus, the Thévenin equivalent resistance is $\frac{R_3 (R_1 + R_2)}{R_1 + R_2 + R_3}$.

To verify, let's find the equivalent resistance by reaching inside the circuit and setting the current source to zero. Because the current is now zero, we can replace the current source by an open circuit. From the viewpoint of the terminals, resistor R_3 is now in parallel with the series combination of R_1 and R_2. Thus, $R_{eq} = (R_3 \parallel R_1 + R_2)$, and we obtain the same result.

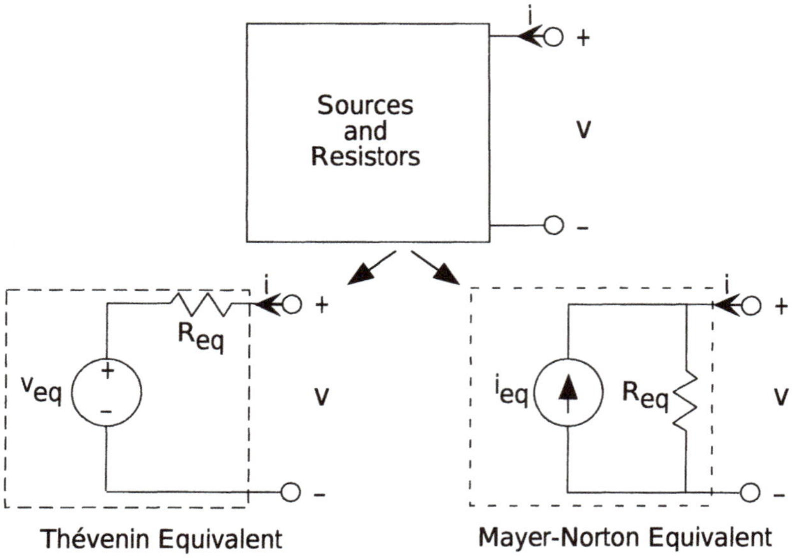

Figure 3.21: *All* circuits containing sources and resistors can be described by simpler equivalent circuits. Choosing the one to use depends on the application, not on what is actually inside the circuit.

As you might expect, equivalent circuits come in two forms: the voltage-source oriented Thévenin equivalent[14] and the current-source oriented **Mayer-Norton equivalent** (Figure 3.21). To derive the latter, the *v-i* relation for the Thévenin equivalent can be written as

$$v = R_{eq}i + v_{eq} \tag{3.9}$$

or

$$i = \frac{v}{R_{eq}} - i_{eq} \tag{3.10}$$

where $i_{eq} = \frac{v_{eq}}{R_{eq}}$ is the Mayer-Norton equivalent source. The Mayer-Norton equivalent shown in Figure 3.21 be easily shown to have this *v-i* relation. Note that both variations have the same equivalent resistance. The short-circuit current equals the negative of the Mayer-Norton equivalent source.

Exercise 3.10 ***(Solution on p. 106.)***

Find the Mayer-Norton equivalent circuit for the circuit below.

[14] "Finding Thévenin Equivalent Circuits" <http://cnx.org/content/m0021/latest/>

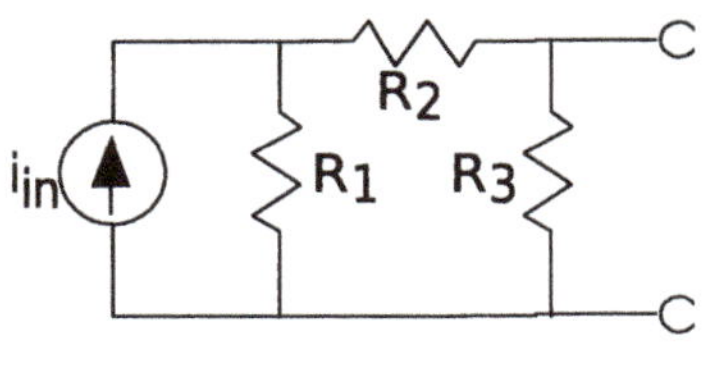

Figure 3.22

Equivalent circuits can be used in two basic ways. The first is to simplify the analysis of a complicated circuit by realizing the *any* portion of a circuit can be described by either a Thévenin or Mayer-Norton equivalent. Which one is used depends on whether what is attached to the terminals is a series configuration (making the Thévenin equivalent the best) or a parallel one (making Mayer-Norton the best).

Another application is modeling. When we buy a flashlight battery, either equivalent circuit can accurately describe it. These models help us understand the limitations of a battery. Since batteries are labeled with a voltage specification, they should serve as voltage sources and the Thévenin equivalent serves as the natural choice. If a load resistance R_L is placed across its terminals, the voltage output can be found using voltage divider: $v = \frac{v_{eq} R_L}{R_L + R_{eq}}$. If we have a load resistance much larger than the battery's equivalent resistance, then, to a good approximation, the battery does serve as a voltage source. If the load resistance is much smaller, we certainly don't have a voltage source (the output voltage depends directly on the load resistance). Consider now the Mayer-Norton equivalent; the current through the load resistance is given by current divider, and equals $i = -\left(\frac{i_{eq} R_{eq}}{R_L + R_{eq}}\right)$. For a current that does not vary with the load resistance, this resistance should be much smaller than the equivalent resistance. If the load resistance is comparable to the equivalent resistance, the battery serves *neither* as a voltage source or a current course. Thus, when you buy a battery, you get a voltage source if its equivalent resistance is much *smaller* than the equivalent resistance of the circuit to which you attach it. On the other hand, if you attach it to a circuit having a small equivalent resistance, you bought a current source.

LÉON CHARLES THÉVENIN: He was an engineer with France's Postes, Télégraphe et Téléphone. In 1883, he published (twice!) a proof of what is now called the Thévenin equivalent while developing ways of teaching electrical engineering concepts at the École Polytechnique. He did not realize that the same result had been published by Hermann Helmholtz[15] , the renowned nineteenth century physicist, thiry years earlier.

HANS FERDINAND MAYER: After earning his doctorate in physics in 1920, he turned to communications engineering when he joined Siemens & Halske in 1922. In 1926, he published in a German technical journal the Mayer-Norton equivalent. During his interesting career, he rose to lead Siemen's Central Laboratory in 1936, surruptiously leaked to the British all he knew of German warfare capabilities a month after the Nazis invaded Poland, was arrested by the Gestapo in 1943 for listening to BBC radio broadcasts, spent two years in Nazi concentration camps, and went to the United States for four years working for the Air Force and Cornell University before returning to Siemens in 1950. He rose to a position on Siemen's Board of Directors before retiring.

EDWARD L. NORTON: Edward Norton[16] was an electrical engineer who worked at Bell Laboratory from its inception in 1922. In the *same* month when Mayer's paper appeared, Norton wrote in an internal technical memorandum a paragraph describing the current-source equivalent. No evidence suggests Norton knew of Mayer's publication.

[15]http://www-gap.dcs.st-and.ac.uk/~history/Mathematicians/Helmholtz.html
[16]http://www.ece.rice.edu/~dhj/norton

3.8 Circuits with Capacitors and Inductors[17]

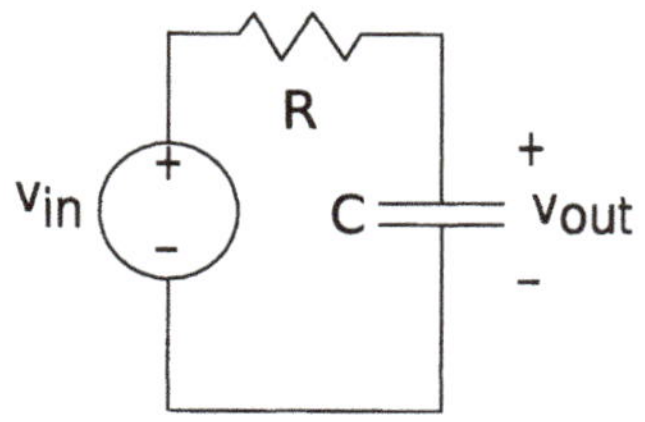

Figure 3.23: A simple *RC* circuit.

Let's consider a circuit having something other than resistors and sources. Because of KVL, we know that $v_{in} = v_R + v_{out}$. The current through the capacitor is given by $i = C\frac{d}{dt}(v_{out})$, and this current equals that passing through the resistor. Substituting $v_R = Ri$ into the KVL equation and using the *v-i* relation for the capacitor, we arrive at

$$RC\frac{d}{dt}(v_{out}) + v_{out} = v_{in} \tag{3.11}$$

The input-output relation for circuits involving energy storage elements takes the form of an ordinary differential equation, which we must solve to determine what the output voltage is for a given input. In contrast to resistive circuits, where we obtain an *explicit* input-output relation, we now have an *implicit* relation that requires more work to obtain answers.

At this point, we could learn how to solve differential equations. Note first that even finding the differential equation relating an output variable to a source is often very tedious. The parallel and series combination rules that apply to resistors don't directly apply when capacitors and inductors occur. We would have to slog our way through the circuit equations, simplifying them until we finally found the equation that related the source(s) to the output. At the turn of the twentieth century, a method was discovered that not only made finding the differential equation easy, but also simplified the solution process in the most common situation. Although not original with him, Charles Steinmetz[18] presented the key paper describing the **impedance** approach in 1893. It allows circuits containing capacitors and inductors to be solved with the *same* methods we have learned to solved resistor circuits. To use impedances, we must master **complex numbers**. Though the arithmetic of complex numbers is mathematically more complicated than with real numbers, the increased insight into circuit behavior and the ease with which circuits are solved with impedances is well worth the diversion. But more importantly, the impedance concept is central to engineering and physics, having a reach far beyond just circuits.

3.9 The Impedance Concept[19]

Rather than solving the differential equation that arises in circuits containing capacitors and inductors, let's pretend that all sources in the circuit are complex exponentials having the *same* frequency. Although this pretense can only be mathematically true, this fiction will greatly ease solving the circuit no matter what the source really is.

[17]This content is available online at <http://cnx.org/content/m0023/2.11/>.
[18]http://www.invent.org/hall_of_fame/139.html
[19]This content is available online at <http://cnx.org/content/m0024/2.22/>.

Simple Circuit

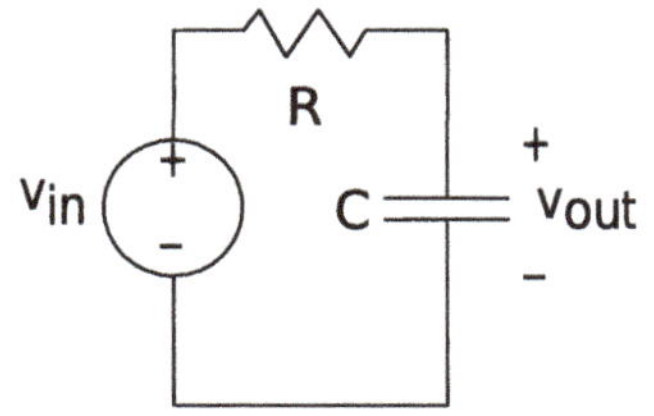

Figure 3.24: A simple *RC* circuit.

Impedance

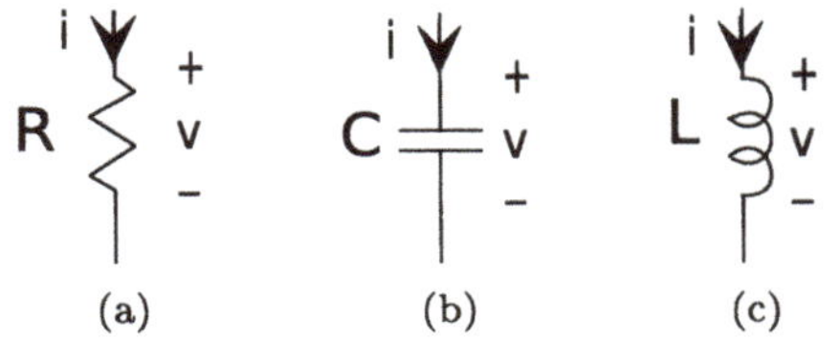

Figure 3.25: (a) Resistor: $Z_R = R$ (b) Capacitor: $Z_C = \frac{1}{j2\pi fC}$ (c) Inductor: $Z_L = j2\pi fL$

For the above example *RC* circuit (Figure 3.24 (Simple Circuit)), let $v_{in} = V_{in}e^{j2\pi ft}$. The complex amplitude V_{in} determines the size of the source and its phase. The critical consequence of assuming that sources have this form is that *all* voltages and currents in the circuit are also complex exponentials, having amplitudes governed by KVL, KCL, and the *v-i* relations and the same frequency as the source. To appreciate why this should be true, let's investigate how each circuit element behaves when either the voltage or current is a complex exponential. For the resistor, $v = Ri$. When $v = Ve^{j2\pi ft}$; then $i = \frac{V}{R}e^{j2\pi ft}$. Thus, if the resistor's voltage is a complex exponential, so is the current, with an amplitude $I = \frac{V}{R}$ (determined by the resistor's *v-i* relation) and a frequency the same as the voltage. Clearly, if the current were assumed to be a complex exponential, so would the voltage. For a capacitor, $i = C\frac{d}{dt}(v)$. Letting the voltage be a complex exponential, we have $i = CVj2\pi fe^{j2\pi ft}$. The amplitude of this complex exponential is $I = CVj2\pi f$. Finally, for the inductor, where $v = L\frac{d}{dt}(i)$, assuming the current to be a complex exponential results in the voltage having the form $v = LIj2\pi fe^{j2\pi ft}$, making its complex amplitude $V = LIj2\pi f$.

The major consequence of assuming complex exponential voltage and currents is that the ratio $Z = \frac{V}{I}$ *for each element does not depend on time, but does depend on source frequency.* This quantity is known as the element's **impedance**.

The impedance is, in general, a complex-valued, frequency-dependent quantity. For example, the magnitude of the capacitor's impedance is inversely related to frequency, and has a phase of $-\left(\frac{\pi}{2}\right)$. This observation means that if the current is a complex exponential and has constant amplitude, the amplitude of the voltage decreases with frequency.

Let's consider Kirchoff's circuit laws. When voltages around a loop are all complex exponentials of the same frequency, we have

$$\begin{aligned} \sum_n (v_n) &= \sum_n \left(V_n e^{j2\pi f t}\right) \\ &= 0 \end{aligned} \quad (3.12)$$

which means

$$\sum_n (V_n) = 0 \quad (3.13)$$

the complex amplitudes of the voltages obey KVL. We can easily imagine that the complex amplitudes of the currents obey KCL.

What we have discovered is that source(s) equaling a complex exponential of the same frequency forces all circuit variables to be complex exponentials of the same frequency. Consequently, the ratio of voltage to current for each element equals the ratio of their complex amplitudes, which depends only on the source's frequency and element values.

This situation occurs because the circuit elements are linear and time-invariant. For example, suppose we had a circuit element where the voltage equaled the square of the current: $v(t) = Ki^2(t)$. If $i(t) = Ie^{j2\pi f t}$, $v(t) = KI^2 e^{j2\pi 2 f t}$, meaning that voltage and current no longer had the same frequency and that their ratio was time-dependent.

Because for linear circuit elements the complex amplitude of voltage is proportional to the complex amplitude of current— $V = ZI$ — assuming complex exponential sources means circuit elements behave as if they were resistors, where instead of resistance, we use impedance. *Because complex amplitudes for voltage and current also obey Kirchoff's laws, we can solve circuits using voltage and current divider and the series and parallel combination rules by considering the elements to be impedances.*

3.10 Time and Frequency Domains[20]

When we find the differential equation relating the source and the output, we are faced with solving the circuit in what is known as the **time domain**. What we emphasize here is that it is often easier to find the output if we use impedances. Because impedances depend only on frequency, we find ourselves in the **frequency domain**. A common error in using impedances is keeping the time-dependent part, the complex exponential, in the fray. The entire point of using impedances is to get rid of time and concentrate on frequency. Only after we find the result in the frequency domain do we go back to the time domain and put things back together again.

To illustrate how the time domain, the frequency domain and impedances fit together, consider the time domain and frequency domain to be two work rooms. Since you can't be two places at the same time, you are faced with solving your circuit problem in one of the two rooms at any point in time. Impedances and complex exponentials are the way you get between the two rooms. Security guards make sure you don't try to sneak time domain variables into the frequency domain room and vice versa. Figure 3.26 (Two Rooms) shows how this works.

As we unfold the impedance story, we'll see that the powerful use of impedances suggested by Steinmetz[21] greatly simplifies solving circuits, alleviates us from solving differential equations, and suggests a general way of thinking about circuits. Because of the importance of this approach, let's go over how it works.

1. Even though it's not, pretend the source is a complex exponential. We do this because the impedance approach simplifies finding how input and output are related. If it were a voltage source having voltage $v_{in} = p(t)$ (a pulse), still let $v_{in} = V_{in} e^{j2\pi f t}$. We'll learn how to "get the pulse back" later.
2. With a source equaling a complex exponential, *all* variables in a linear circuit will also be complex exponentials having the *same* frequency. The circuit's only remaining "mystery" is what each variable's complex amplitude might be. To find these, we consider the source to be a complex number (V_{in} here) and the elements to be impedances.

[20]This content is available online at <http://cnx.org/content/m10708/2.6/>.
[21]http://www.invent.org/hall_of_fame/139.html

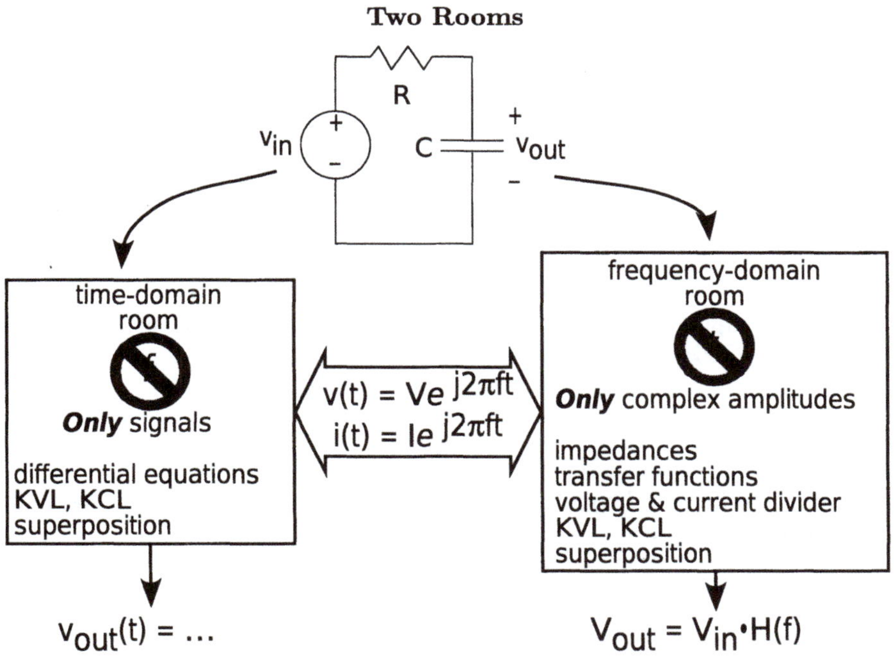

Figure 3.26: The time and frequency domains are linked by assuming signals are complex exponentials. In the time domain, signals can have any form. Passing into the frequency domain 'work room", signals are represented entirely by complex amplitudes.

3. We can now solve using series and parallel combination rules how the complex amplitude of any variable relates to the sources complex amplitude.

Example 3.3

To illustrate the impedance approach, we refer to the *RC* circuit (Figure 3.27 (Simple Circuits)) below, and we assume that $v_{in} = V_{in}e^{j2\pi ft}$.

Using impedances, the complex amplitude of the output voltage V_{out} can be found using voltage divider:

$$V_{out} = \frac{Z_C}{Z_C + Z_R} V_{in}$$

$$V_{out} = \frac{\frac{1}{j2\pi fC}}{\frac{1}{j2\pi fC} + R} V_{in}$$

$$V_{out} = \frac{1}{j2\pi fRC + 1} V_{in}$$

If we refer to the differential equation for this circuit (shown in Circuits with Capacitors and Inductors (Section 3.8) to be $RC\frac{d}{dt}(v_{out}) + v_{out} = v_{in}$), letting the output and input voltages be complex exponentials, we obtain the same relationship between their complex amplitudes. Thus, using impedances is equivalent to using the differential equation and solving it when the source is a complex exponential.

Simple Circuits

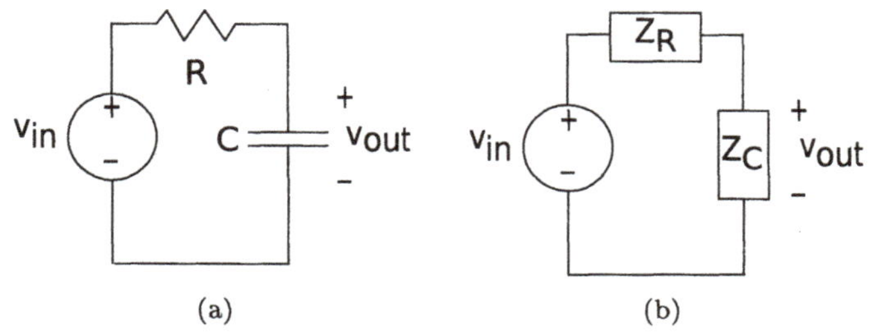

Figure 3.27: (a) A simple *RC* circuit. (b) The impedance counterpart for the *RC* circuit. Note that the source and output voltage are now complex amplitudes.

In fact, we can find the differential equation *directly* using impedances. If we cross-multiply the relation between input and output amplitudes,

$$V_{out}\left(j2\pi fRC+1\right)=V_{in}$$

and then put the complex exponentials back in, we have

$$RCj2\pi fV_{out}e^{j2\pi ft}+V_{out}e^{j2\pi ft}=V_{in}e^{j2\pi ft}$$

In the process of defining impedances, note that the factor $j2\pi f$ arises from the *derivative* of a complex exponential. We can reverse the impedance process, and revert back to the differential equation.

$$RC\frac{d}{dt}\left(v_{out}\right)+v_{out}=v_{in}$$

This is the same equation that was derived much more tediously in Circuits with Capacitors and Inductors (Section 3.8). Finding the differential equation relating output to input is far simpler when we use impedances than with any other technique.

Exercise 3.11 *(Solution on p. 106.)*

Suppose you had an expression where a complex amplitude was divided by $j2\pi f$. What time-domain operation corresponds to this division?

3.11 Power in the Frequency Domain[22]

Recalling that the instantaneous power consumed by a circuit element or an equivalent circuit that represents a collection of elements equals the voltage times the current entering the positive-voltage terminal, $p(t)=v(t)\,i(t)$, what is the equivalent expression using impedances? The resulting calculation reveals more about power consumption in circuits and the introduction of the concept of **average power**.

When all sources produce sinusoids of frequency f, the voltage and current for any circuit element or collection of elements are sinusoids of the same frequency.

$$v(t)=|V|\cos\left(2\pi ft+\phi\right)$$
$$i(t)=|I|\cos\left(2\pi ft+\theta\right)$$

[22]This content is available online at <http://cnx.org/content/m17308/1.2/>.

Here, the complex amplitude of the voltage V equals $|V|e^{j\phi}$ and that of the current is $|I|e^{j\theta}$. We can also write the voltage and current in terms of their complex amplitudes using Euler's formula (Section 2.1.2: Euler's Formula).

$$v(t) = \frac{1}{2}\left(Ve^{j2\pi ft} + V^* e^{-(j2\pi ft)}\right)$$
$$i(t) = \frac{1}{2}\left(Ie^{j2\pi ft} + I^* e^{-(j2\pi ft)}\right)$$

Multiplying these two expressions and simplifying gives

$$\begin{aligned} p(t) &= \frac{1}{4}\left(VI^* + V^*I + VIe^{j4\pi ft} + V^*I^*e^{-(j4\pi ft)}\right) \\ &= \frac{1}{2}\mathrm{Re}\left(VI^*\right) + \frac{1}{2}\mathrm{Re}\left(VIe^{j4\pi ft}\right) \\ &= \frac{1}{2}\mathrm{Re}\left(VI^*\right) + \frac{1}{2}|V||I|\cos\left(4\pi ft + \phi + \theta\right) \end{aligned}$$

We define $\frac{1}{2}VI^*$ to be **complex power**. The real-part of complex power is the first term and since it does not change with time, it represents the power consistently consumed/produced by the circuit. The second term varies with time at a frequency twice that of the source. Conceptually, this term details how power "sloshes" back and forth in the circuit because of the sinusoidal source.

From another viewpoint, the real-part of complex power represents long-term energy consumption/production. Energy is the integral of power and, as the integration interval increases, the first term appreciates while the time-varying term "sloshes." Consequently, the most convenient definition of the **average power** consumed/produced by any circuit is in terms of complex amplitudes.

$$P_{\text{ave}} = \frac{1}{2}\mathrm{Re}\left(VI^*\right) \tag{3.14}$$

Exercise 3.12 *(Solution on p. 107.)*

Suppose the complex amplitudes of the voltage and current have fixed magnitudes. What phase relationship between voltage and current maximizes the average power? In other words, how are ϕ and θ related for maximum power dissipation?

Because the complex amplitudes of the voltage and current are related by the equivalent impedance, average power can also be written as

$$P_{\text{ave}} = \frac{1}{2}\mathrm{Re}\left(Z\right)\left(|I|\right)^2 = \frac{1}{2}\mathrm{Re}\left(\frac{1}{Z}\right)\left(|V|\right)^2$$

These expressions generalize the results (3.3) we obtained for resistor circuits. We have derived a fundamental result: *Only the real part of impedance contributes to long-term power dissipation.* Of the circuit elements, *only* the resistor dissipates power. Capacitors and inductors dissipate no power in the long term. It is important to realize that these statements apply only for sinusoidal sources. If you turn on a constant voltage source in an RC-circuit, charging the capacitor does consume power.

Exercise 3.13 *(Solution on p. 107.)*

In an earlier problem (Section 1.5.1: RMS Values), we found that the rms value of a sinusoid was its amplitude divided by $\sqrt{2}$. What is average power expressed in terms of the rms values of the voltage and current (V_{rms} and I_{rms} respectively)?

3.12 Equivalent Circuits: Impedances and Sources[23]

When we have circuits with capacitors and/or inductors as well as resistors and sources, Thévenin and Mayer-Norton equivalent circuits can still be defined by using impedances and complex amplitudes for voltage and

[23]This content is available online at <http://cnx.org/content/m0030/2.19/>.

currents. For any circuit containing sources, resistors, capacitors, and inductors, the input-output relation for the **complex amplitudes** of the terminal voltage and current is

$$V = Z_{eq}I + V_{eq}$$

$$I = \frac{V}{Z_{eq}} - I_{eq}$$

with $V_{eq} = Z_{eq}I_{eq}$. Thus, we have Thévenin and Mayer-Norton equivalent circuits as shown in Figure 3.28 (Equivalent Circuits).

Example 3.4

Let's find the Thévenin and Mayer-Norton equivalent circuits for Figure 3.29 (Simple RC Circuit). The open-circuit voltage and short-circuit current techniques still work, except we use impedances and complex amplitudes. The open-circuit voltage corresponds to the transfer function we have already found. When we short the terminals, the capacitor no longer has any effect on the circuit, and the short-circuit current I_{sc} equals $\frac{V_{out}}{R}$. The equivalent impedance can be found by setting the source to zero, and finding the impedance using series and parallel combination rules. In our case, the resistor and capacitor are in parallel once the voltage source is removed (setting it to zero amounts to replacing it with a short-circuit). Thus, $Z_{eq} = \left(R \parallel \frac{1}{j2\pi fC}\right) = \frac{R}{1+j2\pi fRC}$. Consequently, we have

$$V_{eq} = \frac{1}{1 + j2\pi fRC} V_{in}$$

$$I_{eq} = \frac{1}{R} V_{in}$$

$$Z_{eq} = \frac{R}{1 + j2\pi fRC}$$

Again, we should check the units of our answer. Note in particular that $j2\pi fRC$ must be dimensionless. Is it?

3.13 Transfer Functions[24]

The ratio of the output and input amplitudes for Figure 3.30 (Simple Circuit), known as the **transfer function** or the **frequency response**, is given by

$$\begin{aligned} \frac{V_{out}}{V_{in}} &= H(f) \\ &= \frac{1}{j2\pi fRC+1} \end{aligned} \quad (3.15)$$

Implicit in using the transfer function is that the input is a complex exponential, and the output is also a complex exponential having the same frequency. The transfer function reveals how the circuit modifies the input amplitude in creating the output amplitude. Thus, the transfer function *completely* describes how the circuit processes the input complex exponential to produce the output complex exponential. The circuit's function is thus summarized by the transfer function. In fact, circuits are often designed to meet transfer function specifications. Because transfer functions are complex-valued, frequency-dependent quantities, we can better appreciate a circuit's function by examining the magnitude and phase of its transfer function (Figure 3.31 (Magnitude and phase of the transfer function)).

[24]This content is available online at <http://cnx.org/content/m0028/2.18/>.

Equivalent Circuits

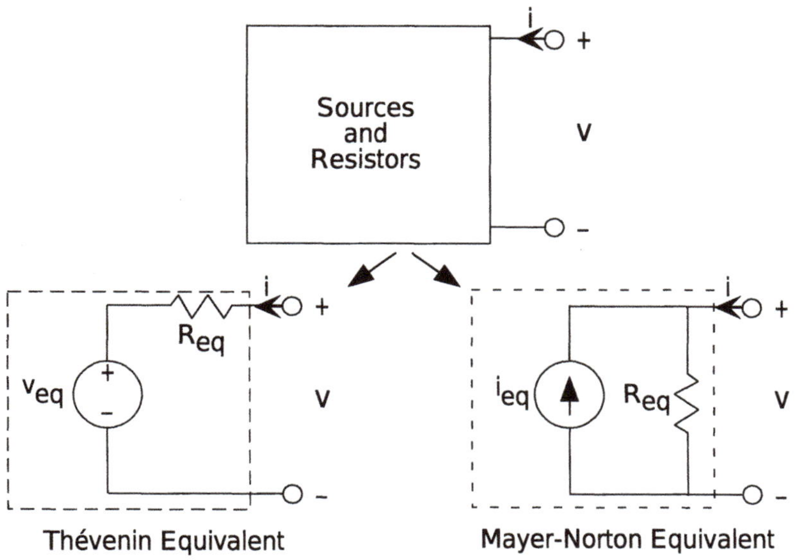

(a) Equivalent circuits with resistors.

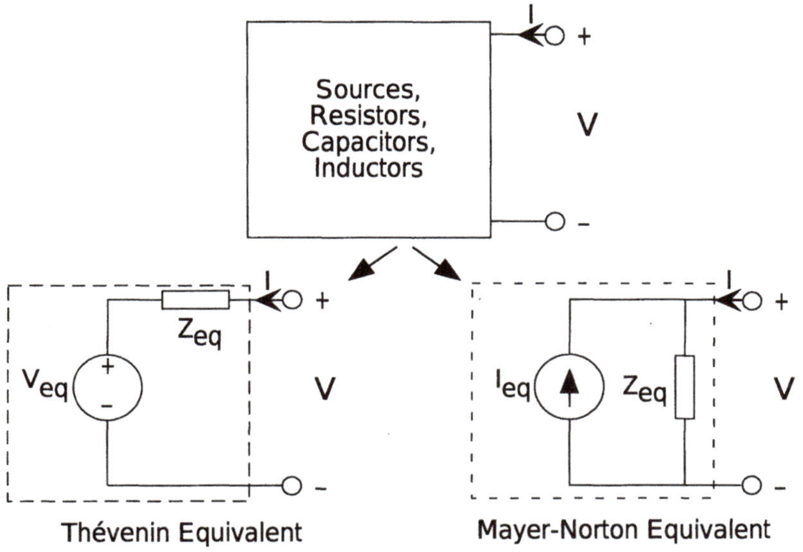

(b) Equivalent circuits with impedances.

Figure 3.28: Comparing the first, simpler, figure with the slightly more complicated second figure, we see two differences. First of all, more circuits (all those containing linear elements in fact) have equivalent circuits that contain equivalents. Secondly, the terminal and source variables are now complex amplitudes, which carries the implicit assumption that the voltages and currents are single complex exponentials, all having the same frequency.

Simple RC Circuit

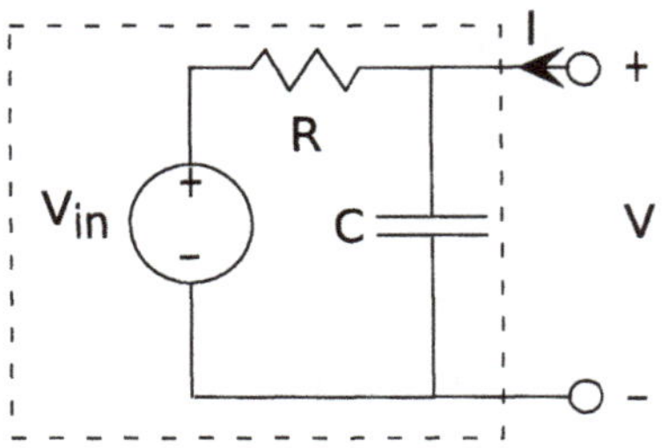

Figure 3.29

Simple Circuit

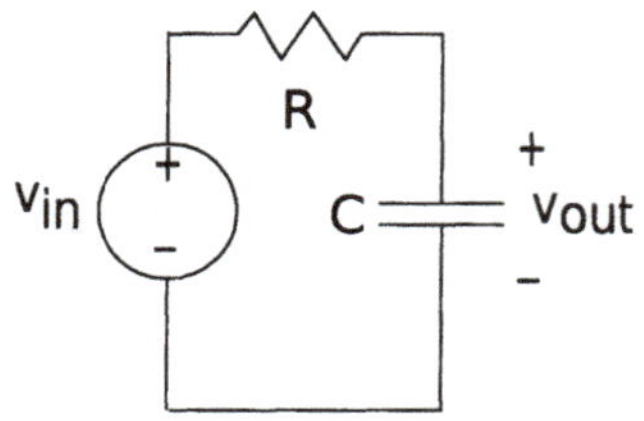

Figure 3.30: A simple *RC* circuit.

Magnitude and phase of the transfer function

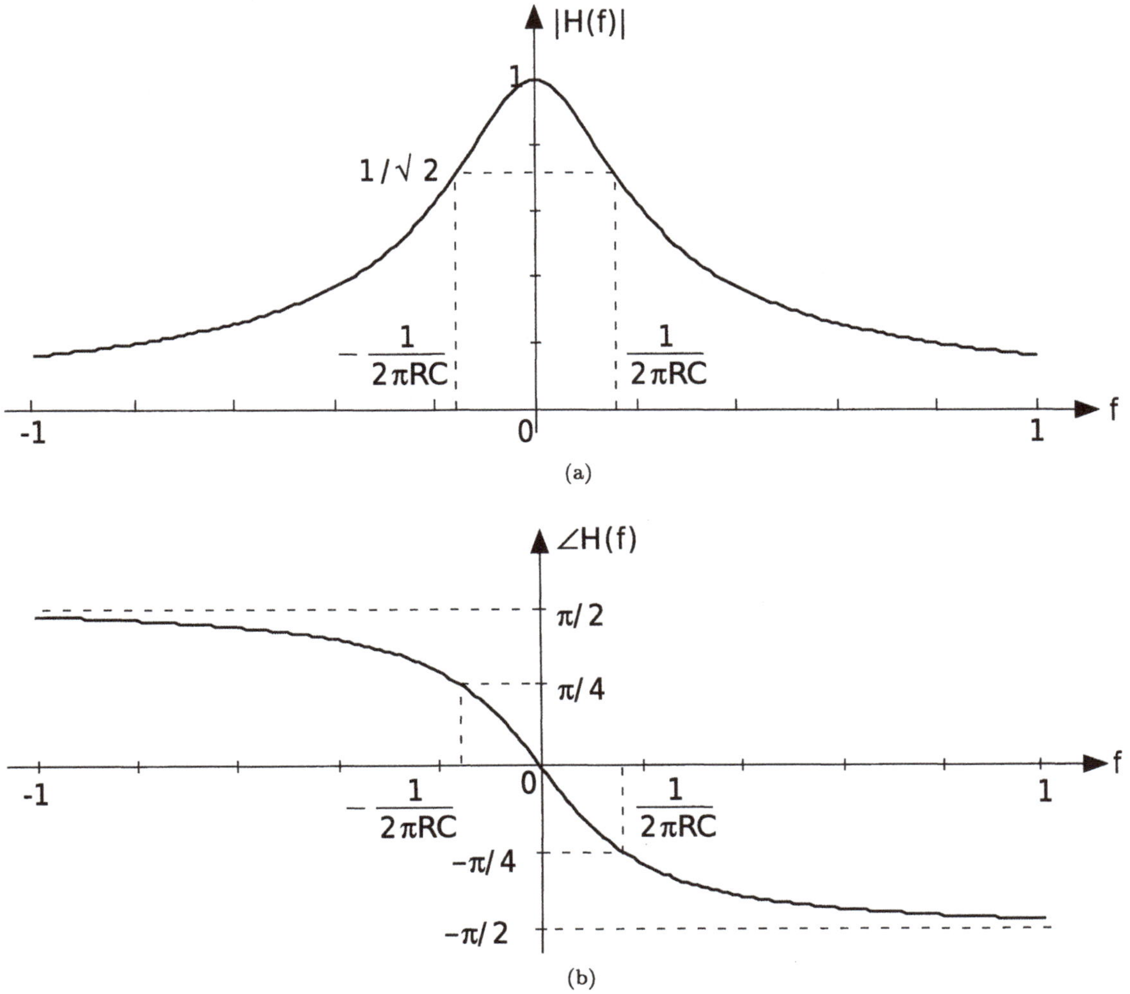

Figure 3.31: Magnitude and phase of the transfer function of the RC circuit shown in Figure 3.30 (Simple Circuit) when $RC = 1$. (a) $|H(f)| = \frac{1}{\sqrt{(2\pi f RC)^2+1}}$ (b) $\angle(H(f)) = -(\arctan(2\pi f RC))$

This transfer function has many important properties and provides }*all* the insights needed to determine how the circuit functions. First of all, note that we can compute the frequency response for both positive and negative frequencies. Recall that sinusoids consist of the sum of two complex exponentials, one having the negative frequency of the other. We will consider how the circuit acts on a sinusoid soon. Do note that the magnitude has *even symmetry*: The negative frequency portion is a mirror image of the positive frequency portion: $|H(-f)| = |H(f)|$. The phase has *odd symmetry*: $\angle(H(-f)) = -(\angle(H(f)))$. These properties of this specific example apply for *all* transfer functions associated with circuits. Consequently, we don't need to plot the negative frequency component; we know what it is from the positive frequency part.

The magnitude equals $\frac{1}{\sqrt{2}}$ of its maximum gain (1 at $f = 0$) when $2\pi f RC = 1$ (the two terms in the denominator of the magnitude are equal). The frequency $f_c = \frac{1}{2\pi RC}$ defines the boundary between two operating ranges.

- For frequencies below this frequency, the circuit does not much alter the amplitude of the complex exponential source.
- For frequencies greater than f_c, the circuit strongly attenuates the amplitude. Thus, when the source frequency is in this range, the circuit's output has a much smaller amplitude than that of the source.

For these reasons, this frequency is known as the **cutoff frequency**. In this circuit the cutoff frequency depends *only* on the product of the resistance and the capacitance. Thus, a cutoff frequency of 1 kHz occurs when $\frac{1}{2\pi RC} = 10^3$ or $RC = \frac{10^{-3}}{2\pi} = 1.59 \times 10^{-4}$. Thus resistance-capacitance combinations of 1.59 kΩ and 100 nF or 10 Ω and 1.59 μF result in the *same* cutoff frequency.

The phase shift caused by the circuit at the cutoff frequency precisely equals $-\left(\frac{\pi}{4}\right)$. Thus, below the cutoff frequency, phase is little affected, but at higher frequencies, the phase shift caused by the circuit becomes $-\left(\frac{\pi}{2}\right)$. This phase shift corresponds to the difference between a cosine and a sine.

We can use the transfer function to find the output when the input voltage is a sinusoid for two reasons. First of all, a sinusoid is the sum of two complex exponentials, each having a frequency equal to the negative of the other. Secondly, because the circuit is linear, superposition applies. If the source is a sine wave, we know that

$$\begin{aligned} v_{in}(t) &= A\sin(2\pi ft) \\ &= \frac{A}{2j}\left(e^{j2\pi ft} - e^{-(j2\pi ft)}\right) \end{aligned} \quad (3.16)$$

Since the input is the sum of two complex exponentials, we know that the output is also a sum of two similar complex exponentials, the only difference being that the complex amplitude of each is multiplied by the transfer function evaluated at each exponential's frequency.

$$v_{out}(t) = \frac{A}{2j}H(f)\,e^{j2\pi ft} - \frac{A}{2j}H(-f)\,e^{-(j2\pi ft)} \quad (3.17)$$

As noted earlier, the transfer function is most conveniently expressed in polar form: $H(f) = |H(f)|e^{j\angle(H(f))}$. Furthermore, $|H(-f)| = |H(f)|$ (even symmetry of the magnitude) and $\angle(H(-f)) = -(\angle(H(f)))$ (odd symmetry of the phase). The output voltage expression simplifies to

$$\begin{aligned} v_{out}(t) &= \frac{A}{2j}|H(f)|e^{j2\pi ft+\angle(H(f))} - \frac{A}{2j}|H(f)|e^{-(j2\pi ft)-\angle(H(f))} \\ &= A|H(f)|\sin(2\pi ft + \angle(H(f))) \end{aligned} \quad (3.18)$$

The circuit's output to a sinusoidal input is also a sinusoid, having a gain equal to the magnitude of the circuit's transfer function evaluated at the source frequency and a phase equal to the phase of the transfer function at the source frequency. It will turn out that this input-output relation description applies to any linear circuit having a sinusoidal source.

Exercise 3.14 *(Solution on p. 107.)*

This input-output property is a special case of a more general result. Show that if the source can be written as the imaginary part of a complex exponential— $v_{in}(t) = \mathrm{Im}\left(Ve^{j2\pi ft}\right)$ — the output is given by $v_{out}(t) = \mathrm{Im}\left(VH(f)\,e^{j2\pi ft}\right)$. Show that a similar result also holds for the real part.

The notion of impedance arises when we assume the sources are complex exponentials. This assumption may seem restrictive; what would we do if the source were a unit step? When we use impedances to find the transfer function between the source and the output variable, we can derive from it the differential equation that relates input and output. The differential equation applies no matter what the source may be. As we have argued, it is far simpler to use impedances to find the differential equation (because we can use series and parallel combination rules) than any other method. In this sense, we have not lost anything by temporarily pretending the source is a complex exponential.

In fact we can also solve the differential equation using impedances! Thus, despite the apparent restrictiveness of impedances, assuming complex exponential sources is actually quite general.

RL circuit

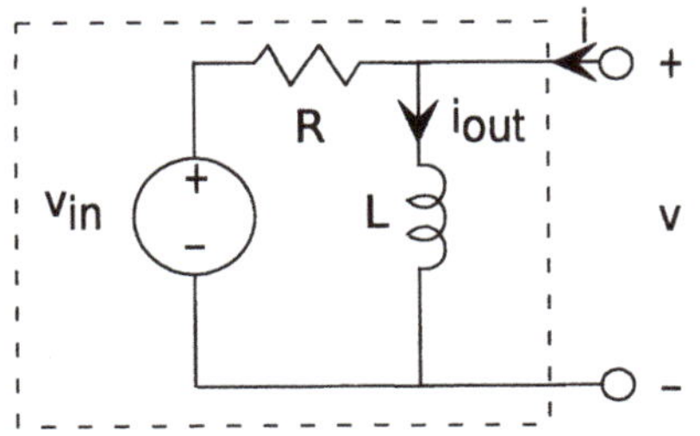

Figure 3.32

3.14 Designing Transfer Functions[25]

If the source consists of two (or more) signals, we know from linear system theory that the output voltage equals the sum of the outputs produced by each signal alone. In short, linear circuits are a special case of linear systems, and therefore superposition applies. In particular, suppose these component signals are complex exponentials, each of which has a frequency different from the others. The transfer function portrays how the circuit affects the amplitude and phase of each component, allowing us to understand how the circuit works on a complicated signal. Those components having a frequency less than the cutoff frequency pass through the circuit with little modification while those having higher frequencies are suppressed. The circuit is said to act as a **filter**, filtering the source signal based on the frequency of each component complex exponential. Because low frequencies pass through the filter, we call it a **lowpass filter** to express more precisely its function.

We have also found the ease of calculating the output for sinusoidal inputs through the use of the transfer function. Once we find the transfer function, we can write the output directly as indicated by the output of a circuit for a sinusoidal input (3.18).

Example 3.5

Let's apply these results to a final example, in which the input is a voltage source and the output is the inductor current. The source voltage equals $V_{in} = 2\cos(2\pi 60t) + 3$. We want the circuit to pass constant (offset) voltage essentially unaltered (save for the fact that the output is a current rather than a voltage) and remove the 60 Hz term. Because the input is the sum of *two* sinusoids–a constant is a zero-frequency cosine–our approach is

1. find the transfer function using impedances;
2. use it to find the output due to each input component;
3. add the results;
4. find element values that accomplish our design criteria.

Because the circuit is a series combination of elements, let's use voltage divider to find the transfer function between V_{in} and V, then use the *v-i* relation of the inductor to find its current.

$$\begin{aligned} \frac{I_{out}}{V_{in}} &= \frac{j2\pi fL}{R + j2\pi fL}\frac{1}{j2\pi fL} \\ &= \frac{1}{j2\pi fL + R} \\ &= H(f) \end{aligned} \tag{3.19}$$

25This content is available online at <http://cnx.org/content/m0031/2.20/>.

where

$$voltage\ divider = \frac{j2\pi fL}{R + j2\pi fL}$$

and

$$inductor\ admittance = \frac{1}{j2\pi fL}$$

[Do the units check?] The form of this transfer function should be familiar; it is a lowpass filter, and it will perform our desired function once we choose element values properly.

The constant term is easiest to handle. The output is given by $3|H(0)| = \frac{3}{R}$. Thus, the value we choose for the resistance will determine the scaling factor of how voltage is converted into current. For the 60 Hz component signal, the output current is $2|H(60)|\cos(2\pi 60t + \angle(H(60)))$. The total output due to our source is

$$i_{out} = 2|H(60)|\cos(2\pi 60t + \angle(H(60))) + 3H(0) \tag{3.20}$$

The cutoff frequency for this filter occurs when the real and imaginary parts of the transfer function's denominator equal each other. Thus, $2\pi f_c L = R$, which gives $f_c = \frac{R}{2\pi L}$. We want this cutoff frequency to be much less than 60 Hz. Suppose we place it at, say, 10 Hz. This specification would require the component values to be related by $\frac{R}{L} = 20\pi = 62.8$. The transfer function at 60 Hz would be

$$\left|\frac{1}{j2\pi 60L + R}\right| = \frac{1}{R}\left|\frac{1}{6j+1}\right| = \frac{1}{R}\frac{1}{\sqrt{37}} \approx 0.16\frac{1}{R} \tag{3.21}$$

which yields an attenuation (relative to the gain at zero frequency) of about 1/6, and result in an output amplitude of $\frac{0.3}{R}$ relative to the constant term's amplitude of $\frac{3}{R}$. A factor of 10 relative size between the two components seems reasonable. Having a 100 mH inductor would require a 6.28 Ω resistor. An easily available resistor value is 6.8 Ω; thus, this choice results in cheaply and easily purchased parts. To make the resistance bigger would require a proportionally larger inductor. Unfortunately, even a 1 H inductor is physically large; consequently low cutoff frequencies require small-valued resistors and large-valued inductors. The choice made here represents only one compromise.

The phase of the 60 Hz component will very nearly be $-\left(\frac{\pi}{2}\right)$, leaving it to be $\frac{0.3}{R}\cos\left(2\pi 60t - \frac{\pi}{2}\right) = \frac{0.3}{R}\sin(2\pi 60t)$. The waveforms for the input and output are shown in Figure 3.33 (Waveforms).

Note that the sinusoid's phase has indeed shifted; the lowpass filter not only reduced the 60 Hz signal's amplitude, but also shifted its phase by 90°.

3.15 Formal Circuit Methods: Node Method[26]

In some (complicated) cases, we cannot use the simplification techniques–such as parallel or series combination rules–to solve for a circuit's input-output relation. In other modules, we wrote *v-i* relations and Kirchoff's laws haphazardly, solving them more on intuition than procedure. We need a formal method that produces a small, easy set of equations that lead directly to the input-output relation we seek. One such technique is the **node method**.

[26]This content is available online at <http://cnx.org/content/m0032/2.18/>.

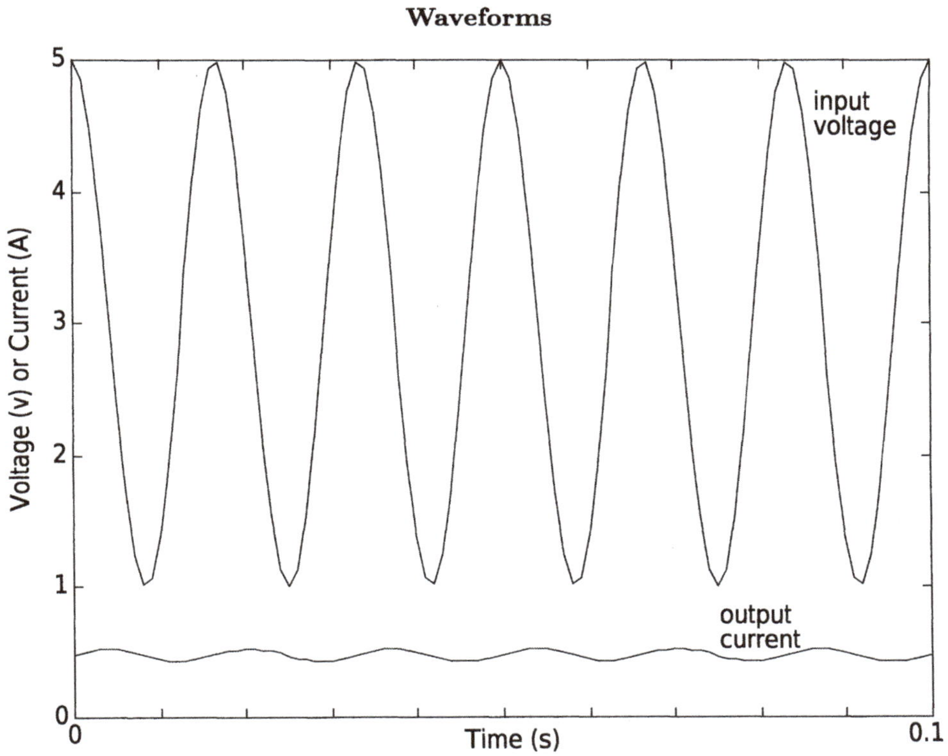

Figure 3.33: Input and output waveforms for the example RL circuit when the element values are $R = 6.28\Omega$ and $L = 100mH$.

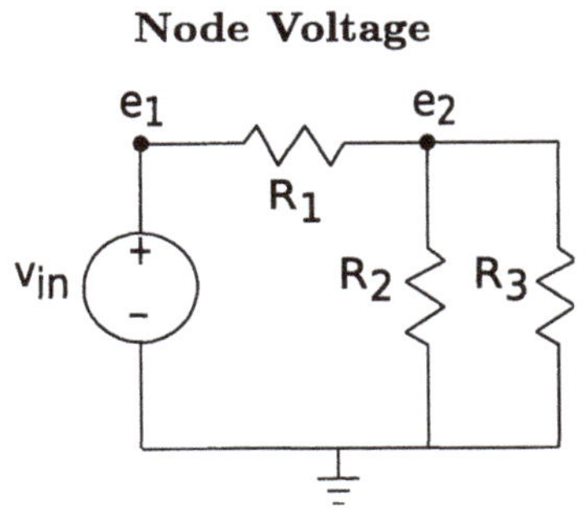

Figure 3.34

The node method begins by finding all nodes–places where circuit elements attach to each other–in the circuit. We call one of the nodes the **reference node**; the choice of reference node is arbitrary, but it is usually chosen to be a point of symmetry or the "bottom" node. For the remaining nodes, we define **node voltages** e_n that represent the voltage between the node and the reference. These node voltages constitute the only unknowns; all we need is a sufficient number of equations to solve for them. In our example, we have two node voltages. *The very act of defining node voltages is equivalent to using all the KVL equations at your disposal.* The reason for this simple, but astounding, fact is that a node voltage is uniquely defined regardless of what path is traced between the node and the reference. Because two paths between a node and reference have the same voltage, the sum of voltages around the loop equals zero.

In some cases, a node voltage corresponds exactly to the voltage across a voltage source. In such cases, the node voltage is specified by the source and is *not* an unknown. For example, in our circuit, $e_1 = v_{in}$; thus, we need only to find one node voltage.

The equations governing the node voltages are obtained by writing KCL equations at each node having an unknown node voltage, using the *v-i* relations for each element. In our example, the only circuit equation is

$$\frac{e_2 - v_{in}}{R_1} + \frac{e_2}{R_2} + \frac{e_2}{R_3} = 0 \tag{3.22}$$

A little reflection reveals that when writing the KCL equations for the sum of currents leaving a node, that node's voltage will *always* appear with a plus sign, and all other node voltages with a minus sign. Systematic application of this procedure makes it easy to write node equations and to check them before solving them. Also remember to check units at this point: Every term should have units of current. In our example, solving for the unknown node voltage is easy:

$$e_2 = \frac{R_2 R_3}{R_1 R_2 + R_1 R_2 + R_2 R_3} v_{in} \tag{3.23}$$

Have we really solved the circuit with the node method? Along the way, we have used KVL, KCL, and the *v-i* relations. Previously, we indicated that the set of equations resulting from applying these laws is necessary and sufficient. This result guarantees that the node method can be used to "solve" *any* circuit. One fallout of this result is that we must be able to find any circuit variable given the node voltages and sources. All circuit variables can be found using the *v-i* relations and voltage divider. For example, the current through R_3 equals $\frac{e_2}{R_3}$.

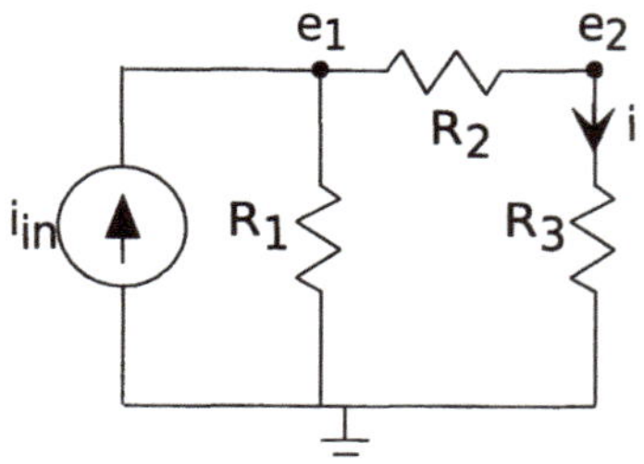

Figure 3.35

The presence of a current source in the circuit does not affect the node method greatly; just include it in writing KCL equations as a current *leaving* the node. The circuit has three nodes, requiring us to define two node voltages. The node equations are

$$\frac{e_1}{R_1} + \frac{e_1 - e_2}{R_2} - i_{in} = 0 \quad \text{(Node 1)}$$

$$\frac{e_2 - e_1}{R_2} + \frac{e_2}{R_3} = 0 \quad \text{(Node 2)}$$

Note that the node voltage corresponding to the node that we are writing KCL for enters with a positive sign, the others with a negative sign, and that the units of each term is given in amperes. Rewrite these equations in the standard set-of-linear-equations form.

$$e_1\left(\frac{1}{R_1}+\frac{1}{R_2}\right)-e_2\frac{1}{R_2}=i_{in}$$

$$(-e_1)\frac{1}{R_2}+e_2\left(\frac{1}{R_2}+\frac{1}{R_3}\right)=0$$

Solving these equations gives

$$e_1=\frac{R_2+R_3}{R_3}e_2$$

$$e_2=\frac{R_1R_3}{R_1+R_2+R_3}i_{in}$$

To find the indicated current, we simply use $i=\frac{e_2}{R_3}$.

Example 3.6: Node Method Example

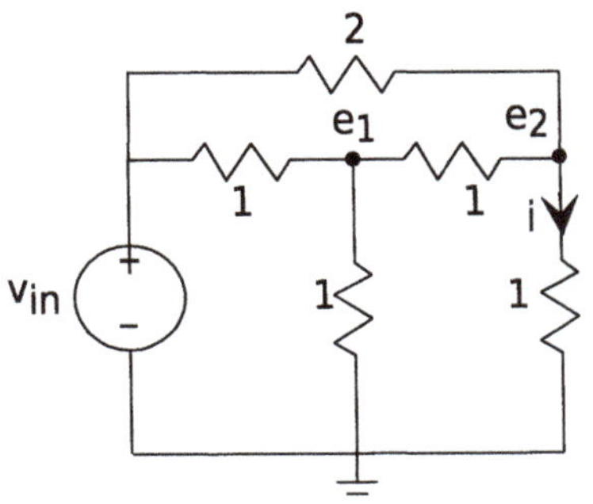

Figure 3.36

In this circuit (Figure 3.36), we cannot use the series/parallel combination rules: The vertical resistor at node 1 keeps the two horizontal 1 Ω resistors from being in series, and the 2 Ω resistor prevents the two 1 Ω resistors at node 2 from being in series. We really do need the node method to solve this circuit! Despite having six elements, we need only define two node voltages. The node equations are

$$\frac{e_1-v_{in}}{1}+\frac{e_1}{1}+\frac{e_1-e_2}{1}=0 \quad \text{(Node 1)}$$

$$\frac{e_2-v_{in}}{2}+\frac{e_2}{1}+\frac{e_2-e_1}{1}=0 \quad \text{(Node 2)}$$

Solving these equations yields $e_1=\frac{2}{5}v_{in}$ and $e_2=\frac{5}{13}v_{in}$. The output current equals $\frac{e_2}{1}=\frac{5}{13}v_{in}$. One unfortunate consequence of using the element's numeric values from the outset is that it becomes impossible to check units while setting up and solving equations.

Exercise 3.15 *(Solution on p. 107.)*

What is the equivalent resistance seen by the voltage source?

Node Method and Impedances

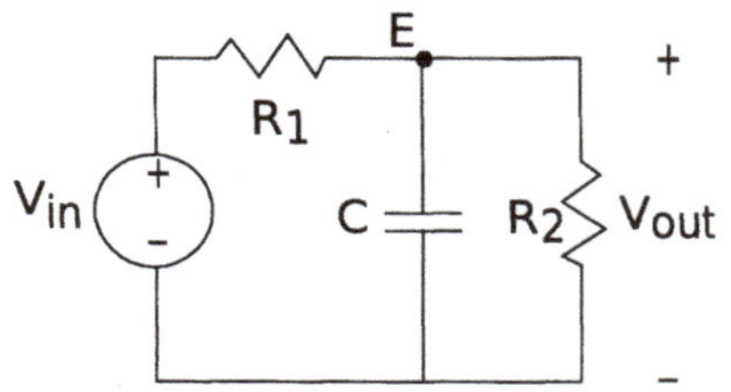

Figure 3.37: Modification of the circuit shown on the left to illustrate the node method and the effect of adding the resistor R_2.

The node method applies to RLC circuits, without significant modification from the methods used on simple resistive circuits, if we use complex amplitudes. We rely on the fact that complex amplitudes satisfy KVL, KCL, and impedance-based *v-i* relations. In the example circuit, we define complex amplitudes for the input and output variables and for the node voltages. We need only one node voltage here, and its KCL equation is

$$\frac{E - V_{in}}{R_1} + Ej2\pi fC + \frac{E}{R_2} = 0$$

with the result

$$E = \frac{R_2}{R_1 + R_2 + j2\pi fR_1R_2C}V_{in}$$

To find the transfer function between input and output voltages, we compute the ratio $\frac{E}{V_{in}}$. The transfer function's magnitude and angle are

$$|H(f)| = \frac{R_2}{\sqrt{(R_1 + R_2)^2 + (2\pi fR_1R_2C)^2}}$$

$$\angle(H(f)) = -\left(\arctan\left(\frac{2\pi fR_1R_2C}{R_1 + R_2}\right)\right)$$

This circuit differs from the one shown previously (Figure 3.30: Simple Circuit) in that the resistor R_2 has been added across the output. What effect has it had on the transfer function, which in the original circuit was a lowpass filter having cutoff frequency $f_c = \frac{1}{2\pi R_1C}$? As shown in Figure 3.38 (Transfer Function), adding the second resistor has two effects: it lowers the gain in the passband (the range of frequencies for which the filter has little effect on the input) and increases the cutoff frequency.

Transfer Function

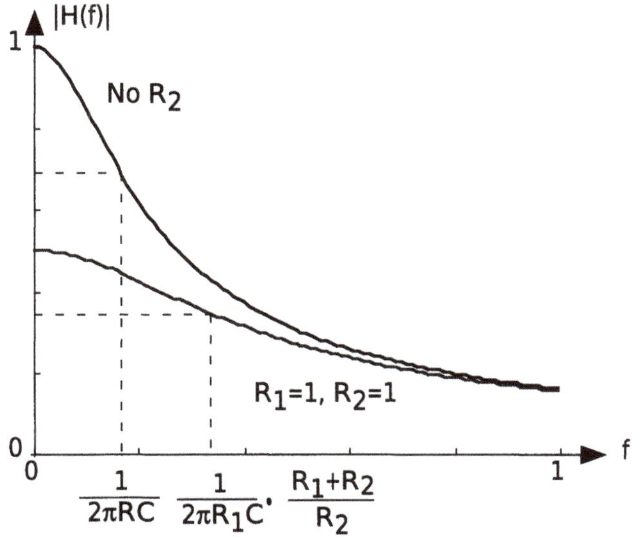

Figure 3.38: Transfer functions of the circuits shown in Figure 3.37 (Node Method and Impedances). Here, $R_1 = 1$, $R_2 = 1$, and $C = 1$.

When $R_2 = R_1$, as shown on the plot, the passband gain becomes half of the original, and the cutoff frequency increases by the same factor. Thus, adding R_2 provides a 'knob' by which we can trade passband gain for cutoff frequency.

Exercise 3.16 ***(Solution on p. 107.)***

We can change the cutoff frequency without affecting passband gain by changing the resistance in the original circuit. Does the addition of the R_2 resistor help in circuit design?

3.16 Power Conservation in Circuits[27]

Now that we have a formal method—the node method—for solving circuits, we can use it to prove a powerful result: KVL and KCL are all that are required to show that *all* circuits conserve power, regardless of what elements are used to build the circuit.

[27]This content is available online at <http://cnx.org/content/m17317/1.1/>.

Part of a general circuit to prove Conservation of Power

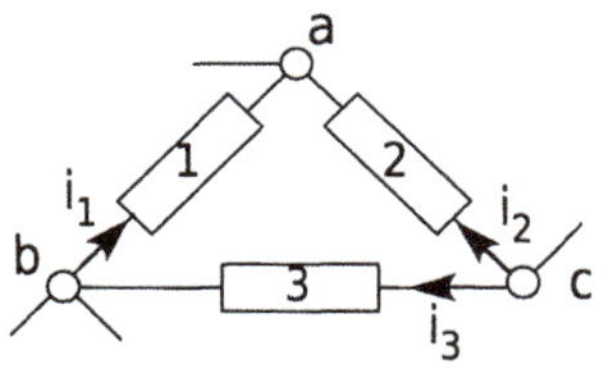

Figure 3.39

First of all, define node voltages for all nodes in a given circuit. Any node chosen as the reference will do. For example, in the portion of a large circuit (Figure 3.39: Part of a general circuit to prove Conservation of Power) depicted here, we define node voltages for nodes a, b and c. With these node voltages, we can express the voltage across any element in terms of them. For example, the voltage across element 1 is given by $v_1 = e_b - e_a$. The instantaneous power for element 1 becomes

$$v_1 i_1 = (e_b - e_a)\, i_1 = e_b i_1 - e_a i_1$$

Writing the power for the other elements, we have

$$\begin{aligned} v_2 i_2 &= e_c i_2 - e_a i_2 \\ v_3 i_3 &= e_c i_3 - e_b i_3 \end{aligned}$$

When we add together the element power terms, we discover that once we collect terms involving a particular node voltage, it is multiplied by the sum of currents leaving the node minus the sum of currents entering. For example, for node b, we have $e_b (i_3 - i_1)$. We see that the currents will obey KCL that multiply each node voltage. Consequently, *we conclude that the sum of element powers must equal zero in any circuit regardless of the elements used to construct the circuit.*

$$\sum_k (v_k i_k) = 0$$

The simplicity and generality with which we proved this results generalizes to other situations as well. In particular, note that the complex amplitudes of voltages and currents obey KVL and KCL, respectively. Consequently, we have that $\sum_k (V_k I_k) = 0$. Furthermore, the complex-conjugate of currents also satisfies KCL, which means we also have $\sum_k (V_k {I_k}^*) = 0$. And finally, we know that evaluating the real-part of an expression is linear. Finding the real-part of this power conservation gives the result that *average power is also conserved in any circuit.*

$$\sum_k \left(\frac{1}{2}\mathrm{Re}\,(V_k {I_k}^*)\right) = 0$$

NOTE: This proof of power conservation can be generalized in another very interesting way. All we need is a set of voltages that obey KVL and a set of currents that obey KCL. Thus, for a given circuit topology (the specific way elements are interconnected), the voltages and currents can be measured at different times and the sum of v-i products is zero.

$$\sum_k (v_k(t_1)\, i_k(t_2)) = 0$$

Even more interesting is the fact that the elements don't matter. We can take a circuit and measure all the voltages. We can then make element-for-element replacements and, if the topology has not changed, we can measure a set of currents. The sum of the product of element voltages and currents will also be zero!

3.17 Electronics[28]

So far we have analyzed *electrical* circuits: The source signal has more power than the output variable, be it a voltage or a current. Power has not been explicitly defined, but no matter. Resistors, inductors, and capacitors as individual elements certainly provide no power gain, and circuits built of them will not magically do so either. Such circuits are termed electrical in distinction to those that do provide power gain: **electronic circuits**. Providing power gain, such as your stereo reading a CD and producing sound, is accomplished by semiconductor circuits that contain transistors. The basic idea of the transistor is to let the weak input signal modulate a strong current provided by a source of electrical power–the power supply–to produce a more powerful signal. A physical analogy is a water faucet: By turning the faucet back and forth, the water flow varies accordingly, and has much more power than expended in turning the handle. The waterpower results from the static pressure of the water in your plumbing created by the water utility pumping the water up to your local water tower. The power supply is like the water tower, and the faucet is the transistor, with the turning achieved by the input signal. Just as in this analogy, a power supply is a source of constant voltage as the water tower is supposed to provide a constant water pressure.

A device that is much more convenient for providing gain (and other useful features as well) than the transistor is the **operational amplifier**, also known as the **op-amp**. An op-amp is an integrated circuit (a complicated circuit involving several transistors constructed on a chip) that provides a large voltage gain *if* you attach the power supply. We can model the op-amp with a new circuit element: the dependent source.

3.18 Dependent Sources[29]

A **dependent source** is either a voltage or current source whose value is proportional to some other voltage or current in the circuit. Thus, there are four different kinds of dependent sources; to describe an op-amp, we need a voltage-dependent voltage source. However, the standard circuit-theoretical model for a transistor[30] contains a current-dependent current source. Dependent sources do not serve as inputs to a circuit like independent sources. They are used to model **active circuits**: those containing electronic elements. The RLC circuits we have been considering so far are known as **passive circuits**.

[28]This content is available online at <http://cnx.org/content/m0035/2.8/>.
[29]This content is available online at <http://cnx.org/content/m0053/2.13/>.
[30]"Small Signal Model for Bipolar Transistor" <http://cnx.org/content/m1019/latest/>

dependent sources

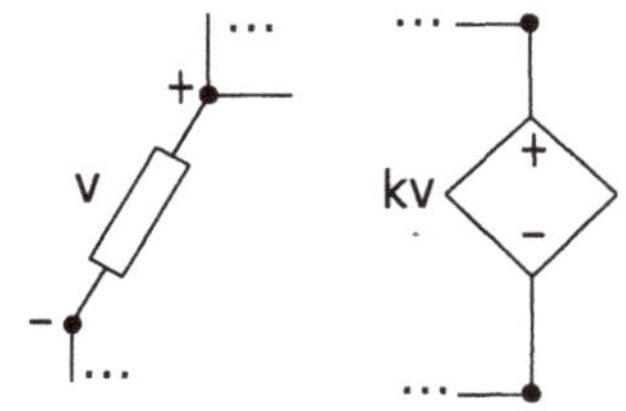

Figure 3.40: Of the four possible dependent sources, depicted is a voltage-dependent voltage source in the context of a generic circuit.

Figure 3.41 (op-amp) shows the circuit symbol for the op-amp and its equivalent circuit in terms of a voltage-dependent voltage source.

op-amp

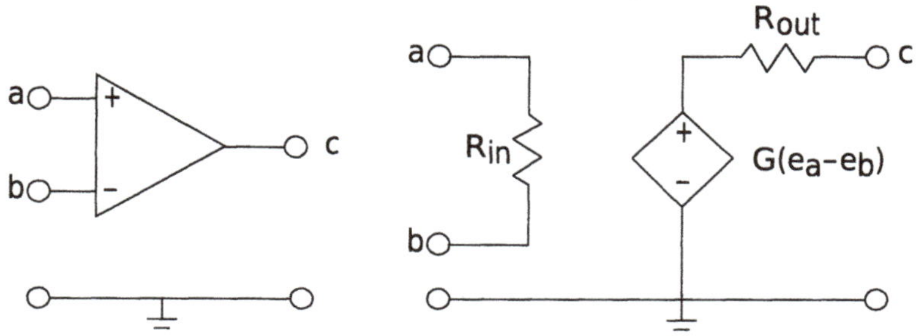

Figure 3.41: The op-amp has four terminals to which connections can be made. Inputs attach to nodes a and b, and the output is node c. As the circuit model on the right shows, the op-amp serves as an amplifier for the difference of the input node voltages.

Here, the output voltage equals an amplified version of the difference of node voltages appearing across its inputs. The dependent source model portrays how the op-amp works quite well. As in most active circuit schematics, the power supply is not shown, but must be present for the circuit model to be accurate. Most operational amplifiers require both positive and negative supply voltages for proper operation.

Because dependent sources cannot be described as impedances, and because the dependent variable cannot "disappear" when you apply parallel/series combining rules, circuit simplifications such as current and voltage divider should not be applied in most cases. Analysis of circuits containing dependent sources essentially requires use of formal methods, like the node method (Section 3.15). Using the node method for such circuits is not difficult, with node voltages defined across the source treated as if they were known (as with independent sources). Consider the circuit shown on the top in Figure 3.42 (feedback op-amp).

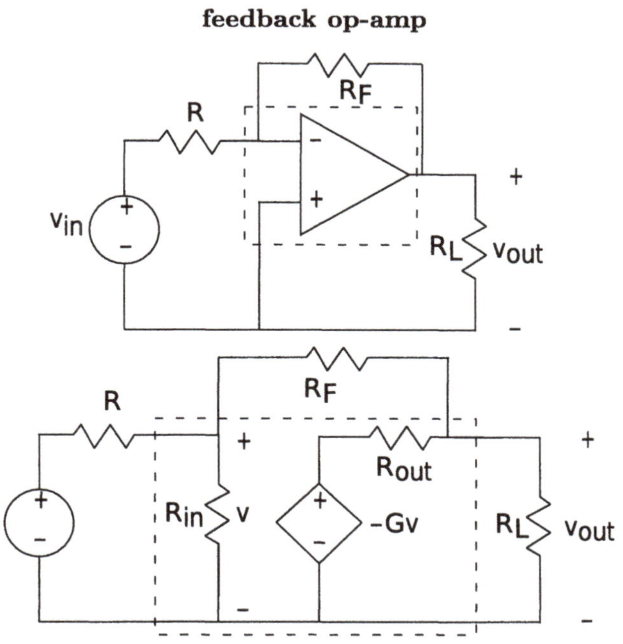

Figure 3.42: The top circuit depicts an op-amp in a feedback amplifier configuration. On the bottom is the equivalent circuit, and integrates the op-amp circuit model into the circuit.

Note that the op-amp is placed in the circuit "upside-down," with its inverting input at the top and serving as the only input. As we explore op-amps in more detail in the next section, this configuration will appear again and again and its usefulness demonstrated. To determine how the output voltage is related to the input voltage, we apply the node method. Only two node voltages— v and v_{out}—need be defined; the remaining nodes are across sources or serve as the reference. The node equations are

$$\frac{v - v_{in}}{R} + \frac{v}{R_{in}} + \frac{v - v_{out}}{R_F} = 0 \tag{3.24}$$

$$\frac{v_{out} - (-G)\, v}{R_{out}} + \frac{v_{out} - v}{R_F} + \frac{v_{out}}{R_L} = 0 \tag{3.25}$$

Note that no special considerations were used in applying the node method to this dependent-source circuit. Solving these to learn how v_{out} relates to v_{in} yields

$$\left(\frac{R_F R_{out}}{R_{out} - G R_F}\left(\frac{1}{R_{out}} + \frac{1}{R_{in}} + \frac{1}{R_L}\right)\left(\frac{1}{R} + \frac{1}{R_{in}} + \frac{1}{R_F}\right) - \frac{1}{R_F}\right) v_{out} = \frac{1}{R} v_{in} \tag{3.26}$$

This expression represents the general input-output relation for this circuit, known as the **standard feedback configuration**. Once we learn more about op-amps (Section 3.19), in particular what its typical element values are, the expression will simplify greatly. Do note that the units check, and that the parameter G of the dependent source is a dimensionless gain.

3.19 Operational Amplifiers[31]

Op-Amp

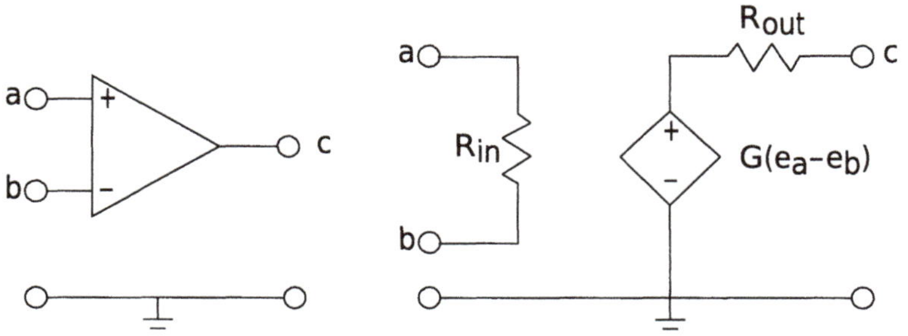

Figure 3.43: The op-amp has four terminals to which connections can be made. Inputs attach to nodes *a* and *b*, and the output is node *c*. As the circuit model on the right shows, the op-amp serves as an amplifier for the difference of the input node voltages.

Op-amps not only have the circuit model shown in Figure 3.43 (Op-Amp), but their element values are very special.

- The **input resistance**, R_{in}, is typically *large*, on the order of 1 MΩ.
- The **output resistance**, R_{out}, is *small*, usually less than 100 Ω.
- The **voltage gain**, G, is *large*, exceeding 10^5.

The large gain catches the eye; it suggests that an op-amp could turn a 1 mV input signal into a 100 V one. If you were to build such a circuit–attaching a voltage source to node *a*, attaching node *b* to the reference, and looking at the output–you would be disappointed. In dealing with electronic components, you cannot forget the unrepresented but needed power supply.

> UNMODELED LIMITATIONS IMPOSED BY POWER SUPPLIES: It is impossible for electronic components to yield voltages that exceed those provided by the power supply or for them to yield currents that exceed the power supply's rating.

Typical power supply voltages required for op-amp circuits are $\pm(15V)$. Attaching the 1 mv signal not only would fail to produce a 100 V signal, the resulting waveform would be severely distorted. While a desirable outcome if you are a rock & roll aficionado, high-quality stereos should not distort signals. Another consideration in designing circuits with op-amps is that these element values are typical: Careful control of the gain can only be obtained by choosing a circuit so that its element values dictate the resulting gain, which must be smaller than that provided by the op-amp.

[31]This content is available online at <http://cnx.org/content/m0036/2.28/>.

opamp

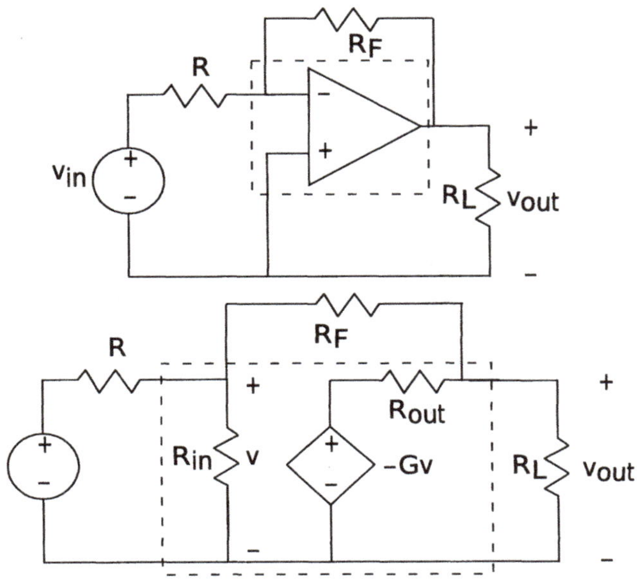

Figure 3.44: The top circuit depicts an op-amp in a feedback amplifier configuration. On the bottom is the equivalent circuit, and integrates the op-amp circuit model into the circuit.

3.19.1 Inverting Amplifier

The feedback configuration shown in Figure 3.44 (opamp) is the most common op-amp circuit for obtaining what is known as an **inverting amplifier**.

$$\left(\frac{R_F R_{out}}{R_{out} - GR_F}\left(\frac{1}{R_{out}} + \frac{1}{R_{in}} + \frac{1}{R_L}\right)\left(\frac{1}{R} + \frac{1}{R_{in}} + \frac{1}{R_F}\right) - \frac{1}{R_F}\right) v_{out} = \frac{1}{R} v_{in} \tag{3.27}$$

provides the exact input-output relationship. In choosing element values with respect to op-amp characteristics, we can simplify the expression dramatically.

- Make the load resistance, R_L, much larger than R_{out}. This situation drops the term $\frac{1}{R_L}$ from the second factor of (3.27).
- Make the resistor, R, smaller than R_{in}, which means that the $\frac{1}{R_{in}}$ term in the third factor is negligible.

With these two design criteria, the expression ((3.27)) becomes

$$\left(\frac{R_F}{R_{out} - GR_F}\left(\frac{1}{R} + \frac{1}{R_F}\right) - \frac{1}{R_F}\right) v_{out} = \frac{1}{R} v_{out} \tag{3.28}$$

Because the gain is large and the resistance R_{out} is small, the first term becomes $-\left(\frac{1}{G}\right)$, leaving us with

$$\left(\left(-\left(\frac{1}{G}\right)\right)\left(\frac{1}{R} + \frac{1}{R_F}\right) - \frac{1}{R_F}\right) v_{out} = \frac{1}{R} v_{in} \tag{3.29}$$

- If we select the values of R_F and R so that $(GR \gg R_F)$, this factor will no longer depend on the op-amp's inherent gain, and it will equal $-\left(\frac{1}{R_F}\right)$.

Under these conditions, we obtain the classic input-output relationship for the op-amp-based inverting amplifier.

$$v_{out} = -\left(\frac{R_F}{R} v_{in}\right) \tag{3.30}$$

Consequently, the gain provided by our circuit is entirely determined by our choice of the feedback resistor R_F and the input resistor R. It is always negative, and can be less than one or greater than one in magnitude. It cannot exceed the op-amp's inherent gain and should not produce such large outputs that distortion results (remember the power supply!). Interestingly, note that this relationship does not depend on the load resistance. This effect occurs because we use load resistances large compared to the op-amp's output resistance. Thus observation means that, if careful, we can place op-amp circuits in cascade, *without* incurring the effect of succeeding circuits changing the behavior (transfer function) of previous ones; see this problem (Problem 3.37).

3.19.2 Active Filters

As long as design requirements are met, the input-output relation for the inverting amplifier also applies when the feedback and input circuit elements are impedances (resistors, capacitors, and inductors).

opamp

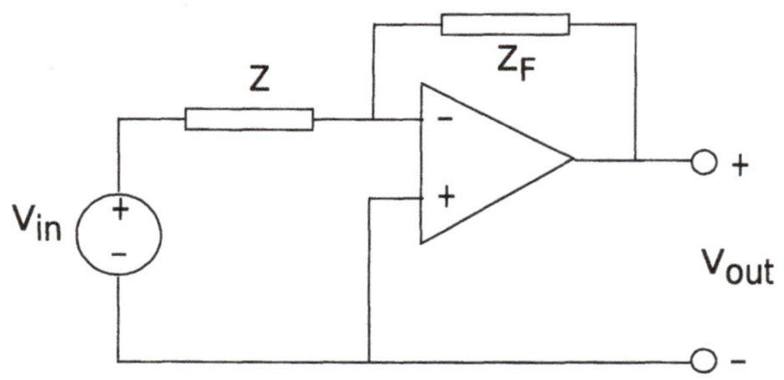

Figure 3.45: $\frac{V_{out}}{V_{in}} = -\left(\frac{Z_F}{Z}\right)$

Example 3.7

Let's design an op-amp circuit that functions as a lowpass filter. We want the transfer function between the output and input voltage to be

$$H(f) = \frac{K}{1 + \frac{jf}{f_c}}$$

where K equals the passband gain and f_c is the cutoff frequency. Let's assume that the inversion (negative gain) does not matter. With the transfer function of the above op-amp circuit in mind, let's consider some choices.

- $Z_F = K$, $Z = 1 + \frac{jf}{f_c}$. This choice means the feedback impedance is a resistor and that the input impedance is a series combination of an inductor and a resistor. In circuit design, we try to avoid inductors because they are physically bulkier than capacitors.

- $Z_F = \frac{1}{1+\frac{jf}{f_c}}$, $Z = \frac{1}{K}$. Consider the reciprocal of the feedback impedance (its admittance): ${Z_F}^{-1} = 1 + \frac{jf}{f_c}$. Since this admittance is a sum of admittances, this expression suggests the parallel combination of a resistor (value $= 1\ \Omega$) and a capacitor (value $= \frac{1}{f_c}F$). We have the right idea, but the values (like $1\ \Omega$) are not right. Consider the general RC parallel combination; its admittance is $\frac{1}{R_F} + j2\pi fC$. Letting the input resistance equal R, the transfer function of the op-amp inverting amplifier now is $H(f) = -\left(\frac{\frac{R_F}{R}}{1+j2\pi fR_FC}\right)$

Thus, we have the gain equal to $\frac{R_F}{R}$ and the cutoff frequency $\frac{1}{R_FC}$.

Creating a specific transfer function with op-amps does not have a unique answer. As opposed to design with passive circuits, electronics is more flexible (a cascade of circuits can be built so that each has little effect on the others; see Problem 3.37) and gain (increase in power and amplitude) can result. To complete our example, let's assume we want a lowpass filter that emulates what the telephone companies do. Signals transmitted over the telephone have an upper frequency limit of about 3 kHz. For the second design choice, we require $R_FC = 3.3 \times 10^{-4}$. Thus, many choices for resistance and capacitance values are possible. A 1 μF capacitor and a 330 Ω resistor, 10 nF and 33 kΩ, and 10 pF and 33 MΩ would all theoretically work. Let's also desire a voltage gain of ten: $\frac{R_F}{R} = 10$, which means $R = \frac{R_F}{10}$. Recall that we must have $R < R_{in}$. As the op-amp's input impedance is about 1 MΩ, we don't want R too large, and this requirement means that the last choice for resistor/capacitor values won't work. We also need to ask for less gain than the op-amp can provide itself. Because the feedback "element" is an impedance (a parallel resistor capacitor combination), we need to examine the gain requirement more carefully. We must have $\frac{|Z_F|}{R} < 10^5$ for all frequencies of interest. Thus, $\frac{\frac{R_F}{|1+j2\pi fR_FC|}}{R} < 10^5$. As this impedance decreases with frequency, the design specification of $\frac{R_F}{R} = 10$ means that this criterion is easily met. Thus, the first two choices for the resistor and capacitor values (as well as many others in this range) will work well. Additional considerations like parts cost might enter into the picture. Unless you have a high-power application (this isn't one) or ask for high-precision components, costs don't depend heavily on component values as long as you stay close to standard values. For resistors, having values $r10^d$, easily obtained values of r are 1, 1.4, 3.3, 4.7, and 6.8, and the decades span 0-8.

Exercise 3.17 *(Solution on p. 107.)*

What is special about the resistor values; why these rather odd-appearing values for r?

3.19.3 Intuitive Way of Solving Op-Amp Circuits

When we meet op-amp design specifications, we can simplify our circuit calculations greatly, so much so that we don't need the op-amp's circuit model to determine the transfer function. Here is our inverting amplifier.

When we take advantage of the op-amp's characteristics—large input impedance, large gain, and small output impedance—we note the two following important facts.

- The current i_{in} must be very small. The voltage produced by the dependent source is 10^5 times the voltage v. Thus, the voltage v must be small, which means that $i_{in} = \frac{v}{R_{in}}$ must be tiny. For example, if the output is about 1 v, the voltage $v = 10^{-5}V$, making the current $i_{in} = 10^{-11}A$. Consequently, we can ignore i_{in} in our calculations and assume it to be zero.
- Because of this assumption—essentially no current flow through R_{in}—the voltage v must also be essentially zero. This means that in op-amp circuits, the voltage across the op-amp's input is basically zero.

Armed with these approximations, let's return to our original circuit as shown in Figure 3.47 (opamp). The node voltage e is essentially zero, meaning that it is essentially tied to the reference node. Thus, the current through the resistor R equals $\frac{v_{in}}{R}$. Furthermore, the feedback resistor appears in parallel with the

opamp

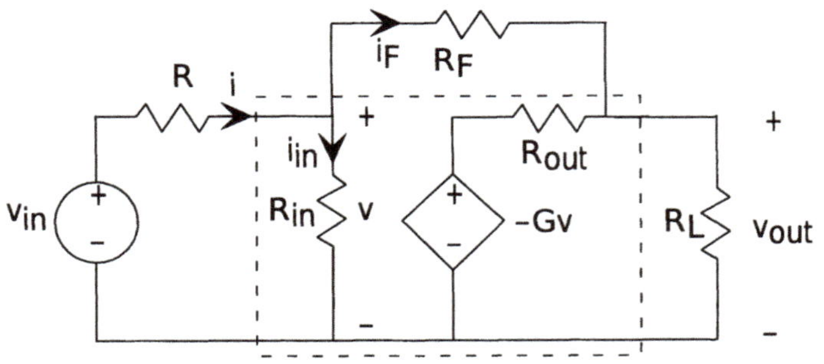

Figure 3.46

opamp

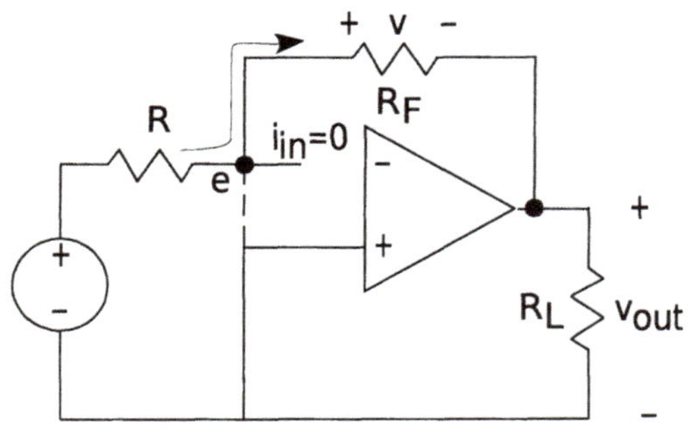

Figure 3.47

load resistor. Because the current going into the op-amp is zero, all of the current flowing through R flows through the feedback resistor ($i_F = i$)! The voltage across the feedback resistor v equals $\frac{v_{in} R_F}{R}$. Because the left end of the feedback resistor is essentially attached to the reference node, the voltage across it equals the negative of that across the output resistor: $v_{out} = -v = -\left(\frac{v_{in} R_F}{R}\right)$. Using this approach makes analyzing new op-amp circuits much easier. When using this technique, check to make sure the results you obtain are consistent with the assumptions of essentially zero current entering the op-amp and nearly zero voltage across the op-amp's inputs.

Example 3.8

Let's try this analysis technique on a simple extension of the inverting amplifier configuration shown in Figure 3.48 (Two Source Circuit). If either of the source-resistor combinations were not present, the inverting amplifier remains, and we know that transfer function. By superposition, we know that the input-output relation is

$$v_{out} = -\left(\frac{R_F}{R_1} v_{in}^{(1)}\right) - \frac{R_F}{R_2} v_{in}^{(2)} \tag{3.31}$$

Two Source Circuit

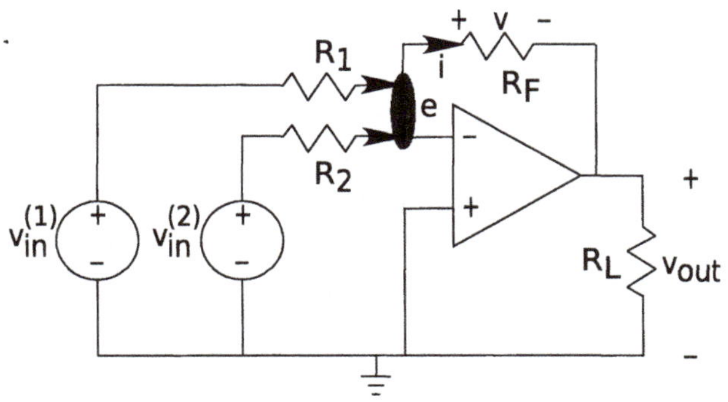

Figure 3.48: Two-source, single-output op-amp circuit example.

Diode

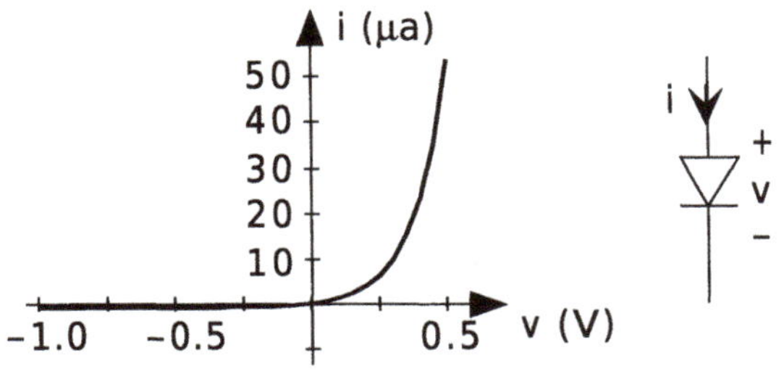

Figure 3.49: *v-i* relation and schematic symbol for the diode. Here, the diode parameters were room temperature and $I_0 = 1\ \mu A$.

When we start from scratch, the node joining the three resistors is at the same potential as the reference, $e \approx 0$, and the sum of currents flowing into that node is zero. Thus, the current i flowing in the resistor R_F equals $\frac{v_{in}^{(1)}}{R_1} + \frac{v_{in}^{(2)}}{R_2}$. Because the feedback resistor is essentially in parallel with the load resistor, the voltages must satisfy $v = -v_{out}$. In this way, we obtain the input-output relation given above.

What utility does this circuit have? Can the basic notion of the circuit be extended without bound?

3.20 The Diode[32]

The resistor, capacitor, and inductor are linear circuit elements in that their *v-i* relations are linear in the mathematical sense. Voltage and current sources are (technically) nonlinear devices: stated simply, doubling the current through a voltage source does not double the voltage. A more blatant, and very useful, nonlinear

[32]This content is available online at <http://cnx.org/content/m0037/2.14/>.

circuit element is the diode (learn more[33]). Its input-output relation has an exponential form.

$$i(t) = I_0 \left(e^{\frac{q}{kT} v(t)} - 1 \right) \tag{3.32}$$

Here, the quantity q represents the charge of a single electron in coulombs, k is Boltzmann's constant, and T is the diode's temperature in K. At room temperature, the ratio $\frac{kT}{q} = 25mv$. The constant I_0 is the leakage current, and is usually very small. Viewing this v-i relation in Figure 3.49 (Diode), the nonlinearity becomes obvious. When the voltage is positive, current flows easily through the diode. This situation is known as **forward biasing**. When we apply a negative voltage, the current is quite small, and equals I_0, known as the **leakage** or **reverse-bias** current. A less detailed model for the diode has any positive current flowing through the diode when it is forward biased, and no current when negative biased. Note that the diode's schematic symbol looks like an arrowhead; the direction of current flow corresponds to the direction the arrowhead points.

diode circuit

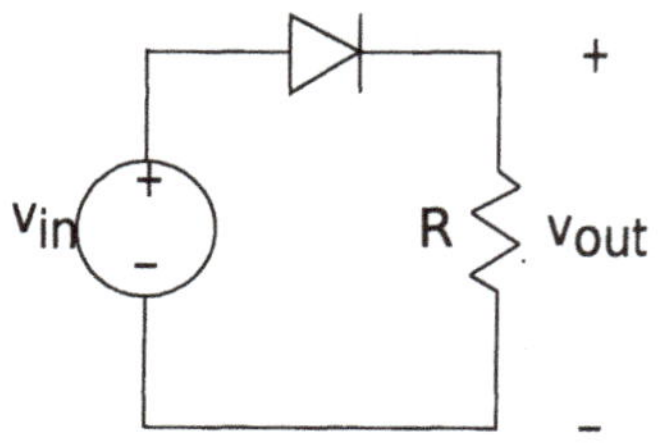

Figure 3.50

Because of the diode's nonlinear nature, we *cannot* use impedances nor series/parallel combination rules analyze circuits containing them. The reliable node method can always be used; it only relies on KVL for its application, and KVL is a statement about voltage drops around a closed path *regardless* of whether the elements are linear or not. Thus, for this simple circuit we have

$$\frac{v_{out}}{R} = I_0 \left(e^{\frac{q}{kT}(v_{in} - v_{out})} - 1 \right) \tag{3.33}$$

This equation *cannot* be solved in closed form. We must understand what is going on from basic principles, using computational and graphical aids. As an approximation, when v_{in} is positive, current flows through the diode so long as the voltage v_{out} is smaller than v_{in} (so the diode is forward biased). If the source is negative or v_{out} "tries" to be bigger than v_{in}, the diode is reverse-biased, and the reverse-bias current flows through the diode. Thus, at this level of analysis, positive input voltages result in positive output voltages with negative ones resulting in $v_{out} = -(RI_0)$.

[33]"P-N Junction: Part II" <http://cnx.org/content/m1004/latest/>

diode circuit

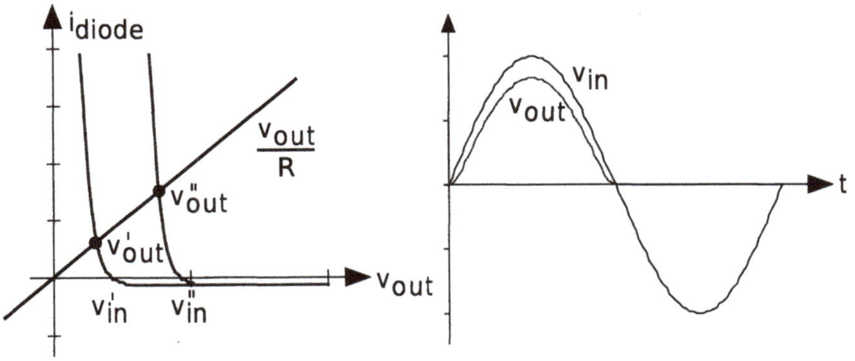

Figure 3.51

We need to detail the exponential nonlinearity to determine how the circuit distorts the input voltage waveform. We can of course numerically solve Figure 3.50 (diode circuit) to determine the output voltage when the input is a sinusoid. To learn more, let's express this equation graphically. We plot each term as a function of v_{out} for various values of the input voltage v_{in}; where they intersect gives us the output voltage. The left side, the current through the output resistor, does not vary itself with v_{in}, and thus we have a fixed straight line. As for the right side, which expresses the diode's *v-i* relation, the point at which the curve crosses the v_{out} axis gives us the value of v_{in}. Clearly, the two curves will always intersect just once for any value of v_{in}, and for positive v_{in} the intersection occurs at a value for v_{out} *smaller* than v_{in}. This reduction is smaller if the straight line has a shallower slope, which corresponds to using a bigger output resistor. For negative v_{in}, the diode is reverse-biased and the output voltage equals $-(RI_0)$.

What utility might this simple circuit have? The diode's nonlinearity cannot be escaped here, and the clearly evident distortion must have some practical application if the circuit were to be useful. This circuit, known as a **half-wave rectifier**, is present in virtually every AM radio *twice* and each serves very different functions! We'll learn what functions later.

diode circuit

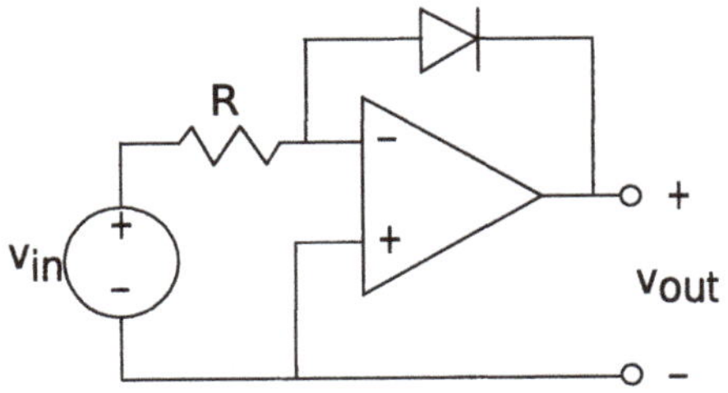

Figure 3.52

Here is a circuit involving a diode that is actually simpler to analyze than the previous one. We know that the current through the resistor must equal that through the diode. Thus, the diode's current is proportional to the input voltage. As the voltage across the diode is related to the logarithm of its current, we see that the input-output relation is

$$v_{out} = -\left(\frac{kT}{q}\ln\left(\frac{v_{in}}{RI_0}+1\right)\right) \qquad (3.34)$$

Clearly, the name **logarithmic amplifier** is justified for this circuit.

3.21 Analog Signal Processing Problems[34]

Problem 3.1: Solving Simple Circuits

a) Write the set of equations that govern Circuit A's (Figure 3.53) behavior.
b) Solve these equations for i_1: In other words, express this current in terms of element and source values by eliminating non-source voltages and currents.
c) For Circuit B, find the value for R_L that results in a current of 5 A passing through it.
d) What is the power dissipated by the load resistor R_L in this case?

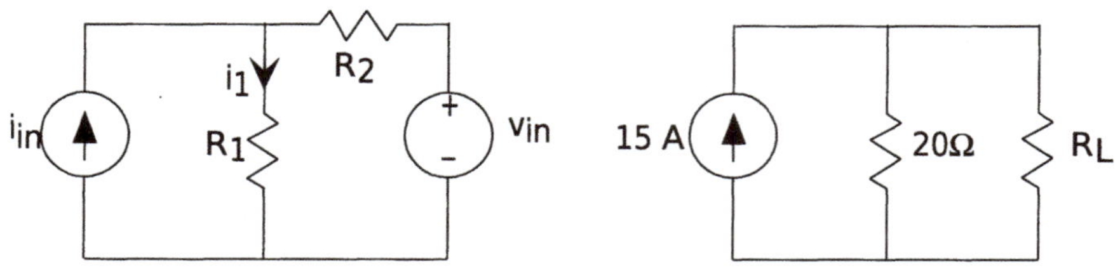

(a) Circuit A (b) Circuit B

Figure 3.53

Problem 3.2: Equivalent Resistance
For each of the following circuits (Figure 3.54), find the equivalent resistance using series and parallel combination rules.

Calculate the conductance seen at the terminals for circuit (c) in terms of each element's conductance. Compare this equivalent conductance formula with the equivalent resistance formula you found for circuit (b). How is the circuit (c) derived from circuit (b)?

Problem 3.3: Superposition Principle
One of the most important consequences of circuit laws is the **Superposition Principle**: The current or voltage defined for any element equals the sum of the currents or voltages produced in the element by the independent sources. This Principle has important consequences in simplifying the calculation of ciruit variables in multiple source circuits.

a) For the depicted circuit (Figure 3.55), find the indicated current using any technique you like (you should use the simplest).

[34]This content is available online at <http://cnx.org/content/m10349/2.37/>.

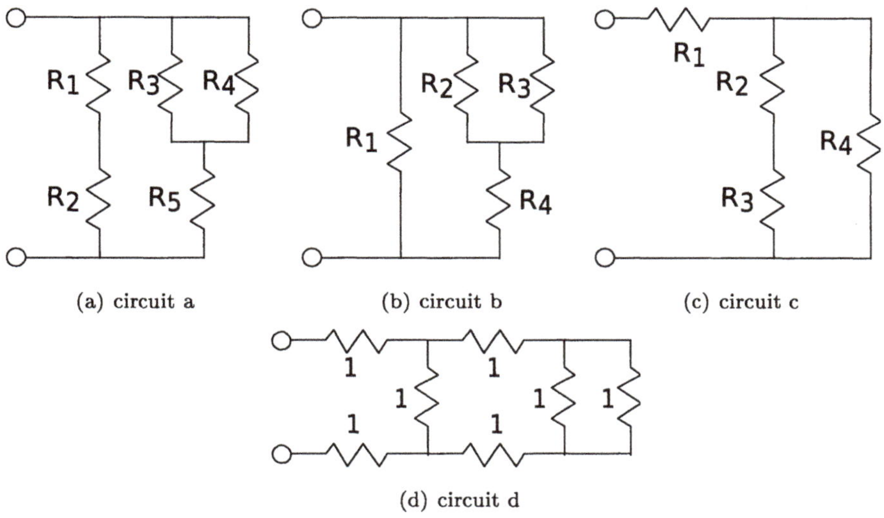

Figure 3.54

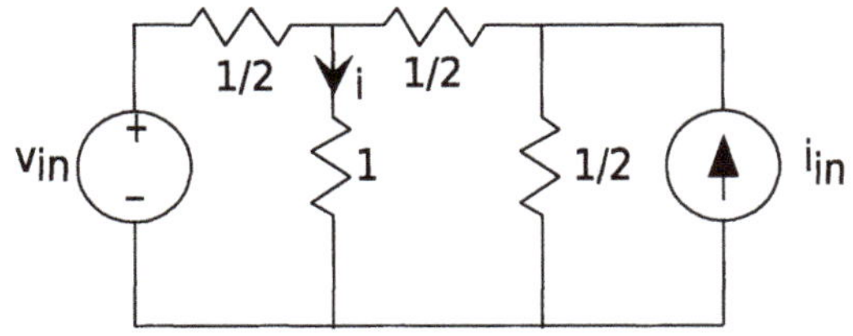

Figure 3.55

b) You should have found that the current i is a linear combination of the two source values: $i = C_1 v_{in} + C_2 i_{in}$. This result means that we can think of the current as a superposition of two components, each of which is due to a source. We can find each component by setting the other sources to zero. Thus, to find the voltage source component, you can set the current source to zero (an open circuit) and use the usual tricks. To find the current source component, you would set the voltage source to zero (a short circuit) and find the resulting current. Calculate the total current i using the Superposition Principle. Is applying the Superposition Principle easier than the technique you used in part (1)?

Problem 3.4: Current and Voltage Divider
Use current of voltage divider rules to calculate the indicated circuit variables in Figure 3.56.

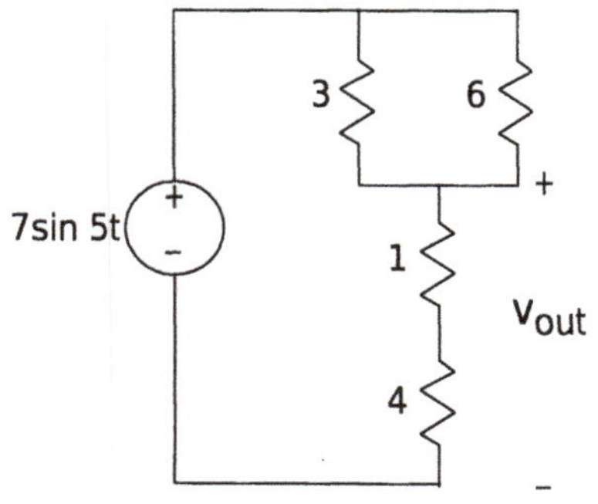

(a) circuit a

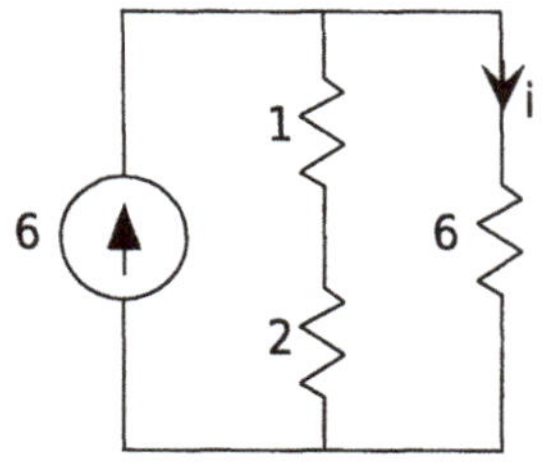

(b) circuit c

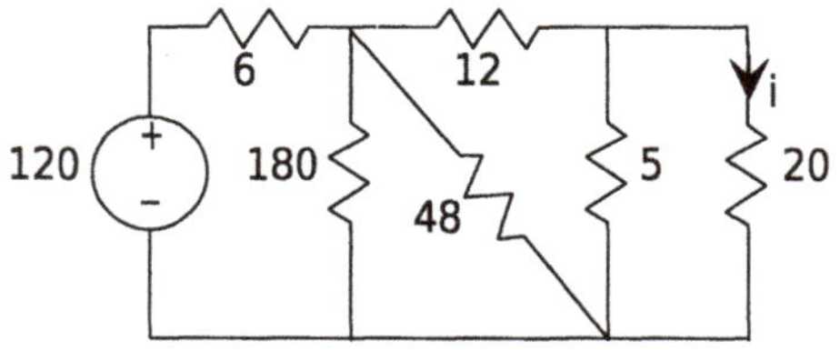

(c) circuit b

Figure 3.56

Problem 3.5: Thévenin and Mayer-Norton Equivalents
Find the Thévenin and Mayer-Norton equivalent circuits for the following circuits (Figure 3.57).

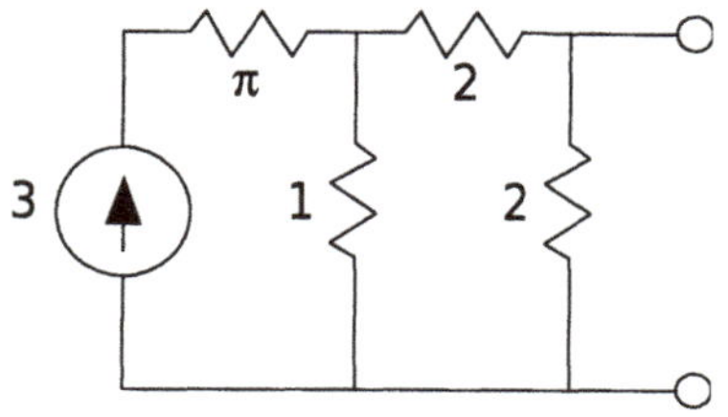

(a) circuit a

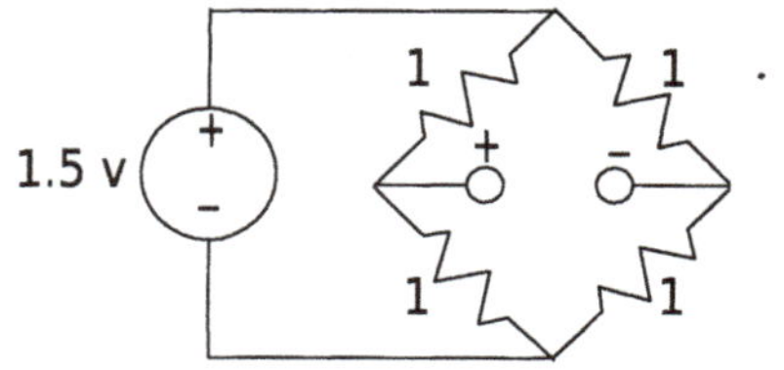

(b) circuit b

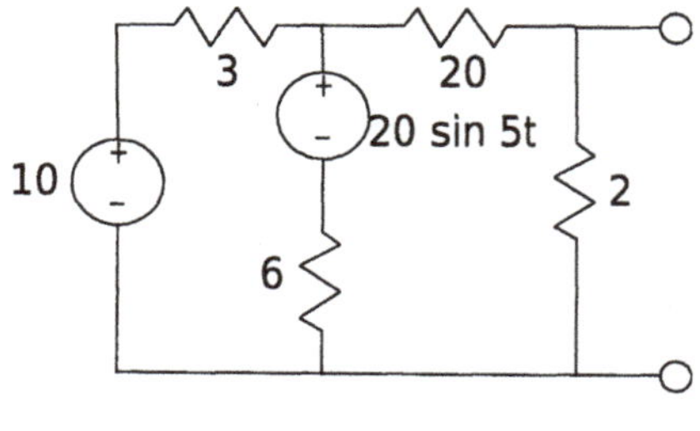

(c) circuit c

Figure 3.57

Problem 3.6: Detective Work
In the depicted circuit (Figure 3.58), the circuit N_1 has the v-i relation $v_1 = 3i_1 + 7$ when $i_s = 2$.

a) Find the Thévenin equivalent circuit for circuit N_2.

b) With $i_s = 2$, determine R such that $i_1 = -1$.

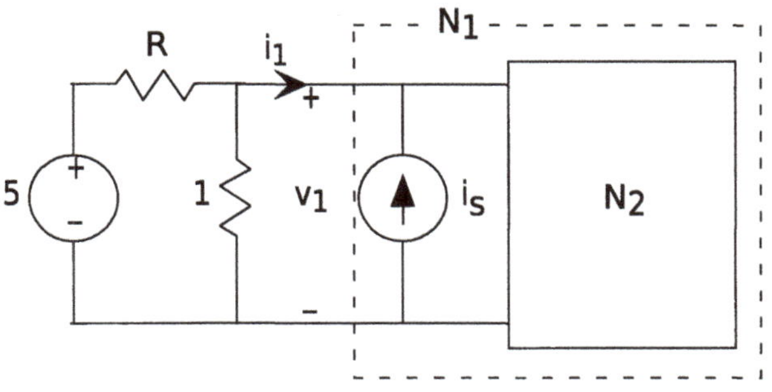

Figure 3.58

Problem 3.7: Cartesian to Polar Conversion
Convert the following expressions into polar form. Plot their location in the complex plane[35].

a) $\left(1+\sqrt{-3}\right)^2$
b) $3+j^4$
c) $\frac{2-j\frac{6}{\sqrt{3}}}{2+j\frac{6}{\sqrt{3}}}$
d) $\left(4-j^3\right)\left(1+j\frac{1}{2}\right)$
e) $3e^{j\pi}+4e^{j\frac{\pi}{2}}$
f) $\left(\sqrt{3}+j\right)2\sqrt{2}e^{-\left(j\frac{\pi}{4}\right)}$
g) $\frac{3}{1+j3\pi}$

Problem 3.8: The Complex Plane
The complex variable z is related to the real variable u according to

$$z = 1 + e^{ju}$$

- Sketch the contour of values z takes on in the complex plane.
- What are the maximum and minimum values attainable by $|z|$?
- Sketch the contour the rational function $\frac{z-1}{z+1}$ traces in the complex plane.

Problem 3.9: Cool Curves
In the following expressions, the variable x runs from zero to infinity. What geometric shapes do the following trace in the complex plane?

a) e^{jx}
b) $1+e^{jx}$
c) $e^{-x}e^{jx}$
d) $e^{jx}+e^{j\left(x+\frac{\pi}{4}\right)}$

Problem 3.10: Trigonometric Identities and Complex Exponentials
Show the following trigonometric identities using complex exponentials. In many cases, they were derived using this approach.

[35]"The Complex Plane" <http://cnx.org/content/m10596/latest/>

a) $\sin(2u) = 2\sin(u)\cos(u)$
b) $\cos^2(u) = \frac{1+\cos(2u)}{2}$
c) $\cos^2(u) + \sin^2(u) = 1$
d) $\frac{d}{du}(\sin(u)) = \cos(u)$

Problem 3.11: Transfer Functions
Find the transfer function relating the complex amplitudes of the indicated variable and the source shown in Figure 3.59. Plot the magnitude and phase of the transfer function.

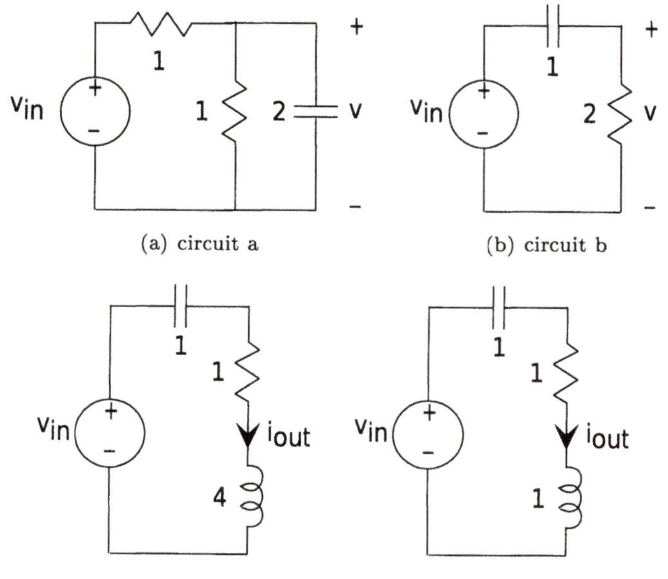

Figure 3.59

Problem 3.12: Using Impedances
Find the differential equation relating the indicated variable to the source(s) using impedances for each circuit shown in Figure 3.60.

Problem 3.13: Transfer Functions
In the following circuit (Figure 3.61), the voltage source equals $v_{in}(t) = 10\sin\left(\frac{t}{2}\right)$.

a) Find the transfer function between the source and the indicated output voltage.
b) For the given source, find the output voltage.

Problem 3.14: A Simple Circuit
You are given this simple circuit (Figure 3.62).

a) What is the transfer function between the source and the indicated output current?
b) If the output current is measured to be $\cos(2t)$, what was the source?

Problem 3.15: Circuit Design

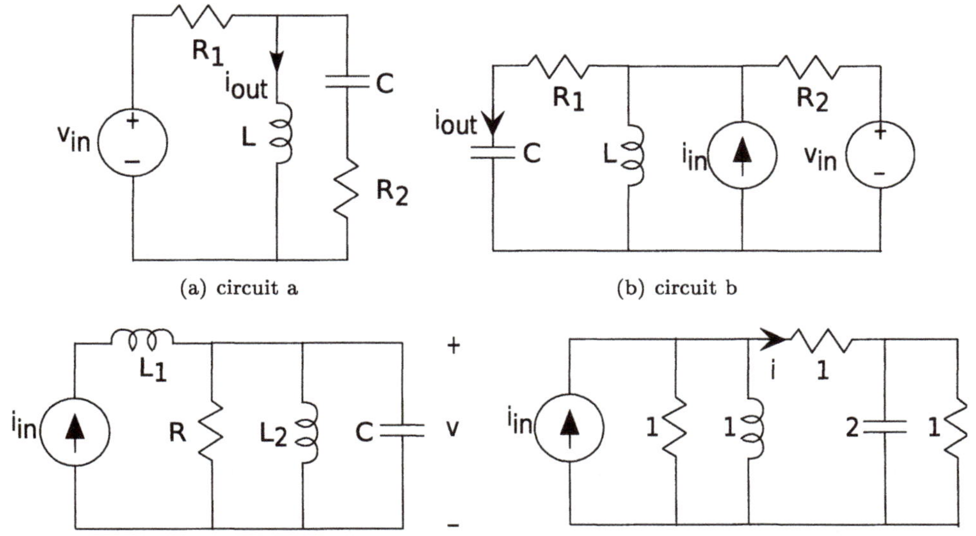

(a) circuit a

(b) circuit b

(c) circuit c

(d) circuit d

Figure 3.60

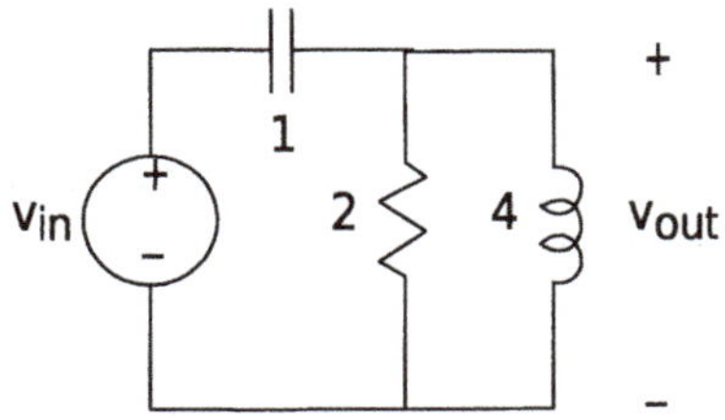

Figure 3.61

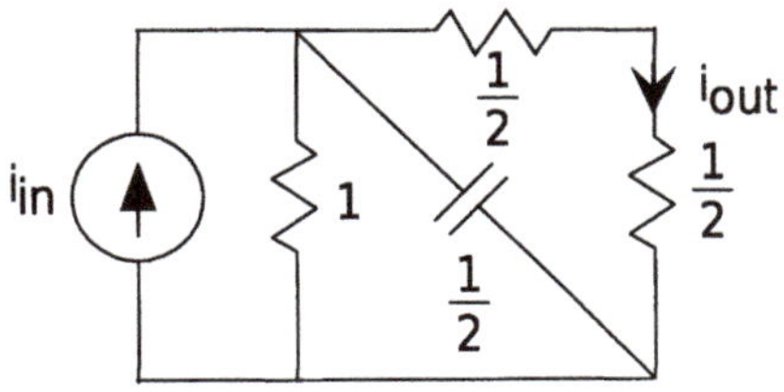

Figure 3.62

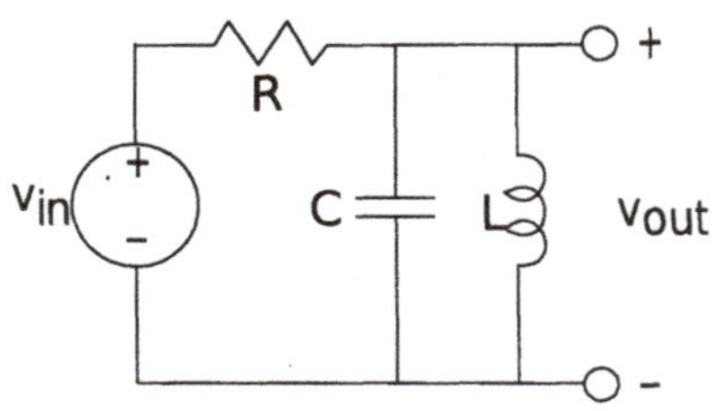

Figure 3.63

a) Find the transfer function between the input and the output voltages for the circuits shown in Figure 3.63.
b) At what frequency does the transfer function have a phase shift of zero? What is the circuit's gain at this frequency?
c) Specifications demand that this circuit have an output impedance (its equivalent impedance) less than 8Ω for frequencies above 1 kHz, the frequency at which the transfer function is maximum. Find element values that satisfy this criterion.

Problem 3.16: Power Transmission

The network shown in Figure 3.64(a) represents a simple power transmission system. The generator produces 60 Hz and is modeled by a simple Thévenin equivalent. The transmission line consists of a long length of copper wire and can be accurately described as a 50Ω resistor.

a) Determine the load current R_L and the average power the generator must produce so that the load receives 1,000 watts of average power. Why does the generator need to generate more than 1,000 watts of average power to meet this requirement?
b) Suppose the load is changed to that shown in Figure 3.64(b). Now how much power must the generator produce to meet the same power requirement? Why is it more than it had to produce to meet the requirement for the resistive load?
c) The load can be *compensated* to have a unity **power factor** (see exercise (Exercise 3.13)) so that the voltage and current are in phase for maximum power efficiency. The compensation technique is to place a circuit in parallel to the load circuit. What element works and what is its value?
d) With this compensated circuit, how much power must the generator produce to deliver 1,000 average power to the load?

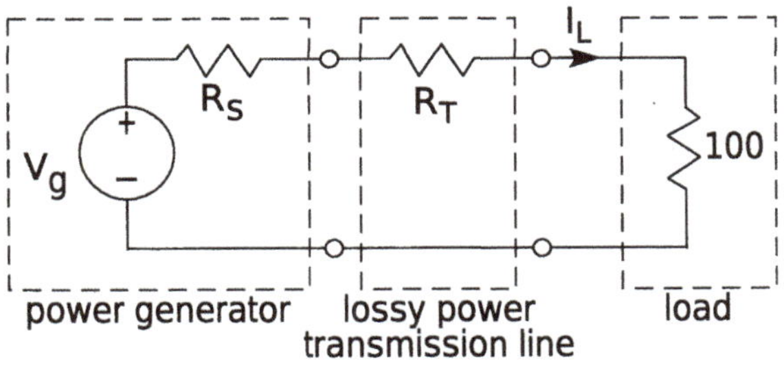

(a) Simple power transmission system

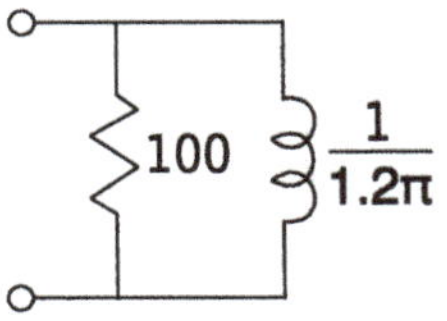

(b) Modified load circuit

Figure 3.64

Problem 3.17: Optimal Power Transmission
The following figure (Figure 3.65) shows a general model for power transmission. The power generator is represented by a Thévinin equivalent and the load by a simple impedance. In most applications, the source components are fixed while there is some latitude in choosing the load.

a) Suppose we wanted the maximize "voltage transmission:" make the voltage across the load as large as possible. What choice of load impedance creates the largest load voltage? What is the largest load voltage?
b) If we wanted the maximum current to pass through the load, what would we choose the load impedance to be? What is this largest current?
c) What choice for the load impedance maximizes the average power dissipated in the load? What is most power the generator can deliver?

NOTE: One way to maximize a function of a complex variable is to write the expression in terms of the variable's real and imaginary parts, evaluate derivatives with respect to each, set both derivatives to zero and solve the two equations simultaneously.

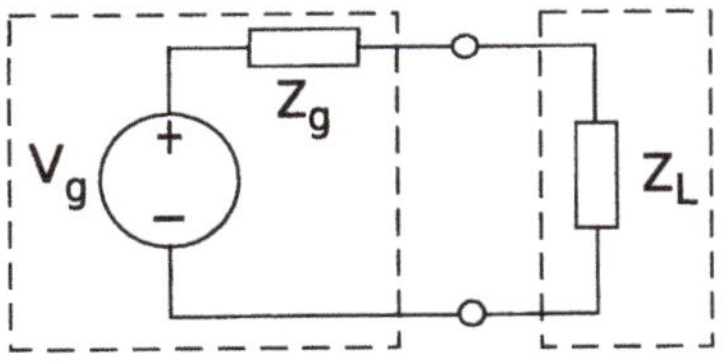

Figure 3.65

Problem 3.18: Sharing a Channel
Two transmitter-receiver pairs want to share the same digital communications channel. The transmitter signals will be added together by the channel. Receiver design is greatly simplified if first we remove the unwanted transmission (as much as possible). Each transmitter signal has the form

$$x_i(t) = A\sin(2\pi f_i t) \quad , \quad 0 \leq t \leq T$$

where the amplitude is either zero or A and each transmitter uses its own frequency f_i. Each frequency is harmonically related to the bit interval duration T, where the transmitter 1 uses the the frequency $\frac{1}{T}$. The datarate is 10Mbps.

a) Draw a block diagram that expresses this communication scenario.
b) Find circuits that the receivers could employ to separate unwanted transmissions. Assume the received signal is a voltage and the output is to be a voltage as well.
c) Find the second transmitter's frequency so that the receivers can suppress the unwanted transmission by at least a factor of ten.

Problem 3.19: Circuit Detective Work
In the lab, the open-circuit voltage measured across an unknown circuit's terminals equals $\sin(t)$. When a $1\ \Omega$ resistor is place across the terminals, a voltage of $\frac{1}{\sqrt{2}}\sin\left(t+\frac{\pi}{4}\right)$ appears.

a) What is the Thévenin equivalent circuit?
b) What voltage will appear if we place a 1 F capacitor across the terminals?

Problem 3.20: More Circuit Detective Work
The left terminal pair of a two terminal-pair circuit is attached to a testing circuit. The test source $v_{in}(t)$ equals $\sin(t)$ (Figure 3.66).

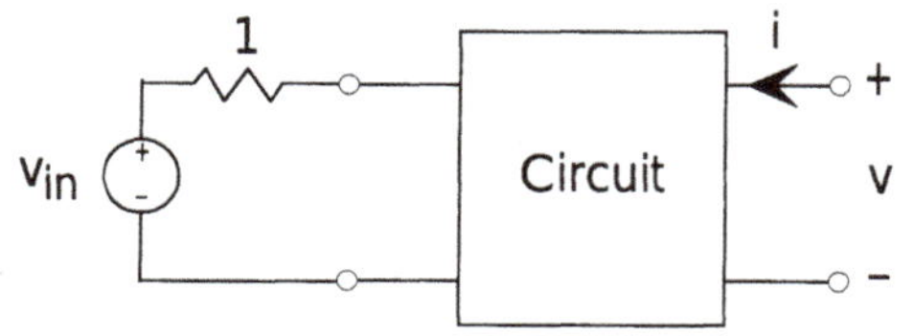

Figure 3.66

We make the following measurements.

- With nothing attached to the terminals on the right, the voltage $v(t)$ equals $\frac{1}{\sqrt{2}}\cos\left(t+\frac{\pi}{4}\right)$.
- When a wire is placed across the terminals on the right, the current $i(t)$ was $-(\sin(t))$.

a) What is the impedance "seen" from the terminals on the right?
b) Find the voltage $v(t)$ if a current source is attached to the terminals on the right so that $i(t)=\sin(t)$.

Problem 3.21: Linear, Time-Invariant Systems
For a system to be completely characterized by a transfer function, it needs not only be linear, but also to be time-invariant. A system is said to be time-invariant if delaying the input delays the output by the same amount. Mathematically, if $S(x(t))=y(t)$, meaning $y(t)$ is the output of a system $S(\bullet)$ when $x(t)$ is the input, $S(\bullet)$ is the time-invariant if $S(x(t-\tau))=y(t-\tau)$ for all delays τ and all inputs $x(t)$. Note that both linear and nonlinear systems have this property. For example, a system that squares its input is time-invariant.

a) Show that if a circuit has fixed circuit elements (their values don't change over time), its input-output relationship is time-invariant. *Hint*: Consider the differential equation that describes a circuit's input-output relationship. What is its general form? Examine the derivative(s) of delayed signals.
b) Show that impedances cannot characterize time-varying circuit elements (R, L, and C). Consequently, show that linear, time-varying systems do not have a transfer function.

c) Determine the linearity and time-invariance of the following. Find the transfer function of the linear, time-invariant (LTI) one(s).

i) diode
ii) $y(t) = x(t)\sin(2\pi f_0 t)$
iii) $y(t) = x(t - \tau_0)$
iv) $y(t) = x(t) + N(t)$

Problem 3.22: Long and Sleepless Nights
Sammy went to lab after a long, sleepless night, and constructed the circuit shown in Figure 3.67.

He cannot remember what the circuit, represented by the impedance Z, was. Clearly, this forgotten circuit is important as the output is the current passing through it.

a) What is the Thévenin equivalent circuit seen by the impedance?
b) In searching his notes, Sammy finds that the circuit is to realize the transfer function

$$H(f) = \frac{1}{j10\pi f + 2}$$

Find the impedance Z as well as values for the other circuit elements.

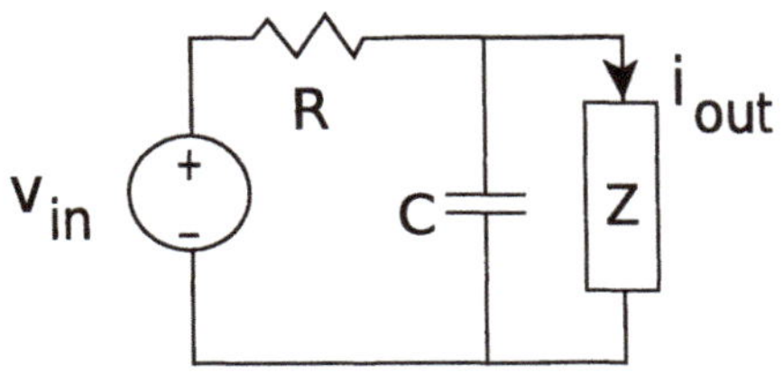

Figure 3.67

Problem 3.23: A Testing Circuit
The simple circuit here (Figure 3.68) was given on a test.

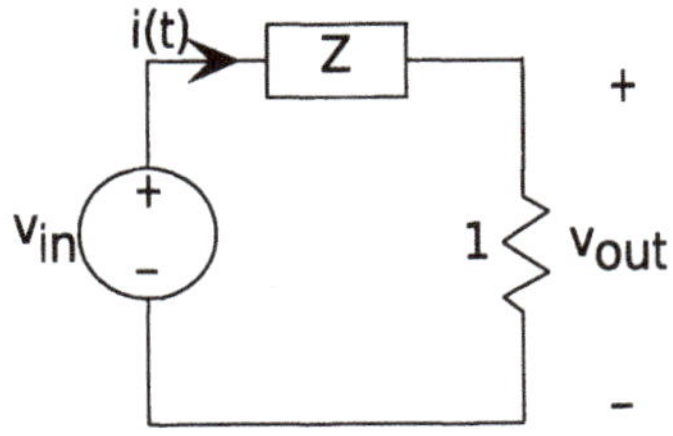

Figure 3.68

When the votlage source is $\sqrt{5}\sin(t)$, the current $i(t) = \sqrt{2}\cos\left(t - \arctan(2) - \frac{\pi}{4}\right)$.

a) What is voltage $v_{out}(t)$?

b) What is the impedance Z at the frequency of the source?

Problem 3.24: Mystery Circuit
You are given a circuit (Figure 3.69) that has two terminals for attaching circuit elements.

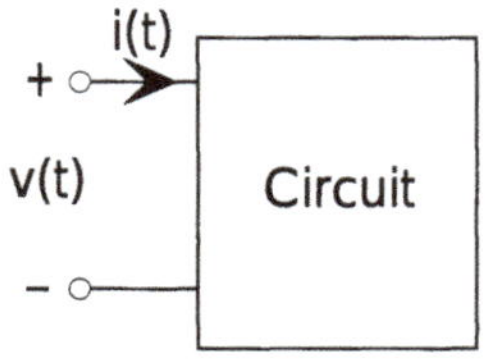

Figure 3.69

When you attach a voltage source equaling $\sin(t)$ to the terminals, the current through the source equals $4\sin\left(t+\frac{\pi}{4}\right)-2\sin(4t)$. When no source is attached (open-circuited terminals), the voltage across the terminals has the form $A\sin(4t+\phi)$.

a) What will the terminal current be when you replace the source by a short circuit?
b) If you were to build a circuit that was identical (from the viewpoint of the terminals) to the given one, what would your circuit be?
c) For your circuit, what are A and ϕ?

Problem 3.25: Mystery Circuit
Sammy must determine as much as he can about a mystery circuit by attaching elements to the terminal and measuring the resulting voltage. When he attaches a $1\,\Omega$ resistor to the circuit's terminals, he measures the voltage across the terminals to be $3\sin(t)$. When he attaches a 1 F capacitor across the terminals, the voltage is now $3\sqrt{2}\sin\left(t-\frac{\pi}{4}\right)$.

a) What voltage should he measure when he attaches nothing to the mystery circuit?
b) What voltage should Sammy measure if he doubled the size of the capacitor to 2 F and attached it to the circuit?

Problem 3.26: Find the Load Impedance
The depicted circuit (Figure 3.70) has a transfer function between the output voltage and the source equal to

$$H(f)=\frac{-8\pi^2f^2}{-8\pi^2f^2+4+j6\pi f}$$

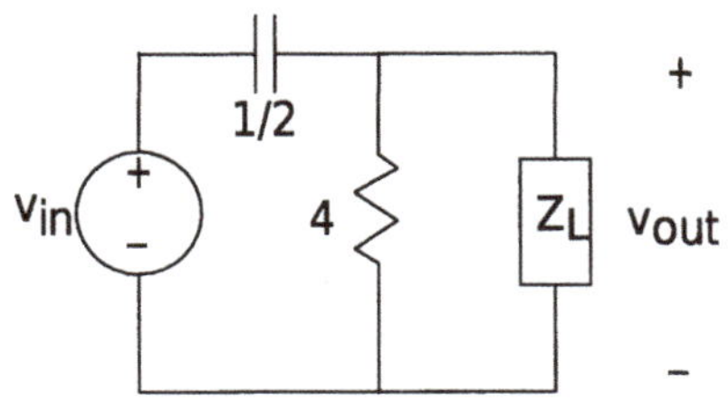

Figure 3.70

a) Sketch the magnitude and phase of the transfer function.
b) At what frequency does the phase equal $\frac{\pi}{2}$?
c) Find a circuit that corresponds to this load impedance. Is your answer unique? If so, show it to be so; if not, give another example.

Problem 3.27: Analog "Hum" Rejection
"Hum" refers to corruption from wall socket power that frequently sneaks into circuits. "Hum" gets its name because it sounds like a persistent humming sound. We want to find a circuit that will remove hum from any signal. A Rice engineer suggests using a simple voltage divider circuit (Figure 3.71) consisting of two series impedances.

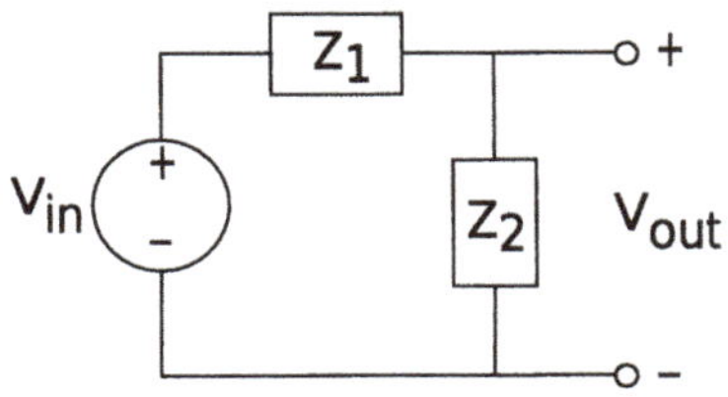

Figure 3.71

a) The impedance Z_1 is a resistor. The Rice engineer must decide between two circuits (Figure 3.72) for the impedance Z_2. Which of these will work?
b) Picking one circuit that works, choose circuit element values that will remove hum.
c) Sketch the magnitude of the resulting frequency response.

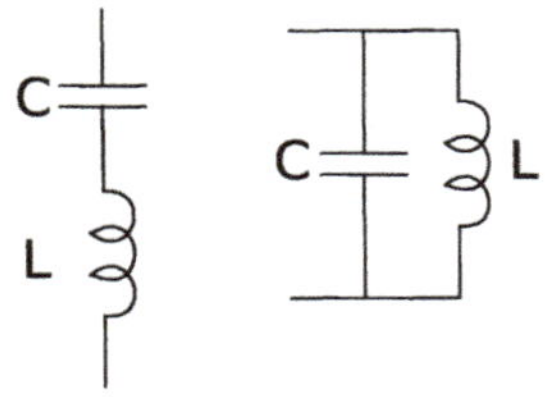

Figure 3.72

Problem 3.28: An Interesting Circuit

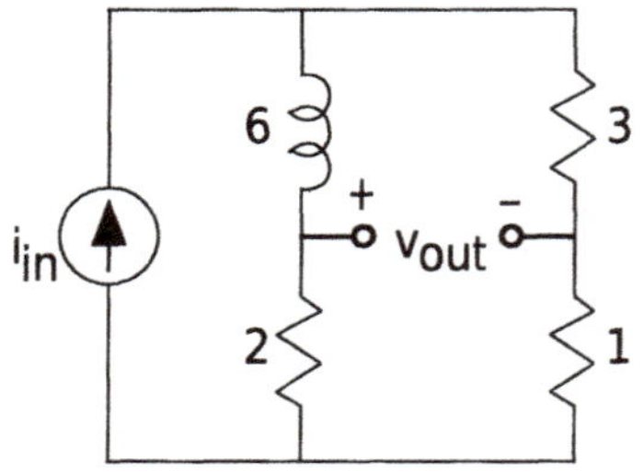

Figure 3.73

a) For the circuit shown in Figure 3.73, find the transfer function.
b) What is the output voltage when the input has the form $i_{in} = 5\sin(2000\pi t)$?

Problem 3.29: A Circuit
You are given the depicted circuit (Figure 3.74).

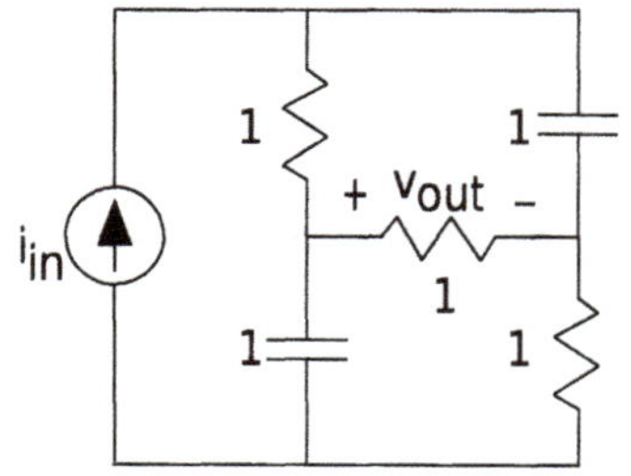

Figure 3.74

a) What is the transfer function between the source and the output voltage?
b) What will the voltage be when the source equals $\sin(t)$?
c) Many function generators produce a constant offset in addition to a sinusoid. If the source equals $1 + \sin(t)$, what is the output voltage?

Problem 3.30: An Interesting and Useful Circuit
The depicted circuit (Figure 3.75) has interesting properties, which are exploited in high-performance oscilloscopes.

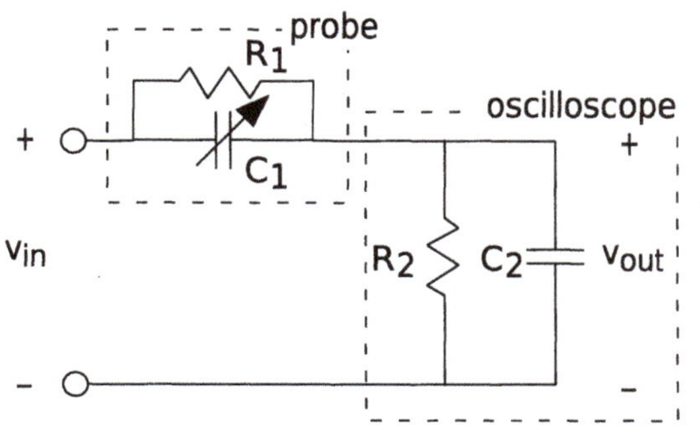

Figure 3.75

The portion of the circuit labeled "Oscilloscope" represents the scope's input impedance. $R_2 = 1M\Omega$ and $C_2 = 30pF$ (note the label under the channel 1 input in the lab's oscilloscopes). A *probe* is a device to attach an oscilloscope to a circuit, and it has the indicated circuit inside it.

a) Suppose for a moment that the probe is merely a wire and that the oscilloscope is attached to a circuit that has a resistive Thévenin equivalent impedance. What would be the effect of the oscilloscope's input impedance on measured voltages?
b) Using the node method, find the transfer function relating the indicated voltage to the source when the probe is used.
c) Plot the magnitude and phase of this transfer function when $R_1 = 9M\Omega$ and $C_1 = 2pF$.
d) For a particular relationship among the element values, the transfer function is quite simple. Find that relationship and describe what is so special about it.
e) The arrow through C_1 indicates that its value can be varied. Select the value for this capacitor to make the special relationship valid. What is the impedance seen by the circuit being measured for this special value?

Problem 3.31: A Circuit Problem
You are given the depicted circuit (Figure 3.76).

a) Find the differential equation relating the output voltage to the source.
b) What is the impedance "seen" by the capacitor?

Problem 3.32: Analog Computers
Because the differential equations arising in circuits resemble those that describe mechanical motion, we can use circuit models to describe mechanical systems. An ELEC 241 student wants to understand the suspension system on his car. Without a suspension, the car's body moves in concert with the bumps in the raod. A well-designed suspension system will smooth out bumpy roads, reducing the car's vertical motion.

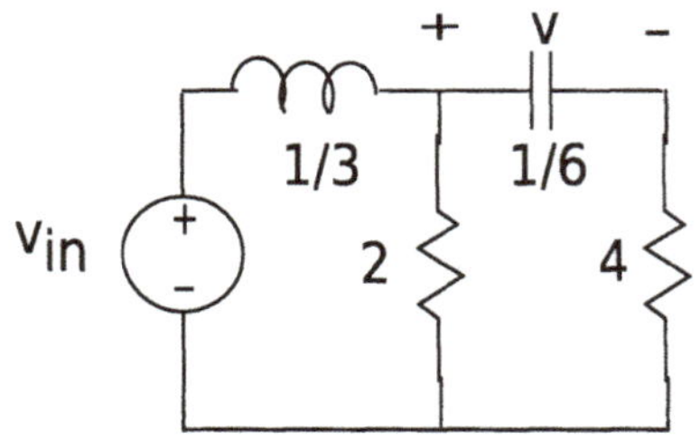

Figure 3.76

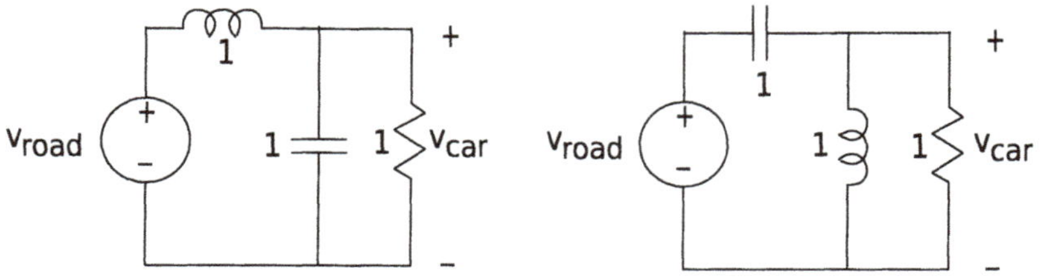

Figure 3.77

If the bumps are very gradual (think of a hill as a large but very gradual bump), the car's vertical motion should follow that of the road. The student wants to find a simple circuit that will model the car's motion. He is trying to decide between two circuit models (Figure 3.77).

Here, road and car displacements are represented by the voltages $v_{road}(t)$ and $v_{car}(t)$, respectively.

a) Which circuit would you pick? Why?
b) For the circuit you picked, what will be the amplitude of the car's motion if the road has a displacement given by $v_{road}(t) = 1 + \sin(2t)$?

Problem 3.33: Dependent Sources
Find the voltage v_{out} in each of the depicted circuits (Figure 3.78).
Problem 3.34: Transfer Functions and Circuits
You are given the depicted network (Figure 3.79).

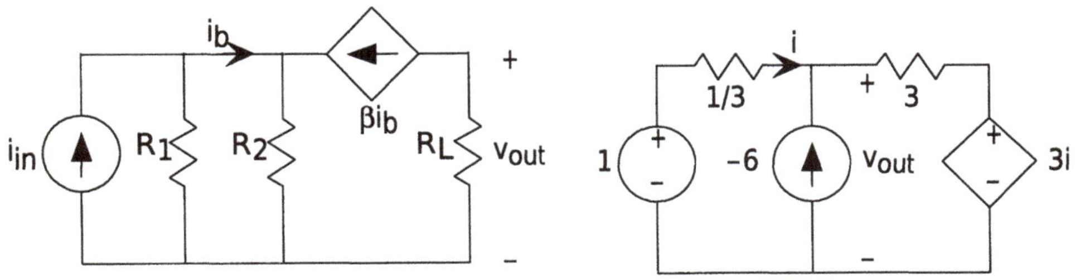

(a) circuit a

(b) circuit b

Figure 3.78

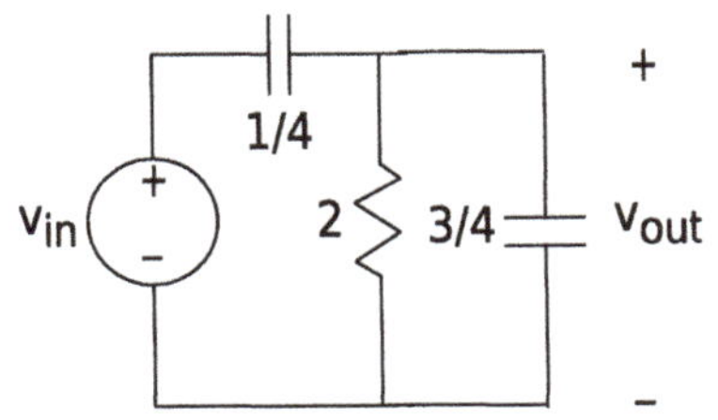

Figure 3.79

a) Find the transfer function between V_{in} and V_{out}.
b) Sketch the magnitude and phase of your transfer function. Label important frequency, amplitude and phase values.
c) Find $v_{out}(t)$ when $v_{in}(t) = \sin\left(\frac{t}{2} + \frac{\pi}{4}\right)$.

Problem 3.35: Fun in the Lab
You are given an unopenable box that has two terminals sticking out. You assume the box contains a circuit. You measure the voltage $\sin\left(t + \frac{\pi}{4}\right)$ across the terminals when nothing is connected to them and the current $\sqrt{2}\cos(t)$ when you place a wire across the terminals.

a) Find a circuit that has these characteristics.
b) You attach a 1 H inductor across the terminals. What voltage do you measure?

Problem 3.36: Operational Amplifiers
Find the transfer function between the source voltage(s) and the indicated output voltage for the circuits shown in Figure 3.80.

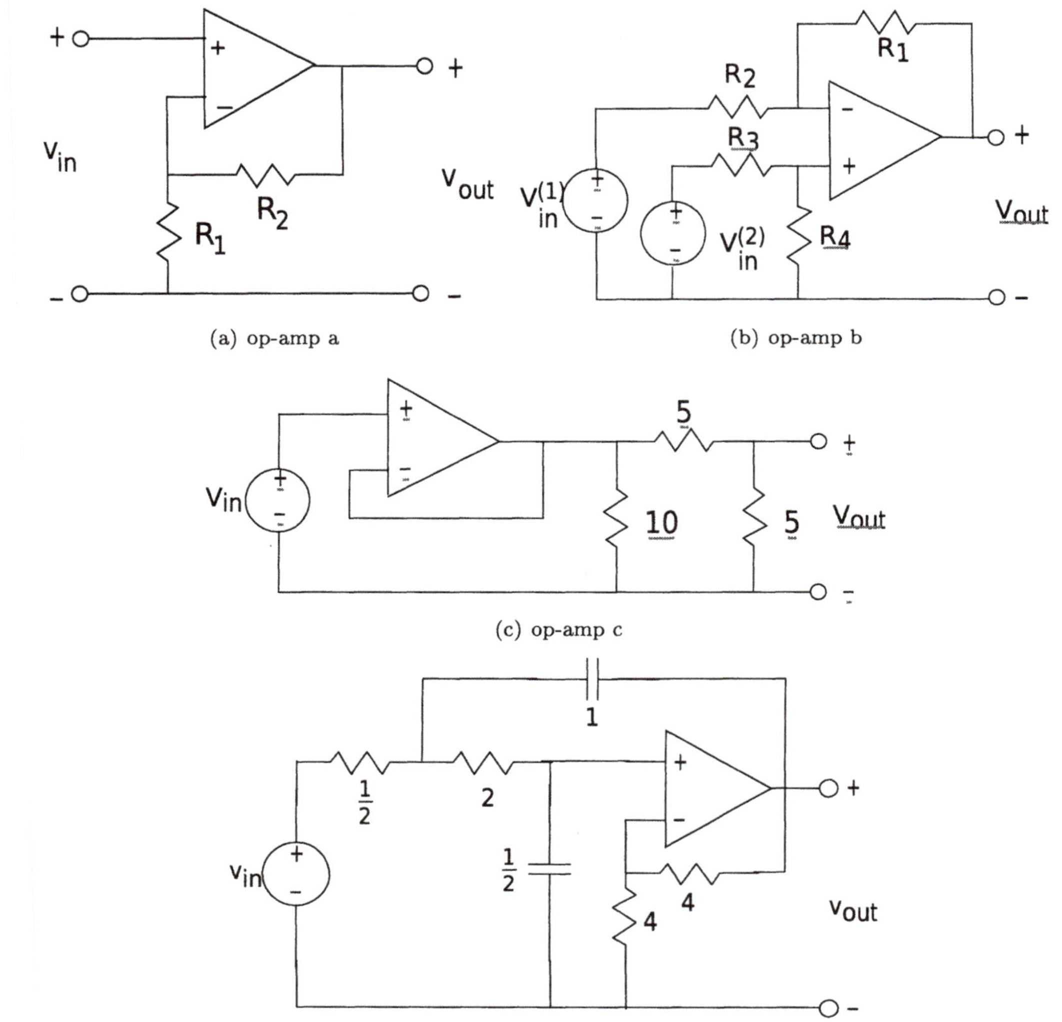

(a) op-amp a (b) op-amp b

(c) op-amp c

(d) op-amp d

Figure 3.80

Problem 3.37: Why Op-Amps are Useful
The circuit (Figure 3.81) of a cascade of op-amp circuits illustrate the reason why op-amp realizations of transfer functions are so useful.

a) Find the transfer function relating the complex amplitude of the voltage $v_{out}(t)$ to the source. Show that this transfer function equals the product of each stage's transfer function.
b) What is the load impedance appearing across the first op-amp's output?
c) Figure 3.82 illustrates that sometimes "designs" can go wrong. Find the transfer function for this op-amp circuit (Figure 3.82), and then show that it can't work! Why can't it?

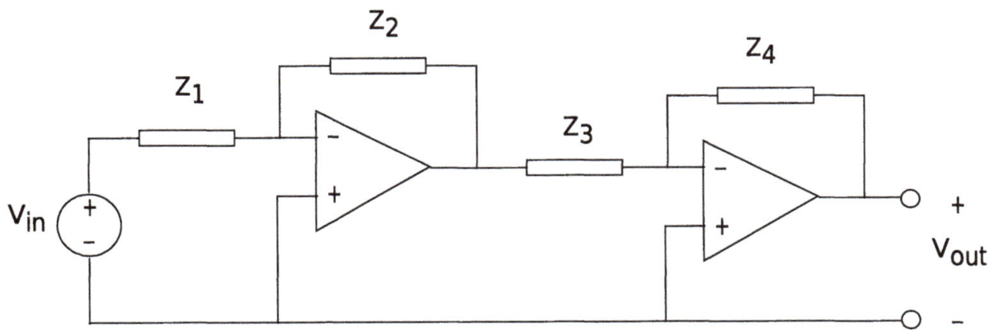

Figure 3.81

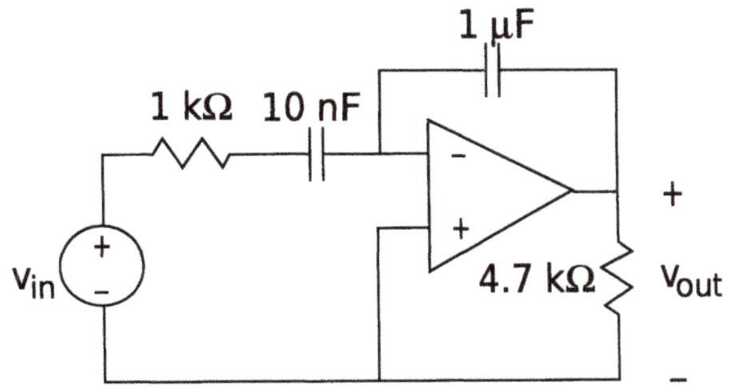

Figure 3.82

Problem 3.38: Operational Amplifiers
Consider the depicted circuit (Figure 3.83).

a) Find the transfer function relating the voltage $v_{out}(t)$ to the source.
b) In particular, $R_1 = 530\Omega$, $C_1 = 1\mu F$, $R_2 = 5.3k\Omega$, $C_2 = 0.01\mu F$, and $R_3 = R_4 = 5.3k\Omega$. Characterize the resulting transfer function and determine what use this circuit might have.

Problem 3.39: Designing a Bandpass Filter
We want to design a bandpass filter that has transfer the function

$$H(f) = 10\frac{j2\pi f}{\left(j\frac{f}{f_l}+1\right)\left(j\frac{f}{f_h}+1\right)}$$

Here, f_l is the cutoff frequency of the low-frequency edge of the passband and f_h is the cutoff frequency of the high-frequency edge. We want $f_l = 1kHz$ and $f_h = 10kHz$.

a) Plot the magnitude and phase of this frequency response. Label important amplitude and phase values and the frequencies at which they occur.

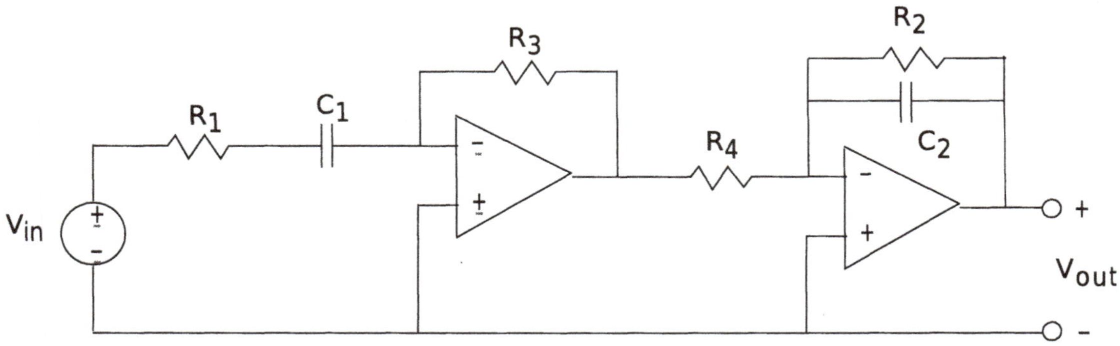

Figure 3.83

b) Design a bandpass filter that meets these specifications. Specify component values.

Problem 3.40: Pre-emphasis or De-emphasis?
In audio applications, prior to analog-to-digital conversion signals are passed through what is known as a **pre-emphasis circuit** that leaves the low frequencies alone but provides increasing gain at increasingly higher frequencies beyond some frequency f_0. **De-emphasis circuits** do the opposite and are applied after digital-to-analog conversion. After pre-emphasis, digitization, conversion back to analog and de-emphasis, the signal's spectrum should be what it was.

The op-amp circuit here (Figure 3.84) has been designed for pre-emphasis or de-emphasis (Samantha can't recall which).

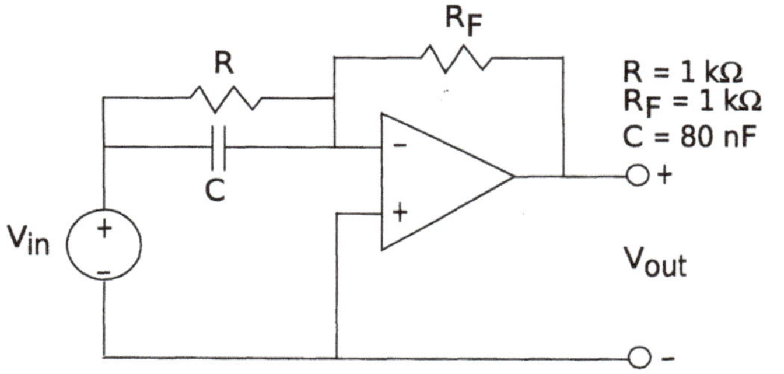

Figure 3.84

a) Is this a pre-emphasis or de-emphasis circuit? Find the frequency f_0 that defines the transition from low to high frequencies.
b) What is the circuit's output when the input voltage is $\sin(2\pi ft)$, with $f = 4kHz$?
c) What circuit could perform the opposite function to your answer for the first part?

Problem 3.41: Active Filter
Find the transfer function of the depicted active filter (Figure 3.85).

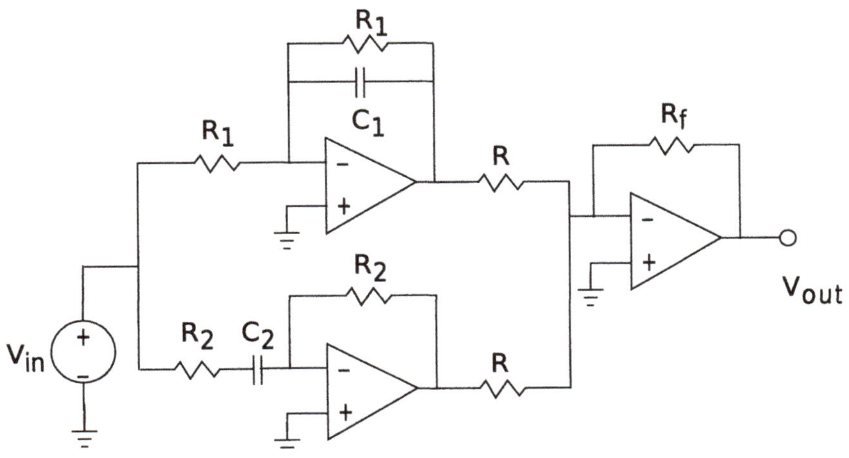

Figure 3.85

Problem 3.42: This is a filter?
You are given a circuit (Figure 3.86).

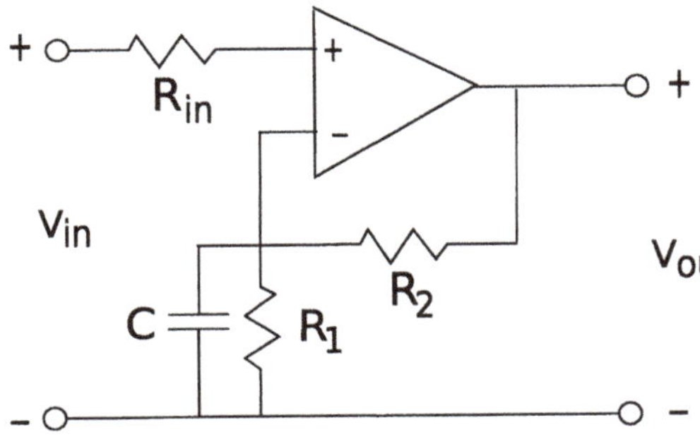

Figure 3.86

a) What is this circuit's transfer function? Plot the magnitude and phase.
b) If the input signal is the sinusoid $\sin(2\pi f_0 t)$, what will the output be when f_0 is larger than the filter's "cutoff frequency"?

Problem 3.43: Optical Receivers
In your optical telephone, the receiver circuit had the form shown (Figure 3.87).

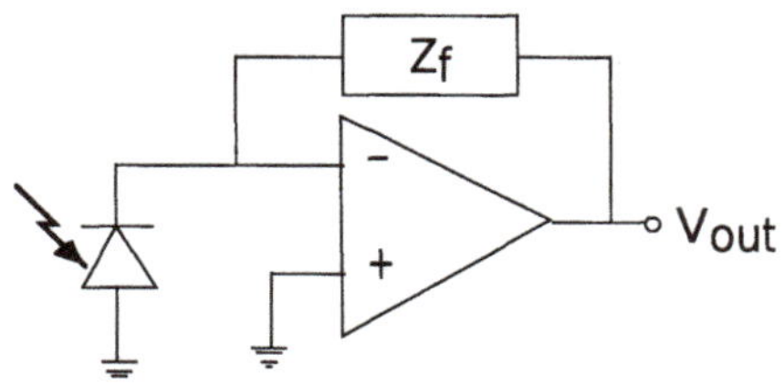

Figure 3.87

This circuit served as a transducer, converting light energy into a voltage v_{out}. The photodiode acts as a current source, producing a current proportional to the light intesity falling upon it. As is often the case in this crucial stage, the signals are small and noise can be a problem. Thus, the op-amp stage serves to boost the signal and to filter out-of-band noise.

a) Find the transfer function relating light intensity to v_{out}.
b) What should the circuit realizing the feedback impedance Z_f be so that the transducer acts as a 5 kHz lowpass filter?
c) A clever engineer suggests an alternative circuit (Figure 3.88) to accomplish the same task. Determine whether the idea works or not. If it does, find the impedance Z_{in} that accomplishes the lowpass filtering task. If not, show why it does not work.

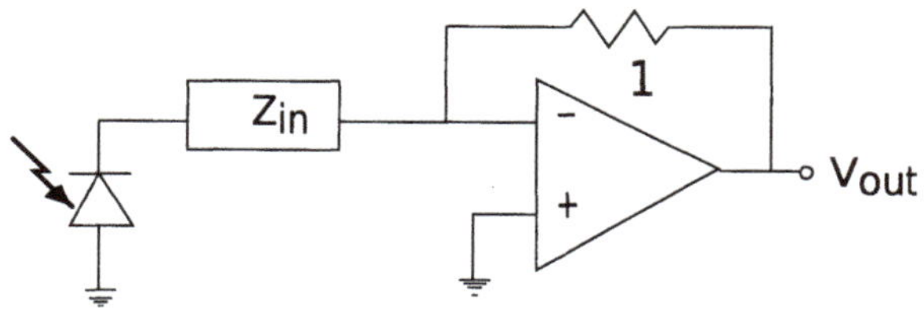

Figure 3.88

Problem 3.44: Reverse Engineering
The depicted circuit (Figure 3.89) has been developed by the TBBG Electronics design group. They are trying to keep its use secret; we, representing RU Electronics, have discovered the schematic and want to figure out the intended application. Assume the diode is ideal.

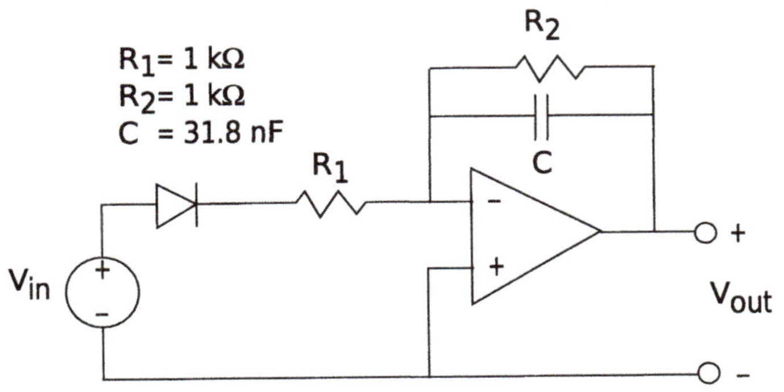

Figure 3.89

a) Assuming the diode is a short-circuit (it has been removed from the circuit), what is the circuit's transfer function?
b) With the diode in place, what is the circuit's output when the input voltage is $\sin(2\pi f_0 t)$?
c) What function might this circuit have?

Solutions to Exercises in Chapter 3

Solution to Exercise 3.1 (p. 36)
One kilowatt-hour equals 360,000 watt-seconds, which indeed directly corresponds to 360,000 joules.

Solution to Exercise 3.2 (p. 41)
KCL says that the sum of currents entering or leaving a node must be zero. If we consider two nodes together as a "supernode", KCL applies as well to currents entering the combination. Since no currents enter an entire circuit, the sum of currents must be zero. If we had a two-node circuit, the KCL equation of one *must* be the negative of the other, We can combine all but one node in a circuit into a supernode; KCL for the supernode must be the negative of the remaining node's KCL equation. Consequently, specifying $n-1$ KCL equations always specifies the remaining one.

Solution to Exercise 3.3 (p. 43)
The circuit serves as an amplifier having a gain of $\frac{R_2}{R_1+R_2}$.

Solution to Exercise 3.4 (p. 44)
The power consumed by the resistor R_1 can be expressed as

$$(v_{\text{in}} - v_{\text{out}})\, i_{\text{out}} = \frac{R_1}{(R_1+R_2)^2} {v_{\text{in}}}^2$$

Solution to Exercise 3.5 (p. 44)

$$\frac{1}{R_1+R_2} {v_{\text{in}}}^2 = \frac{R_1}{(R_1+R_2)^2} {v_{\text{in}}}^2 + + \frac{R_2}{(R_1+R_2)^2} {v_{\text{in}}}^2$$

Solution to Exercise 3.6 (p. 46)
Replacing the current source by a voltage source does not change the fact that the voltages are identical. Consequently, $v_{in} = R_2 i_{out}$ or $i_{out} = \frac{v_{in}}{R_2}$. This result does not depend on the resistor R_1, which means that we simply have a resistor (R_2) across a voltage source. The two-resistor circuit has no apparent use.

Solution to Exercise 3.7 (p. 47)
$R_{eq} = \frac{R_2}{1+\frac{R_2}{R_L}}$. Thus, a 10% change means that the ratio $\frac{R_2}{R_L}$ must be less than 0.1. A 1% change means that $\frac{R_2}{R_L} < 0.01$.

Solution to Exercise 3.8 (p. 49)
In a series combination of resistors, the current is the same in each; in a parallel combination, the voltage is the same. For a series combination, the equivalent resistance is the sum of the resistances, which will be larger than any component resistor's value; for a parallel combination, the equivalent conductance is the sum of the component conductances, which is larger than any component conductance. The equivalent resistance is therefore smaller than any component resistance.

Solution to Exercise 3.9 (p. 51)
$v_{oc} = \frac{R_2}{R_1+R_2} v_{in}$ and $i_{sc} = -\left(\frac{v_{in}}{R_1}\right)$ (resistor R_2 is shorted out in this case). Thus, $v_{eq} = \frac{R_2}{R_1+R_2} v_{in}$ and $R_{eq} = \frac{R_1 R_2}{R_1+R_2}$.

Solution to Exercise 3.10 (p. 52)
$i_{eq} = \frac{R_1}{R_1+R_2} i_{in}$ and $R_{eq} = (R_3 \parallel R_1 + R_2)$.

Solution to Exercise 3.11 (p. 58)
Division by $j2\pi f$ arises from integrating a complex exponential. Consequently,

$$\left(\frac{1}{j2\pi f} V \Leftrightarrow \int V e^{j2\pi f t} dt\right)$$

Solution to Exercise 3.12 (p. 59)

For maximum power dissipation, the imaginary part of complex power should be zero. As the complex power is given by $VI^* = |V||I|e^{j(\phi-\theta)}$, zero imaginary part occurs when the phases of the voltage and currents agree.

Solution to Exercise 3.13 (p. 59)

$P_{\text{ave}} = V_{\text{rms}} I_{\text{rms}} \cos(\phi - \theta)$. The cosine term is known as the **power factor**.

Solution to Exercise 3.14 (p. 64)

The key notion is writing the imaginary part as the difference between a complex exponential and its complex conjugate:

$$\text{Im}\left(Ve^{j2\pi ft}\right) = \frac{Ve^{j2\pi ft} - V^* e^{-(j2\pi ft)}}{2j} \quad (3.35)$$

The response to $Ve^{j2\pi ft}$ is $VH(f)\, e^{j2\pi ft}$, which means the response to $V^* e^{-(j2\pi ft)}$ is $V^* H(-f)\, e^{-(j2\pi ft)}$. As $H(-f) = \left(H(f)^*\right)$, the Superposition Principle says that the output to the imaginary part is $\text{Im}\left(VH(f)\, e^{j2\pi ft}\right)$. The same argument holds for the real part: $\text{Re}\left(Ve^{j2\pi ft}\right) \to \text{Re}\left(VH(f)\, e^{j2\pi ft}\right)$.

Solution to Exercise 3.15 (p. 69)

To find the equivalent resistance, we need to find the current flowing through the voltage source. This current equals the current we have just found plus the current flowing through the other vertical $1\,\Omega$ resistor. This current equals $\frac{e_1}{1} = \frac{6}{13} v_{in}$, making the total current through the voltage source (flowing out of it) $\frac{11}{13} v_{in}$. Thus, the equivalent resistance is $\frac{13}{11}\Omega$.

Solution to Exercise 3.16 (p. 71)

Not necessarily, especially if we desire individual knobs for adjusting the gain and the cutoff frequency.

Solution to Exercise 3.17 (p. 79)

The ratio between adjacent values is about $\sqrt{2}$.

Chapter 4

Frequency Domain

4.1 Introduction to the Frequency Domain[1]

In developing ways of analyzing linear circuits, we invented the impedance method because it made solving circuits easier. Along the way, we developed the notion of a circuit's frequency response or transfer function. This notion, which also applies to all linear, time-invariant systems, describes how the circuit responds to a sinusoidal input when we express it in terms of a complex exponential. We also learned the Superposition Principle for linear systems: The system's output to an input consisting of a sum of two signals is the sum of the system's outputs to each individual component.

The study of the frequency domain combines these two notions–a system's sinusoidal response is easy to find and a linear system's output to a sum of inputs is the sum of the individual outputs–to develop the crucial idea of a signal's **spectrum**. We begin by finding that those signals that can be represented as a sum of sinusoids is very large. In fact, *all signals can be expressed as a superposition of sinusoids*.

As this story unfolds, we'll see that information systems rely heavily on spectral ideas. For example, radio, television, and cellular telephones transmit over different portions of the spectrum. In fact, spectrum is so important that communications systems are regulated as to which portions of the spectrum they can use by the Federal Communications Commission in the United States and by International Treaty for the world (see Frequency Allocations (Section 7.3)). Calculating the spectrum is easy: The **Fourier transform** defines how we can find a signal's spectrum.

4.2 Complex Fourier Series[2]

In an earlier module (Exercise 2.4), we showed that a square wave could be expressed as a superposition of pulses. As useful as this decomposition was in this example, it does not generalize well to other periodic signals: How can a superposition of pulses equal a smooth signal like a sinusoid? Because of the importance of sinusoids to linear systems, you might wonder whether they could be added together to represent a large number of periodic signals. You would be right and in good company as well. Euler[3] and Gauss[4] in particular worried about this problem, and Jean Baptiste Fourier[5] got the credit even though tough mathematical issues were not settled until later. They worked on what is now known as the **Fourier series**: representing *any* periodic signal as a superposition of sinusoids.

But the Fourier series goes well beyond being another signal decomposition method. Rather, the Fourier series begins our journey to appreciate how a signal can be described in either the time-domain or the

[1] This content is available online at <http://cnx.org/content/m0038/2.10/>.
[2] This content is available online at <http://cnx.org/content/m0042/2.24/>.
[3] http://www-groups.dcs.st-and.ac.uk/~history/Mathematicians/Euler.html
[4] http://www-groups.dcs.st-and.ac.uk/~history/Mathematicians/Guass.html
[5] http://www-groups.dcs.st-and.ac.uk/~history/Mathematicians/Fourier.html

frequency-domain with *no* compromise. Let $s(t)$ be a *periodic* signal with period T. We want to show that periodic signals, even those that have constant-valued segments like a square wave, can be expressed as sum of **harmonically** related sine waves: sinusoids having frequencies that are integer multiples of the **fundamental frequency**. Because the signal has period T, the fundamental frequency is $\frac{1}{T}$. The complex Fourier series expresses the signal as a superposition of complex exponentials having frequencies $\frac{k}{T}$, $k = \{\dots, -1, 0, 1, \dots\}$.

$$s(t) = \sum_{k=-\infty}^{\infty} \left(c_k e^{j\frac{2\pi kt}{T}}\right) \tag{4.1}$$

with $c_k = \frac{1}{2}(a_k + (-(jb_k)))$. The real and imaginary parts of the **Fourier coefficients** c_k are written in this unusual way for convenience in defining the classic Fourier series. The zeroth coefficient equals the signal's average value and is real-valued for real-valued signals: $c_0 = a_0$. The family of functions $\left\{e^{j\frac{2\pi kt}{T}}\right\}$ are called **basis functions** and form the foundation of the Fourier series. No matter what the periodic signal might be, these functions are always present and form the representation's building blocks. They depend on the signal period T, and are indexed by k.

KEY POINT: Assuming we know the period, knowing the Fourier coefficients is equivalent to knowing the signal. Thus, it makes not difference if we have a time-domain or a frequency-domain characterization of the signal.

Exercise 4.1 ***(Solution on p. 152.)***

What is the complex Fourier series for a sinusoid?

To find the Fourier coefficients, we note the orthogonality property

$$\int_0^T e^{j\frac{2\pi kt}{T}} e^{(-j)\frac{2\pi lt}{T}} \, dt = \begin{cases} T \text{ if } k = l \\ 0 \text{ if } k \neq l \end{cases} \tag{4.2}$$

Assuming for the moment that the complex Fourier series "works," we can find a signal's complex Fourier coefficients, its **spectrum**, by exploiting the orthogonality properties of harmonically related complex exponentials. Simply multiply each side of (4.1) by $e^{-(j2\pi lt)}$ and integrate over the interval $[0, T]$.

$$\begin{aligned} c_k &= \tfrac{1}{T}\int_0^T s(t)\, e^{-(j\frac{2\pi kt}{T})}\, dt \\ c_0 &= \tfrac{1}{T}\int_0^T s(t)\, dt \end{aligned} \tag{4.3}$$

Example 4.1

Finding the Fourier series coefficients for the square wave is very simple. $sq_T(t)$. Mathematically, this signal can be expressed as

$$sq_T(t) = \begin{cases} 1 \text{ if } 0 \leq t < \frac{T}{2} \\ -1 \text{ if } \frac{T}{2} \leq t < T \end{cases}$$

The expression for the Fourier coefficients has the form

$$c_k = \frac{1}{T}\int_0^{\frac{T}{2}} e^{-(j\frac{2\pi kt}{T})}\, dt - \frac{1}{T}\int_{\frac{T}{2}}^{T} e^{-(j\frac{2\pi kt}{T})}\, dt \tag{4.4}$$

NOTE: When integrating an expression containing j, treat it just like any other constant.

The two integrals are very similar, one equaling the negative of the other. The final expression becomes

$$\begin{aligned} b_k &= 2\left(-\left(\tfrac{1}{j2\pi k}\left((-1)^k - 1\right)\right)\right) \\ &= \begin{cases} \frac{2}{j\pi k} \text{ if k odd} \\ 0 \text{ if k even} \end{cases} \end{aligned} \tag{4.5}$$

Thus, the complex Fourier series for the square wave is

$$\mathrm{sq}(t) = \sum_{k \in \{\dots,-3,-1,1,3,\dots\}} \left(\frac{2}{j\pi k} e^{(+j)\frac{2\pi kt}{T}}\right) \tag{4.6}$$

Consequently, the square wave equals a sum of complex exponentials, but only those having frequencies equal to odd multiples of the fundamental frequency $\frac{1}{T}$. The coefficients decay slowly as the frequency index k increases. This index corresponds to the k-th harmonic of the signal's period.

A signal's Fourier series spectrum c_k has interesting properties.

Property 4.1:
If $s(t)$ is real, $c_{-k} = c_k{}^*$ (real-valued periodic signals have conjugate-symmetric spectra).

This result follows from the integral that calculates the c_k from the signal. Furthermore, this result means that $\mathrm{Re}(c_k) = \mathrm{Re}(c_{-k})$: The real part of the Fourier coefficients for real-valued signals is even. Similarly, $\mathrm{Im}(c_k) = -(\mathrm{Im}(c_{-k}))$: The imaginary parts of the Fourier coefficients have odd symmetry. Consequently, if you are given the Fourier coefficients for positive indices and zero and are told the signal is real-valued, you can find the negative-indexed coefficients, hence the entire spectrum. This kind of symmetry, $c_k = c_k{}^*$, is known as **conjugate symmetry**.

Property 4.2:
If $s(-t) = s(t)$, which says the signal has even symmetry about the origin, $c_{-k} = c_k$.

Given the previous property for real-valued signals, the Fourier coefficients of even signals are real-valued. A real-valued Fourier expansion amounts to an expansion in terms of only cosines, which is the simplest example of an even signal.

Property 4.3:
If $s(-t) = -(s(t))$, which says the signal has odd symmetry, $c_{-k} = -c_k$.

Therefore, the Fourier coefficients are purely imaginary. The square wave is a great example of an odd-symmetric signal.

Property 4.4:
The spectral coefficients for a periodic signal delayed by τ $s(t-\tau)$ are $c_k e^{-\left(\frac{j2\pi k\tau}{T}\right)}$, where c_k denotes the spectrum of $s(t)$. Delaying a signal by τ seconds results in a spectrum having a **linear phase shift** of $-\left(\frac{2\pi k\tau}{T}\right)$ in comparison to the spectrum of the undelayed signal. Note that the spectral magnitude is unaffected. Showing this property is easy.

Proof:

$$\begin{aligned} \frac{1}{T}\int_0^T s(t-\tau)\, e^{(-j)\frac{2\pi kt}{T}}\, dt &= \frac{1}{T}\int_{-\tau}^{T-\tau} s(t)\, e^{(-j)\frac{2\pi k(t+\tau)}{T}}\, dt \\ &= \frac{1}{T} e^{(-j)\frac{2\pi k\tau}{T}} \int_{-\tau}^{T-\tau} s(t)\, e^{(-j)\frac{2\pi kt}{T}}\, dt \end{aligned} \tag{4.7}$$

Note that the range of integration extends over a period of the integrand. Consequently, it should not matter how we integrate over a period, which means that $\int_{-\tau}^{T-\tau} (\cdot)\, dt = \int_0^T (\cdot)\, dt$, and we have our result.

The complex Fourier series obeys **Parseval's Theorem**, one of the most important results in signal analysis. This general mathematical result says you can calculate a signal's power in either the time domain or the frequency domain.

Theorem 4.1: Parseval's Theorem
Average power calculated in the time domain equals the power calculated in the frequency domain.

$$\frac{1}{T}\int_0^T s^2(t)\, dt = \sum_{k=-\infty}^{\infty} \left((|c_k|)^2\right) \tag{4.8}$$

This result is a (simpler) re-expression of how to calculate a signal's power than with the real-valued Fourier series expression for power (4.22).

Let's calculate the Fourier coefficients of the periodic pulse signal shown here (Figure 4.1).

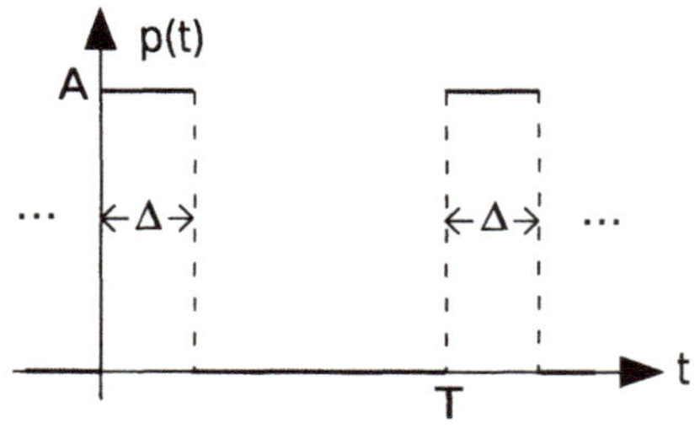

Figure 4.1: Periodic pulse signal.

The pulse width is Δ, the period T, and the amplitude A. The complex Fourier spectrum of this signal is given by

$$c_k = \frac{1}{T}\int_0^{\Delta} Ae^{-\left(\frac{j2\pi k\Delta}{T}\right)}dt = \left(-\left(\frac{A}{j2\pi k}\right)\right)\left(e^{-\left(\frac{j2\pi k\Delta}{T}\right)}-1\right)$$

At this point, simplifying this expression requires knowing an interesting property.

$$1-e^{-(j\theta)} = e^{-\left(\frac{j\theta}{2}\right)}\left(e^{+\frac{j\theta}{2}}-e^{-\left(\frac{j\theta}{2}\right)}\right) = e^{-\left(\frac{j\theta}{2}\right)}2j\sin\left(\frac{\theta}{2}\right)$$

Armed with this result, we can simply express the Fourier series coefficients for our pulse sequence.

$$c_k = Ae^{-\left(\frac{j\pi k\Delta}{T}\right)}\frac{\sin\left(\frac{\pi k\Delta}{T}\right)}{\pi k} \tag{4.9}$$

Because this signal is real-valued, we find that the coefficients do indeed have conjugate symmetry: $c_k = {c_{-k}}^*$. The periodic pulse signal has neither even nor odd symmetry; consequently, no additional symmetry exists in the spectrum. Because the spectrum is complex valued, to plot it we need to calculate its magnitude and phase.

$$|c_k| = A\left|\frac{\sin\left(\frac{\pi k\Delta}{T}\right)}{\pi k}\right| \tag{4.10}$$

$$\angle(c_k) = -\left(\frac{\pi k\Delta}{T}\right) + \pi\mathrm{neg}\left(\frac{\sin\left(\frac{\pi k\Delta}{T}\right)}{\pi k}\right)\mathrm{sign}(k)$$

The function $\mathrm{neg}(\cdot)$ equals -1 if its argument is negative and zero otherwise. The somewhat complicated expression for the phase results because the sine term can be negative; magnitudes must be positive, leaving the occasional negative values to be accounted for as a phase shift of π.

Periodic Pulse Sequence

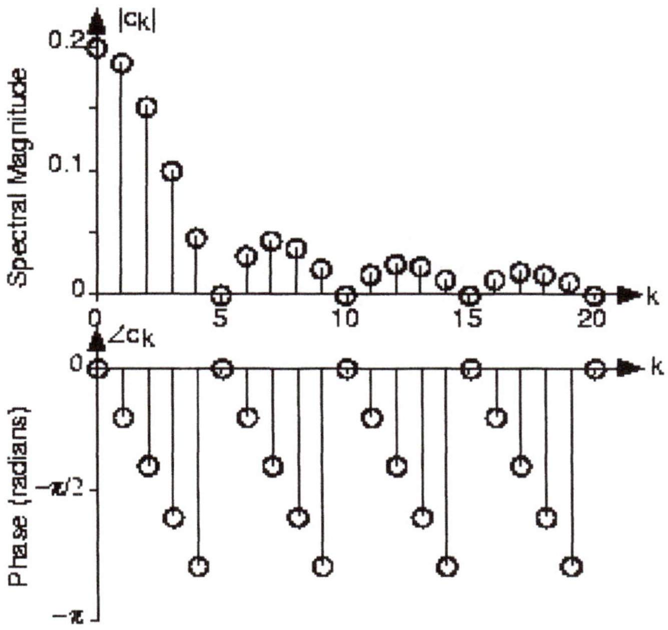

Figure 4.2: The magnitude and phase of the periodic pulse sequence's spectrum is shown for positive-frequency indices. Here $\frac{\Delta}{T} = 0.2$ and $A = 1$.

Also note the presence of a linear phase term (the first term in $\angle(c_k)$ is proportional to frequency $\frac{k}{T}$). Comparing this term with that predicted from delaying a signal, a delay of $\frac{\Delta}{2}$ is present in our signal. Advancing the signal by this amount centers the pulse about the origin, leaving an even signal, which in turn means that its spectrum is real-valued. Thus, our calculated spectrum is consistent with the properties of the Fourier spectrum.

Exercise 4.2 ***(Solution on p. 152.)***

What is the value of c_0? Recalling that this spectral coefficient corresponds to the signal's average value, does your answer make sense?

The phase plot shown in Figure 4.2 (Periodic Pulse Sequence) requires some explanation as it does not seem to agree with what (4.10) suggests. There, the phase has a linear component, with a jump of π every time the sinusoidal term changes sign. We must realize that any integer multiple of 2π can be added to a phase at each frequency *without* affecting the value of the complex spectrum. We see that at frequency index 4 the phase is nearly $-\pi$. The phase at index 5 is undefined because the magnitude is zero in this example. At index 6, the formula suggests that the phase of the linear term should be less than (more negative) than $-\pi$. In addition, we expect a shift of $-\pi$ in the phase between indices 4 and 6. Thus, the phase value predicted by the formula is a little less than $-(2\pi)$. Because we can add 2π without affecting the value of the spectrum at index 6, the result is a slightly negative number as shown. Thus, the formula and the plot do agree. In phase calculations like those made in MATLAB, values are usually confined to the range $[-\pi, \pi)$ by adding some (possibly negative) multiple of 2π to each phase value.

4.3 Classic Fourier Series[6]

The classic Fourier series as derived originally expressed a periodic signal (period T) in terms of harmonically related sines and cosines.

$$s(t) = a_0 + \sum_{k=1}^{\infty} \left(a_k \cos\left(\frac{2\pi kt}{T}\right)\right) + \sum_{k=1}^{\infty} \left(b_k \sin\left(\frac{2\pi kt}{T}\right)\right) \tag{4.11}$$

The complex Fourier series and the sine-cosine series are identical, each representing a signal's spectrum. The **Fourier coefficients**, a_k and b_k, express the real and imaginary parts respectively of the spectrum while the coefficients c_k of the complex Fourier series express the spectrum as a magnitude and phase. Equating the classic Fourier series (4.11) to the complex Fourier series (4.1), an extra factor of two and complex conjugate become necessary to relate the Fourier coefficients in each.

$$c_k = \frac{1}{2}(a_k - jb_k)$$

Exercise 4.3 *(Solution on p. 152.)*

Derive this relationship between the coefficients of the two Fourier series.

Just as with the complex Fourier series, we can find the Fourier coefficients using the **orthogonality** properties of sinusoids. Note that the cosine and sine of harmonically related frequencies, even the *same* frequency, are orthogonal.

$$\int_0^T \sin\left(\frac{2\pi kt}{T}\right)\cos\left(\frac{2\pi lt}{T}\right) dt = 0 \;,\quad k \in \mathbb{Z} \;\; l \in \mathbb{Z} \tag{4.12}$$

$$\int_0^T \sin\left(\frac{2\pi kt}{T}\right)\sin\left(\frac{2\pi lt}{T}\right) dt = \begin{cases} \frac{T}{2} \text{ if } k = l \text{ and } k \neq 0 \text{ and } l \neq 0 \\ 0 \text{ if } k \neq l \text{ or } k = 0 = l \end{cases}$$

$$\int_0^T \cos\left(\frac{2\pi kt}{T}\right)\cos\left(\frac{2\pi lt}{T}\right) dt = \begin{cases} \frac{T}{2} \text{ if } k = l \text{ and } k \neq 0 \text{ and } l \neq 0 \\ T \text{ if } k = 0 = l \\ 0 \text{ if } k \neq l \end{cases}$$

These orthogonality relations follow from the following important trigonometric identities.

$$\begin{aligned} \sin(\alpha)\sin(\beta) &= \tfrac{1}{2}(\cos(\alpha-\beta) - \cos(\alpha+\beta)) \\ \cos(\alpha)\cos(\beta) &= \tfrac{1}{2}(\cos(\alpha+\beta) + \cos(\alpha-\beta)) \\ \sin(\alpha)\cos(\beta) &= \tfrac{1}{2}(\sin(\alpha+\beta) + \sin(\alpha-\beta)) \end{aligned} \tag{4.13}$$

These identities allow you to substitute a sum of sines and/or cosines for a product of them. Each term in the sum can be integrating by noticing one of two important properties of sinusoids.

- The integral of a sinusoid over an *integer* number of periods equals zero.
- The integral of the *square* of a unit-amplitude sinusoid over a period T equals $\frac{T}{2}$.

[6]This content is available online at <http://cnx.org/content/m0039/2.22/>.

To use these, let's, for example, multiply the Fourier series for a signal by the cosine of the lth harmonic $\cos\left(\frac{2\pi lt}{T}\right)$ and integrate. The idea is that, because integration is linear, the integration will sift out all but the term involving a_l.

$$\int_0^T s(t)\cos\left(\frac{2\pi lt}{T}\right)dt = \int_0^T a_0\cos\left(\frac{2\pi lt}{T}\right)dt + \sum_{k=1}^{\infty}\left(a_k\int_0^T\cos\left(\frac{2\pi kt}{T}\right)\cos\left(\frac{2\pi lt}{T}\right)dt\right) + \sum_{k=1}^{\infty}\left(b_k\int_0^T\sin\left(\frac{2\pi kt}{T}\right)\cos\left(\frac{2\pi lt}{T}\right)dt\right) \quad (4.14)$$

The first and third terms are zero; in the second, the only non-zero term in the sum results when the indices k and l are equal (but not zero), in which case we obtain $\frac{a_l T}{2}$. If $k = 0 = l$, we obtain $a_0 T$. Consequently,

$$a_l = \frac{2}{T}\int_0^T s(t)\cos\left(\frac{2\pi lt}{T}\right)dt \ , \ l \neq 0$$

All of the Fourier coefficients can be found similarly.

$$\begin{aligned} a_0 &= \frac{1}{T}\int_0^T s(t)\,dt \\ a_k &= \frac{2}{T}\int_0^T s(t)\cos\left(\frac{2\pi kt}{T}\right)dt \ , \ k \neq 0 \\ b_k &= \frac{2}{T}\int_0^T s(t)\sin\left(\frac{2\pi kt}{T}\right)dt \end{aligned} \quad (4.15)$$

Exercise 4.4 *(Solution on p. 152.)*

The expression for a_0 is referred to as the *average value* of $s(t)$. Why?

Exercise 4.5 *(Solution on p. 152.)*

What is the Fourier series for a unit-amplitude square wave?

Example 4.2

Let's find the Fourier series representation for the half-wave rectified sinusoid.

$$s(t) = \begin{cases} \sin\left(\frac{2\pi t}{T}\right) \text{ if } 0 \leq t < \frac{T}{2} \\ 0 \text{ if } \frac{T}{2} \leq t < T \end{cases} \quad (4.16)$$

Begin with the sine terms in the series; to find b_k we must calculate the integral

$$b_k = \frac{2}{T}\int_0^{\frac{T}{2}}\sin\left(\frac{2\pi t}{T}\right)\sin\left(\frac{2\pi kt}{T}\right)dt \quad (4.17)$$

Using our trigonometric identities turns our integral of a product of sinusoids into a sum of integrals of individual sinusoids, which are much easier to evaluate.

$$\begin{aligned} \int_0^{\frac{T}{2}}\sin\left(\frac{2\pi t}{T}\right)\sin\left(\frac{2\pi kt}{T}\right)dt &= \frac{1}{2}\int_0^{\frac{T}{2}}\cos\left(\frac{2\pi(k-1)t}{T}\right) - \cos\left(\frac{2\pi(k+1)t}{T}\right)dt \\ &= \begin{cases} \frac{1}{2} \text{ if } k = 1 \\ 0 \text{ otherwise} \end{cases} \end{aligned} \quad (4.18)$$

Thus,

$$b_1 = \frac{1}{2}$$

$$b_2 = b_3 = \cdots = 0$$

On to the cosine terms. The average value, which corresponds to a_0, equals $\frac{1}{\pi}$. The remainder of the cosine coefficients are easy to find, but yield the complicated result

$$a_k = \begin{cases} -\left(\frac{2}{\pi}\frac{1}{k^2-1}\right) \text{ if } k \in \{2, 4, \dots\} \\ 0 \text{ if k odd} \end{cases} \tag{4.19}$$

Thus, the Fourier series for the half-wave rectified sinusoid has non-zero terms for the average, the fundamental, and the even harmonics.

4.4 A Signal's Spectrum[7]

A periodic signal, such as the half-wave rectified sinusoid, consists of a sum of elemental sinusoids. A plot of the Fourier coefficients as a function of the frequency index, such as shown in Figure 4.3 (Fourier Series spectrum of a half-wave rectified sine wave), displays the signal's **spectrum**. The word "spectrum" implies that the independent variable, here k, corresponds somehow to frequency. Each coefficient is directly related to a sinusoid having a frequency of $\frac{k}{T}$. Thus, if we half-wave rectified a 1 kHz sinusoid, $k = 1$ corresponds to 1 kHz, $k = 2$ to 2 kHz, etc.

Fourier Series spectrum of a half-wave rectified sine wave

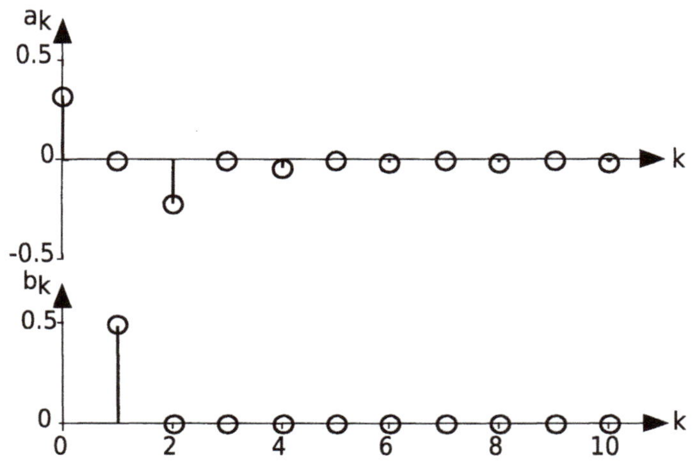

Figure 4.3: The Fourier series spectrum of a half-wave rectified sinusoid is shown. The index indicates the multiple of the fundamental frequency at which the signal has energy.

A subtle, but very important, aspect of the Fourier spectrum is its *uniqueness*: You can unambiguously find the spectrum from the signal (decomposition (4.15)) and the signal from the spectrum (composition). Thus, any aspect of the signal can be found from the spectrum and vice versa. *A signal's frequency domain expression is its spectrum.* A periodic signal can be defined either in the time domain (as a function) or in the frequency domain (as a spectrum).

[7]This content is available online at <http://cnx.org/content/m0040/2.19/>.

A fundamental aspect of solving electrical engineering problems is whether the time or frequency domain provides the most understanding of a signal's properties and the simplest way of manipulating it. The uniqueness property says that either domain can provide the right answer. As a simple example, suppose we want to know the (periodic) signal's maximum value. Clearly the time domain provides the answer directly. To use a frequency domain approach would require us to find the spectrum, form the signal from the spectrum and calculate the maximum; we're back in the time domain!

Another feature of a signal is its average **power**. A signal's instantaneous power is defined to be its square. The average power is the average of the instantaneous power over some time interval. For a periodic signal, the natural time interval is clearly its period; for nonperiodic signals, a better choice would be entire time or time from onset. For a periodic signal, the average power is the square of its root-mean-squared (rms) value. We define the **rms** value of a periodic signal to be

$$rms(s) = \sqrt{\frac{1}{T}\int_0^T s^2(t)\,dt} \tag{4.20}$$

and thus its average power is

$$\begin{aligned} power(s) &= rms^2(s) \\ &= \frac{1}{T}\int_0^T s^2(t)\,dt \end{aligned} \tag{4.21}$$

Exercise 4.6 ***(Solution on p. 152.)***

What is the rms value of the half-wave rectified sinusoid?

To find the average power in the frequency domain, we need to substitute the spectral representation of the signal into this expression.

$$power(s) = \frac{1}{T}\int_0^T \left(a_0 + \sum_{k=1}^{\infty}\left(a_k \cos\left(\frac{2\pi kt}{T}\right)\right) + \sum_{k=1}^{\infty}\left(b_k \sin\left(\frac{2\pi kt}{T}\right)\right)\right)^2 dt$$

The square inside the integral will contain all possible pairwise products. However, the orthogonality properties (4.12) say that most of these crossterms integrate to zero. The survivors leave a rather simple expression for the power we seek.

$$power(s) = {a_0}^2 + \frac{1}{2}\sum_{k=1}^{\infty}\left({a_k}^2 + {b_k}^2\right) \tag{4.22}$$

It could well be that computing this sum is easier than integrating the signal's square. Furthermore, the contribution of each term in the Fourier series toward representing the signal can be measured by its contribution to the signal's average power. Thus, the power contained in a signal at its kth harmonic is $\frac{{a_k}^2+{b_k}^2}{2}$. The **power spectrum**, $P_s(k)$, such as shown in Figure 4.4 (Power Spectrum of a Half-Wave Rectified Sinusoid), plots each harmonic's contribution to the total power.

Exercise 4.7 ***(Solution on p. 152.)***

In high-end audio, deviation of a sine wave from the ideal is measured by the **total harmonic distortion**, which equals the total power in the harmonics higher than the first compared to power in the fundamental. Find an expression for the total harmonic distortion for any periodic signal. Is this calculation most easily performed in the time or frequency domain?

4.5 Fourier Series Approximation of Signals[8]

It is interesting to consider the sequence of signals that we obtain as we incorporate more terms into the Fourier series approximation of the half-wave rectified sine wave (Example 4.2). Define $s_K(t)$ to be the

[8]This content is available online at <http://cnx.org/content/m10687/2.8/>.

Power Spectrum of a Half-Wave Rectified Sinusoid

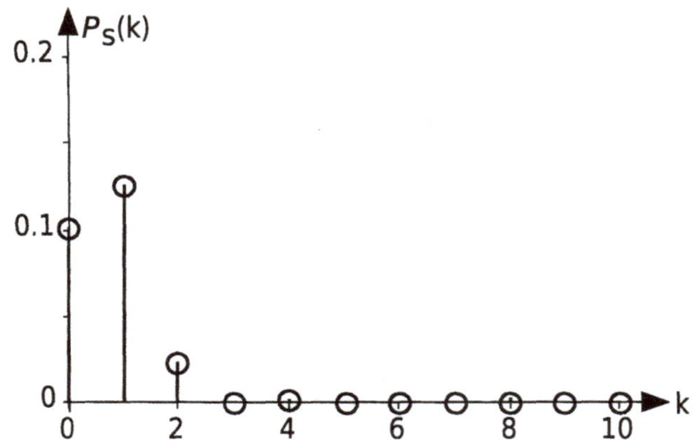

Figure 4.4: Power spectrum of a half-wave rectified sinusoid.

signal containing $K+1$ Fourier terms.

$$s_K(t) = a_0 + \sum_{k=1}^{K} \left(a_k \cos \left(\frac{2\pi kt}{T} \right) \right) + \sum_{k=1}^{K} \left(b_k \sin \left(\frac{2\pi kt}{T} \right) \right) \tag{4.23}$$

Figure 4.5 (Fourier Series spectrum of a half-wave rectified sine wave) shows how this sequence of signals portrays the signal more accurately as more terms are added.

We need to assess quantitatively the accuracy of the Fourier series approximation so that we can judge how rapidly the series approaches the signal. When we use a $K+1$-term series, the error—the difference between the signal and the $K+1$-term series—corresponds to the unused terms from the series.

$$\epsilon_K(t) = \sum_{k=K+1}^{\infty} \left(a_k \cos \left(\frac{2\pi kt}{T} \right) \right) + \sum_{k=K+1}^{\infty} \left(b_k \sin \left(\frac{2\pi kt}{T} \right) \right) \tag{4.24}$$

To find the rms error, we must square this expression and integrate it over a period. Again, the integral of most cross-terms is zero, leaving

$$rms(\epsilon_K) = \sqrt{\frac{1}{2} \sum_{k=K+1}^{\infty} \left({a_k}^2 + {b_k}^2 \right)} \tag{4.25}$$

Figure 4.6 (Approximation error for a half-wave rectified sinusoid) shows how the error in the Fourier series for the half-wave rectified sinusoid decreases as more terms are incorporated. In particular, the use of four terms, as shown in the bottom plot of Figure 4.5 (Fourier Series spectrum of a half-wave rectified sine wave), has a rms error (relative to the rms value of the signal) of about 3%. The Fourier series in this case converges quickly to the signal.

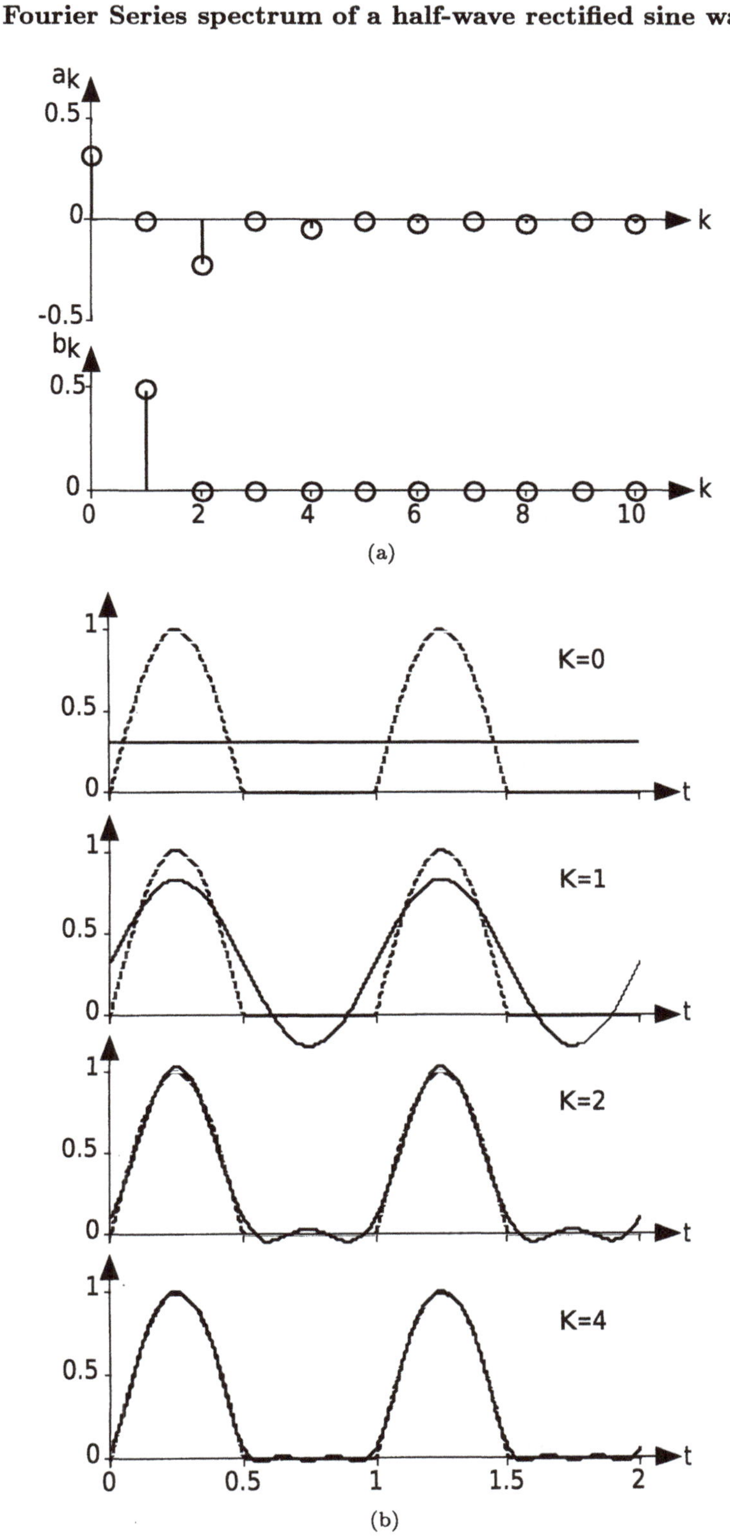

Figure 4.5: The Fourier series spectrum of a half-wave rectified sinusoid is shown in the upper portion. The index indicates the multiple of the fundamental frequency at which the signal has energy. The cumulative effect of adding terms to the Fourier series for the half-wave rectified sine wave is shown in the bottom portion. The dashed line is the actual signal, with the solid line showing the finite series approximation to the indicated number of terms, $K + 1$.

Approximation error for a half-wave rectified sinusoid

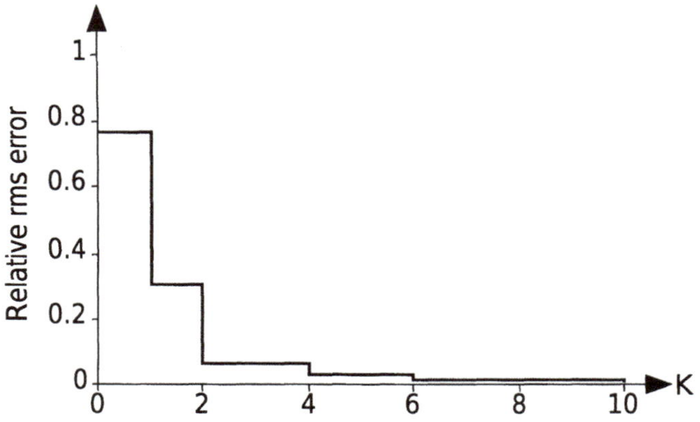

Figure 4.6: The rms error calculated according to (4.25) is shown as a function of the number of terms in the series for the half-wave rectified sinusoid. The error has been normalized by the rms value of the signal.

We can look at Figure 4.7 (Power spectrum and approximation error for a square wave) to see the power spectrum and the rms approximation error for the square wave. Because the Fourier coefficients decay more slowly here than for the half-wave rectified sinusoid, the rms error is not decreasing quickly. Said another way, the square-wave's spectrum contains more power at higher frequencies than does the half-wave-rectified sinusoid. This difference between the two Fourier series results because the half-wave rectified sinusoid's Fourier coefficients are proportional to $\frac{1}{k^2}$ while those of the square wave are proportional to $\frac{1}{k}$. If fact, after 99 terms of the square wave's approximation, the error is bigger than 10 terms of the approximation for the half-wave rectified sinusoid. Mathematicians have shown that no signal has an rms approximation error that decays more slowly than it does for the square wave.

Exercise 4.8 *(Solution on p. 153.)*

Calculate the harmonic distortion for the square wave.

More than just decaying slowly, Fourier series approximation shown in Figure 4.8 (Fourier series approximation of a square wave) exhibits interesting behavior. Although the square wave's Fourier series requires more terms for a given representation accuracy, when comparing plots it is not clear that the two are equal. Does the Fourier series really equal the square wave at *all* values of t? In particular, at each step-change in the square wave, the Fourier series exhibits a peak followed by rapid oscillations. As more terms are added to the series, the oscillations seem to become more rapid and smaller, but the peaks are not decreasing. For the Fourier series approximation for the half-wave rectified sinusoid (Figure 4.5: Fourier Series spectrum of a half-wave rectified sine wave), no such behavior occurs. What is happening?

Consider this mathematical question intuitively: Can a discontinuous function, like the square wave, be expressed as a sum, even an infinite one, of continuous signals? One should at least be suspicious, and in fact, it can't be thus expressed. This issue brought Fourier[9] much criticism from the French Academy of Science (Laplace, Lagrange, Monge and LaCroix comprised the review committee) for several years after its presentation on 1807. It was not resolved for also a century, and its resolution is interesting and important to understand from a practical viewpoint.

The extraneous peaks in the square wave's Fourier series *never* disappear; they are termed **Gibb's**

[9] http://www-groups.dcs.st-and.ac.uk/~history/Mathematicians/Fourier.html

Power spectrum and approximation error for a square wave

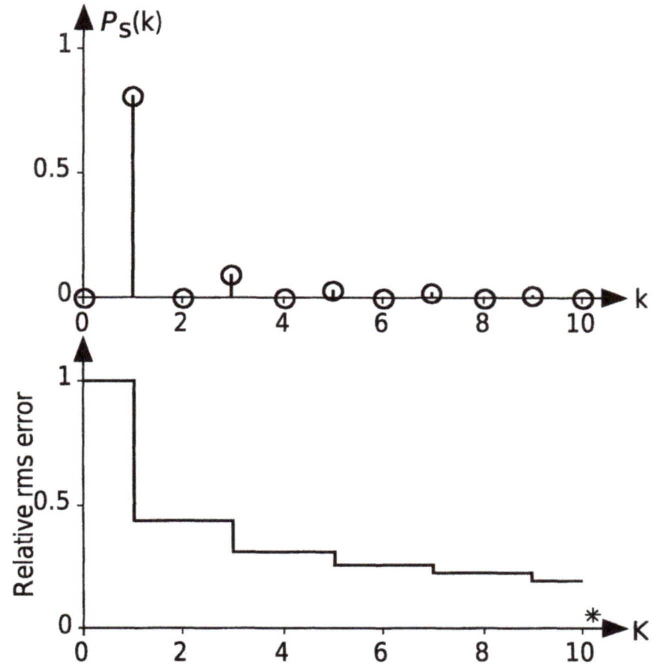

Figure 4.7: The upper plot shows the power spectrum of the square wave, and the lower plot the rms error of the finite-length Fourier series approximation to the square wave. The asterisk denotes the rms error when the number of terms K in the Fourier series equals 99.

phenomenon after the American physicist Josiah Willard Gibbs. They occur whenever the signal is discontinuous, and will always be present whenever the signal has jumps.

Let's return to the question of equality; how can the equal sign in the definition of the Fourier series be justified? The partial answer is that *pointwise*—each and every value of t—equality is *not* guaranteed. However, mathematicians later in the nineteenth century showed that the rms error of the Fourier series was always zero.

$$\lim_{K\to\infty} \text{rms}\,(\epsilon_K) = 0$$

What this means is that the error between a signal and its Fourier series approximation may not be zero, but that its rms value will be zero! It is through the eyes of the rms value that we redefine equality: The usual definition of equality is called **pointwise equality**: Two signals $s_1(t)$, $s_2(t)$ are said to be equal pointwise if $s_1(t) = s_2(t)$ for all values of t. A new definition of equality is **mean-square equality**: Two signals are said to be equal in the mean square if $rms\,(s_1 - s_2) = 0$. For Fourier series, Gibb's phenomenon peaks have finite height and zero width. The error differs from zero only at isolated points—whenever the periodic signal contains discontinuities—and equals about 9% of the size of the discontinuity. The value of a function at a finite set of points does not affect its integral. This effect underlies the reason why defining the value of a discontinuous function, like we refrained from doing in defining the step function (Section 2.2.4: Unit Step), at its discontinuity is meaningless. Whatever you pick for a value has no practical relevance for either the signal's spectrum or for how a system responds to the signal. The Fourier series value "at" the discontinuity is the average of the values on either side of the jump.

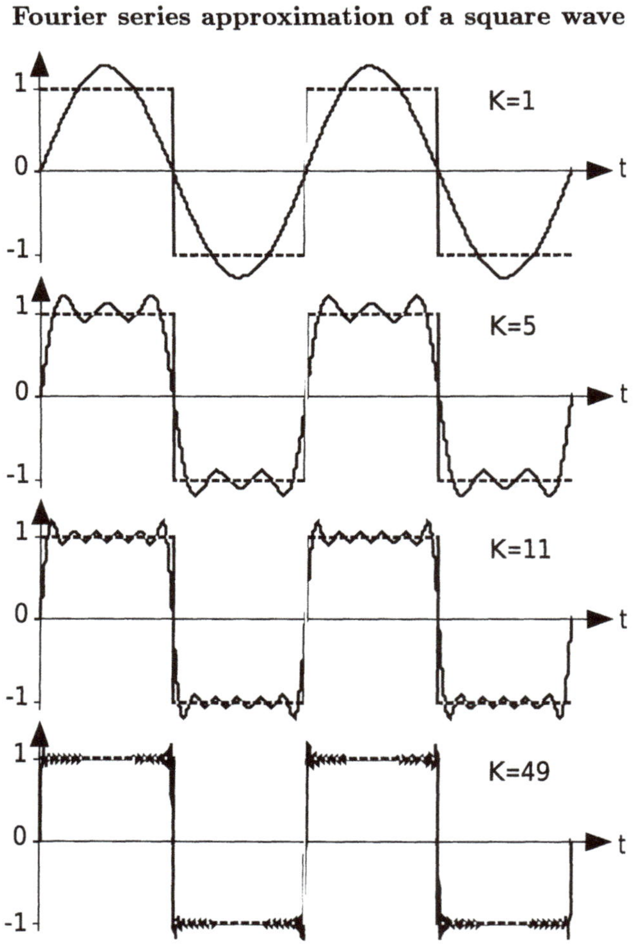

Figure 4.8: Fourier series approximation to sq (t). The number of terms in the Fourier sum is indicated in each plot, and the square wave is shown as a dashed line over two periods.

4.6 Encoding Information in the Frequency Domain[10]

To emphasize the fact that every periodic signal has both a time and frequency domain representation, we can exploit both to *encode information* into a signal. Refer to the Fundamental Model of Communication (Figure 1.4: Fundamental model of communication). We have an information source, and want to construct a transmitter that produces a signal $x(t)$. For the source, let's assume we have information to encode every T seconds. For example, we want to represent typed letters produced by an extremely good typist (a key is struck every T seconds). Let's consider the complex Fourier series formula in the light of trying to encode information.

$$x(t) = \sum_{k=-K}^{K} \left(c_k e^{j\frac{2\pi kt}{T}} \right) \tag{4.26}$$

We use a finite sum here merely for simplicity (fewer parameters to determine). An important aspect of the spectrum is that each frequency component c_k can be manipulated separately: Instead of finding the

[10]This content is available online at <http://cnx.org/content/m0043/2.15/>.

Fourier spectrum from a time-domain specification, let's construct it in the frequency domain by selecting the c_k according to some rule that relates coefficient values to the alphabet. In defining this rule, we want to always create a real-valued signal $x(t)$. Because of the Fourier spectrum's properties (Property 4.1, p. 111), the spectrum must have conjugate symmetry. This requirement means that we can only assign positive-indexed coefficients (positive frequencies), with negative-indexed ones equaling the complex conjugate of the corresponding positive-indexed ones.

Assume we have N letters to encode: $\{a_1, \dots, a_N\}$. One simple encoding rule could be to make a single Fourier coefficient be non-zero and all others zero for each letter. For example, if a_n occurs, we make $c_n = 1$ and $c_k = 0$, $k \neq n$. In this way, the nth harmonic of the frequency $\frac{1}{T}$ is used to represent a letter. Note that the **bandwidth**—the range of frequencies required for the encoding—equals $\frac{N}{T}$. Another possibility is to consider the binary representation of the letter's index. For example, if the letter a_{13} occurs, converting 13 to its base 2 representation, we have $13 = 1101_2$. We can use the pattern of zeros and ones to represent directly which Fourier coefficients we "turn on" (set equal to one) and which we "turn off."

Exercise 4.9 ***(Solution on p. 153.)***

Compare the bandwidth required for the direct encoding scheme (one nonzero Fourier coefficient for each letter) to the binary number scheme. Compare the bandwidths for a 128-letter alphabet. Since both schemes represent information without loss – we can determine the typed letter uniquely from the signal's spectrum – both are viable. Which makes more efficient use of bandwidth and thus might be preferred?

Exercise 4.10 ***(Solution on p. 153.)***

Can you think of an information-encoding scheme that makes even more efficient use of the spectrum? In particular, can we use only one Fourier coefficient to represent N letters uniquely?

We can create an encoding scheme in the frequency domain (p. 123) to represent an alphabet of letters. But, as this information-encoding scheme stands, we can represent one letter for all time. However, we note that the Fourier coefficients depend *only* on the signal's characteristics over a single period. We could change the signal's spectrum every T as each letter is typed. In this way, we turn spectral coefficients on and off as letters are typed, thereby encoding the entire typed document. For the receiver (see the Fundamental Model of Communication (Figure 1.4: Fundamental model of communication)) to retrieve the typed letter, it would simply use the Fourier formula for the complex Fourier spectrum[11] for each T-second interval to determine what each typed letter was. Figure 4.9 (Encoding Signals) shows such a signal in the time-domain.

[11] "Complex Fourier Series and Their Properties", (2) <http://cnx.org/content/m0065/latest/#complex>

Encoding Signals

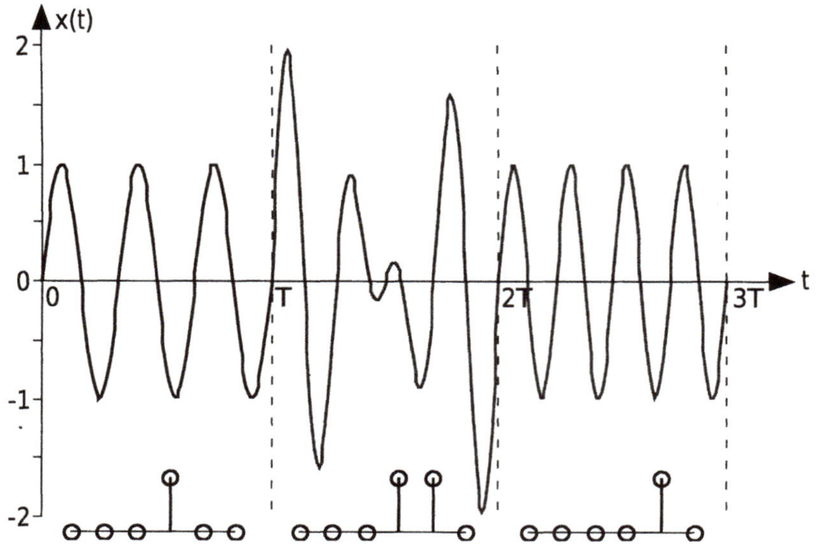

Figure 4.9: The encoding of signals via the Fourier spectrum is shown over three "periods." In this example, only the third and fourth harmonics are used, as shown by the spectral magnitudes corresponding to each T-second interval plotted below the waveforms. Can you determine the phase of the harmonics from the waveform?

In this Fourier-series encoding scheme, we have used the fact that spectral coefficients can be independently specified and that they can be uniquely recovered from the time-domain signal over one "period." Do note that the signal representing the entire document is no longer periodic. By understanding the Fourier series' properties (in particular that coefficients are determined only over a T-second interval, we can construct a communications system. This approach represents a simplification of how modern modems represent text that they transmit over telephone lines.

4.7 Filtering Periodic Signals[12]

The Fourier series representation of a periodic signal makes it easy to determine how a linear, time-invariant filter reshapes such signals *in general.* The fundamental property of a linear system is that its input-output relation obeys superposition: $L(a_1 s_1(t) + a_2 s_2(t)) = a_1 L(s_1(t)) + a_2 L(s_2(t))$. Because the Fourier series represents a periodic signal as a linear combination of complex exponentials, we can exploit the superposition property. Furthermore, we found for linear circuits that their output to a complex exponential input is just the frequency response evaluated at the signal's frequency times the complex exponential. Said mathematically, if $x(t) = e^{j\frac{2\pi kt}{T}}$, then the output $y(t) = H\left(\frac{k}{T}\right) e^{j\frac{2\pi kt}{T}}$ because $f = \frac{k}{T}$. Thus, if $x(t)$ is periodic thereby having a Fourier series, a linear circuit's output to this signal will be the superposition of the output to each component.

$$y(t) = \sum_{k=-\infty}^{\infty} \left(c_k H\left(\frac{k}{T}\right) e^{j\frac{2\pi kt}{T}} \right) \tag{4.27}$$

[12] This content is available online at <http://cnx.org/content/m0044/2.9/>.

Thus, the output has a Fourier series, which means that it too is periodic. Its Fourier coefficients equal $c_k H\left(\frac{k}{T}\right)$. *To obtain the spectrum of the output, we simply multiply the input spectrum by the frequency response* . The circuit modifies the magnitude and phase of each Fourier coefficient. Note especially that while the Fourier coefficients do not depend on the signal's period, the circuit's transfer function does depend on frequency, which means that the circuit's output will differ as the period varies.

Filtering a period signal

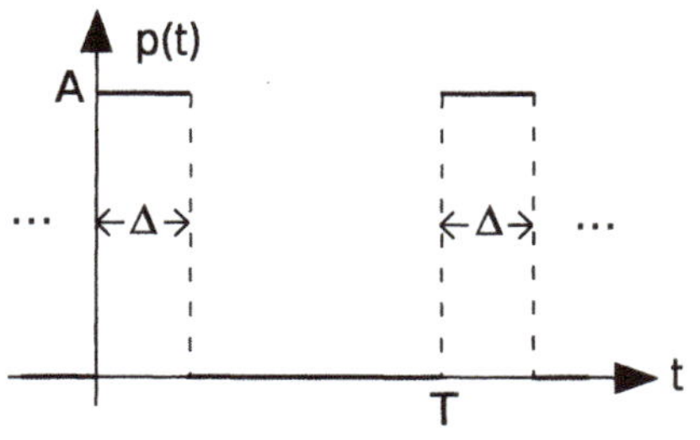

(a)

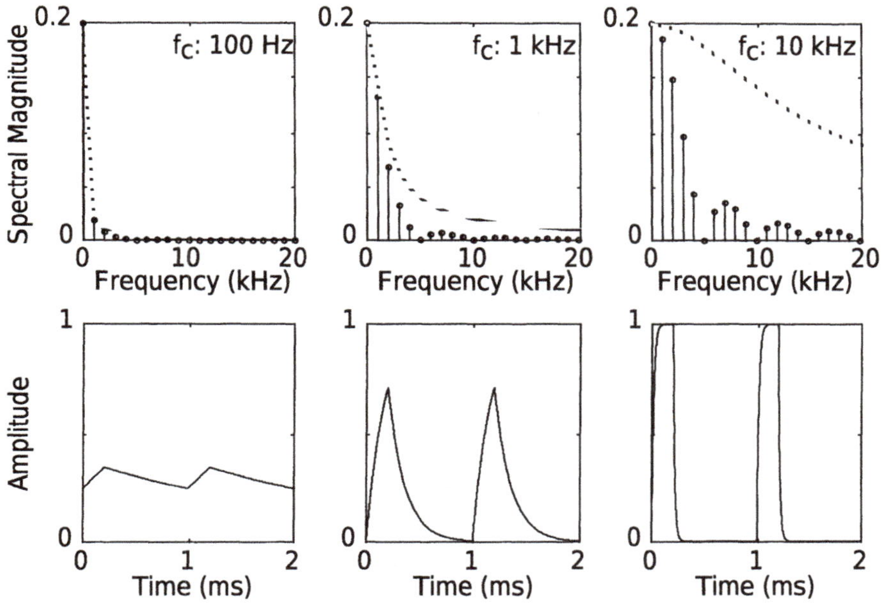

(b)

Figure 4.10: A periodic pulse signal, such as shown on the left part ($\frac{\Delta}{T} = 0.2$), serves as the input to an *RC* lowpass filter. The input's period was 1 ms (millisecond). The filter's cutoff frequency was set to the various values indicated in the top row, which display the output signal's spectrum and the filter's transfer function. The bottom row shows the output signal derived from the Fourier series coefficients shown in the top row. (a) Periodic pulse signal (b) Top plots show the pulse signal's spectrum for various cutoff frequencies. Bottom plots show the filter's output signals.

Example 4.3

The periodic pulse signal shown on the left above serves as the input to a *RC*-circuit that has the transfer function (calculated elsewhere (Figure 3.31: Magnitude and phase of the transfer function))

$$H(f) = \frac{1}{1 + j2\pi fRC} \tag{4.28}$$

Figure 4.10 (Filtering a period signal) shows the output changes as we vary the filter's cutoff frequency. Note how the signal's spectrum extends well above its fundamental frequency. Having a cutoff frequency ten times higher than the fundamental does perceptibly change the output waveform, rounding the leading and trailing edges. As the cutoff frequency decreases (center, then left), the rounding becomes more prominent, with the leftmost waveform showing a small ripple.

Exercise 4.11 ***(Solution on p. 153.)***

What is the average value of each output waveform? The correct answer may surprise you.

This example also illustrates the impact a lowpass filter can have on a waveform. The simple *RC* filter used here has a rather gradual frequency response, which means that higher harmonics are smoothly suppressed. Later, we will describe filters that have much more rapidly varying frequency responses, allowing a much more dramatic selection of the input's Fourier coefficients.

More importantly, we have calculated the output of a circuit to a periodic input *without* writing, much less solving, the differential equation governing the circuit's behavior. Furthermore, we made these calculations entirely in the frequency domain. Using Fourier series, we can calculate how *any* linear circuit will respond to a periodic input.

4.8 Derivation of the Fourier Transform[13]

Fourier series clearly open the frequency domain as an interesting and useful way of determining how circuits and systems respond to *periodic* input signals. Can we use similar techniques for nonperiodic signals? What is the response of the filter to a single pulse? Addressing these issues requires us to find the Fourier spectrum of all signals, both periodic and nonperiodic ones. We need a definition for *the* Fourier spectrum of a signal, periodic or not. This spectrum is calculated by what is known as the **Fourier transform**.

Let $s_T(t)$ be a periodic signal having period T. We want to consider what happens to this signal's spectrum as we let the period become longer and longer. We denote the spectrum for any assumed value of the period by $c_k(T)$. We calculate the spectrum according to the familiar formula

$$c_k(T) = \frac{1}{T}\int_{-\left(\frac{T}{2}\right)}^{\frac{T}{2}} s_T(t)\, e^{-\left(\frac{j2\pi kt}{T}\right)}\, dt \tag{4.29}$$

where we have used a symmetric placement of the integration interval about the origin for subsequent derivational convenience. Let f be a *fixed* frequency equaling $\frac{k}{T}$; we vary the frequency index k proportionally as we increase the period. Define

$$S_T(f) \equiv Tc_k(T) = \int_{-\left(\frac{T}{2}\right)}^{\frac{T}{2}} s_T(t)\, e^{-(j2\pi ft)}\, dt \tag{4.30}$$

making the corresponding Fourier series

$$s_T(t) = \sum_{k=-\infty}^{\infty} \left(S_T(f)\, e^{j2\pi ft} \frac{1}{T} \right) \tag{4.31}$$

[13]This content is available online at <http://cnx.org/content/m0046/2.19/>.

As the period increases, the spectral lines become closer together, becoming a continuum. Therefore,

$$\lim_{T\to\infty} s_T(t) \equiv s(t) = \int_{-\infty}^{\infty} S(f)\, e^{j2\pi ft} df \tag{4.32}$$

with

$$S(f) = \int_{-\infty}^{\infty} s(t)\, e^{-(j2\pi ft)} dt \tag{4.33}$$

$S(f)$ is the Fourier transform of $s(t)$ (the Fourier transform is symbolically denoted by the uppercase version of the signal's symbol) and is defined for *any* signal for which the integral ((4.33)) converges.

Example 4.4

Let's calculate the Fourier transform of the pulse signal (Section 2.2.5: Pulse), $p(t)$.

$$P(f) = \int_{-\infty}^{\infty} p(t)\, e^{-(j2\pi ft)} dt = \int_{0}^{\Delta} e^{-(j2\pi ft)} dt = \frac{1}{-(j2\pi f)} \left(e^{-(j2\pi f\Delta)} - 1\right)$$

$$P(f) = e^{-(j2\pi f\Delta)} \frac{\sin(\pi f\Delta)}{\pi f}$$

Note how closely this result resembles the expression for Fourier series coefficients of the periodic pulse signal (4.10).

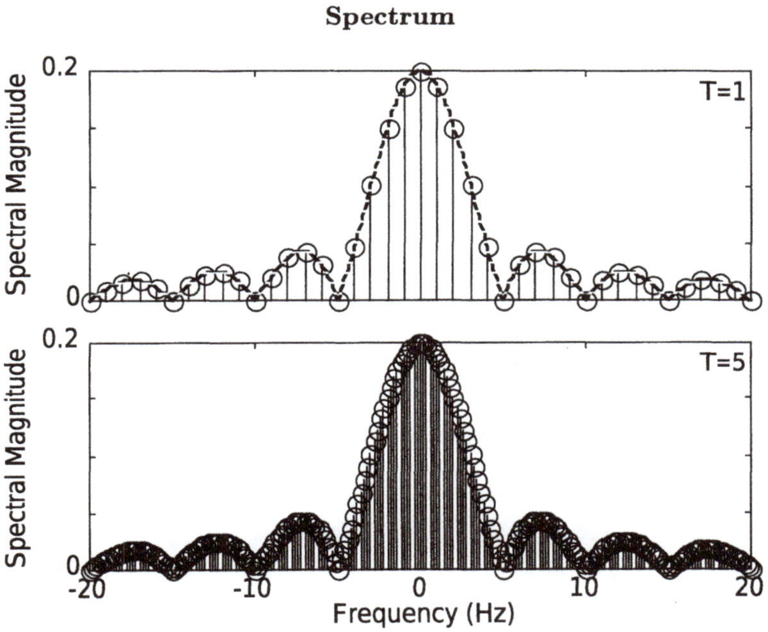

Figure 4.11: The upper plot shows the magnitude of the Fourier series spectrum for the case of $T = 1$ with the Fourier transform of $p(t)$ shown as a dashed line. For the bottom panel, we expanded the period to $T = 5$, keeping the pulse's duration fixed at 0.2, and computed its Fourier series coefficients.

Figure 4.11 (Spectrum) shows how increasing the period does indeed lead to a continuum of coefficients, and that the Fourier transform does correspond to what the continuum becomes. The quantity $\frac{\sin(t)}{t}$ has a special name, the **sinc** (pronounced "sink") function, and is denoted by $sinc(t)$. Thus, the magnitude of the pulse's Fourier transform equals $|\Delta sinc(\pi f \Delta)|$.

The Fourier transform relates a signal's time and frequency domain representations to each other. The direct Fourier transform (or simply the Fourier transform) calculates a signal's frequency domain representation from its time-domain variant ((4.34)). The inverse Fourier transform ((4.35)) finds the time-domain representation from the frequency domain. Rather than explicitly writing the required integral, we often symbolically express these transform calculations as $\mathcal{F}(s)$ and $\mathcal{F}^{-1}(S)$, respectively.

$$\begin{aligned} \mathcal{F}(s) &= S(f) \\ &= \int_{-\infty}^{\infty} s(t)\, e^{-(j2\pi ft)} dt \end{aligned} \tag{4.34}$$

$$\begin{aligned} \mathcal{F}^{-1}(S) &= s(t) \\ &= \int_{-\infty}^{\infty} S(f)\, e^{+j2\pi ft} df \end{aligned} \tag{4.35}$$

We must have $s(t) = \mathcal{F}^{-1}(\mathcal{F}(s(t)))$ and $S(f) = \mathcal{F}\left(\mathcal{F}^{-1}(S(f))\right)$, and these results are indeed valid with minor exceptions.

> NOTE: Recall that the Fourier series for a square wave gives a value for the signal at the discontinuities equal to the average value of the jump. This value may differ from how the signal is *defined* in the time domain, but being unequal at a point is indeed minor.

Showing that you "get back to where you started" is difficult from an analytic viewpoint, and we won't try here. Note that the direct and inverse transforms differ only in the sign of the exponent.

Exercise 4.12 ***(Solution on p. 153.)***

The differing exponent signs means that some curious results occur when we use the wrong sign. What is $\mathcal{F}(S(f))$? In other words, use the wrong exponent sign in evaluating the inverse Fourier transform.

Properties of the Fourier transform and some useful transform pairs are provided in the accompanying tables (Short Table of Fourier Transform Pairs, p. 129 and Fourier Transform Properties, p. 129). Especially important among these properties is **Parseval's Theorem**, which states that power computed in either domain equals the power in the other.

$$\int_{-\infty}^{\infty} s^2(t)\, dt = \int_{-\infty}^{\infty} (|S(f)|)^2 df \tag{4.36}$$

Of practical importance is the conjugate symmetry property: When $s(t)$ is real-valued, the spectrum at negative frequencies equals the complex conjugate of the spectrum at the corresponding positive frequencies. Consequently, we need only plot the positive frequency portion of the spectrum (we can easily determine the remainder of the spectrum).

Exercise 4.13 ***(Solution on p. 153.)***

How many Fourier transform operations need to be applied to get the original signal back: $\mathcal{F}(\cdots(\mathcal{F}(s))) = s(t)$?

Note that the mathematical relationships between the time domain and frequency domain versions of the same signal are termed **transforms**. We are transforming (in the nontechnical meaning of the word) a signal from one representation to another. We express Fourier transform *pairs* as $(s(t) \leftrightarrow S(f))$. A signal's time and frequency domain representations are uniquely related to each other. A signal thus "exists" in both the time and frequency domains, with the Fourier transform bridging between the two. We can define an information carrying signal in either the time or frequency domains; it behooves the wise engineer to use the simpler of the two.

A common misunderstanding is that while a signal exists in both the time and frequency domains, a single formula expressing a signal must contain *only* time or frequency: Both cannot be present simultaneously. This situation mirrors what happens with complex amplitudes in circuits: As we reveal how communications systems work and are designed, we will define signals entirely in the frequency domain without explicitly finding their time domain variants. This idea is shown in another module (Section 4.6) where we define Fourier series coefficients according to letter to be transmitted. Thus, a signal, though most familiarly defined in the time-domain, really can be defined equally as well (and sometimes more easily) in the frequency domain. For example, impedances depend on frequency and the time variable cannot appear.

We will learn (Section 4.9) that finding a linear, time-invariant system's output in the time domain can be most easily calculated by determining the input signal's spectrum, performing a simple calculation in the frequency domain, and inverse transforming the result. Furthermore, understanding communications and information processing systems requires a thorough understanding of signal structure and of how systems work in *both* the time and frequency domains.

The only difficulty in calculating the Fourier transform of any signal occurs when we have periodic signals (in either domain). Realizing that the Fourier series is a special case of the Fourier transform, we simply calculate the Fourier series coefficients instead, and plot them along with the spectra of nonperiodic signals on the same frequency axis.

Short Table of Fourier Transform Pairs

$s(t)$	$S(f)$
$e^{-(at)}u(t)$	$\frac{1}{j2\pi f+a}$
$e^{(-a)\|t\|}$	$\frac{2a}{4\pi^2 f^2+a^2}$
$p(t) = \begin{cases} 1 \text{ if } \|t\| < \frac{\Delta}{2} \\ 0 \text{ if } \|t\| > \frac{\Delta}{2} \end{cases}$	$\frac{\sin(\pi f \Delta)}{\pi f}$
$\frac{\sin(2\pi W t)}{\pi t}$	$S(f) = \begin{cases} 1 \text{ if } \|f\| < W \\ 0 \text{ if } \|f\| > W \end{cases}$

Fourier Transform Properties

	Time-Domain	**Frequency Domain**
Linearity	$a_1 s_1(t) + a_2 s_2(t)$	$a_1 S_1(f) + a_2 S_2(f)$
Conjugate Symmetry	$s(t) \in \mathbb{R}$	$S(f) = S(-f)^*$
Even Symmetry	$s(t) = s(-t)$	$S(f) = S(-f)$
Odd Symmetry	$s(t) = -(s(-t))$	$S(f) = -(S(-f))$
Scale Change	$s(at)$	$\frac{1}{\lvert a\rvert} S\left(\frac{f}{a}\right)$
Time Delay	$s(t-\tau)$	$e^{-(j2\pi f\tau)} S(f)$
Complex Modulation	$e^{j2\pi f_0 t} s(t)$	$S(f - f_0)$
Amplitude Modulation by Cosine	$s(t)\cos(2\pi f_0 t)$	$\frac{S(f-f_0)+S(f+f_0)}{2}$
Amplitude Modulation by Sine	$s(t)\sin(2\pi f_0 t)$	$\frac{S(f-f_0)-S(f+f_0)}{2j}$
Differentiation	$\frac{d}{dt} s(t)$	$j2\pi f S(f)$
Integration	$\int_{-\infty}^{t} s(\alpha)\, d\alpha$	$\frac{1}{j2\pi f} S(f)$ if $S(0) = 0$
Multiplication by t	$ts(t)$	$\frac{1}{-(j2\pi)} \frac{d}{df} S(f)$
Area	$\int_{-\infty}^{\infty} s(t)\, dt$	$S(0)$
Value at Origin	$s(0)$	$\int_{-\infty}^{\infty} S(f)\, df$
Parseval's Theorem	$\int_{-\infty}^{\infty} (\lvert s(t)\rvert)^2 dt$	$\int_{-\infty}^{\infty} (\lvert S(f)\rvert)^2 df$

Example 4.5

In communications, a very important operation on a signal $s(t)$ is to **amplitude modulate** it. Using this operation more as an example rather than elaborating the communications aspects here, we want to compute the Fourier transform — the spectrum — of

$$(1 + s(t))\cos(2\pi f_c t)$$

Thus,

$$(1 + s(t))\cos(2\pi f_c t) = \cos(2\pi f_c t) + s(t)\cos(2\pi f_c t)$$

For the spectrum of $\cos(2\pi f_c t)$, we use the Fourier series. Its period is $\frac{1}{f_c}$, and its only nonzero Fourier coefficients are $c_{\pm 1} = \frac{1}{2}$. The second term is *not* periodic unless $s(t)$ has the same period as the sinusoid. Using Euler's relation, the spectrum of the second term can be derived as

$$s(t)\cos(2\pi f_c t) = \int_{-\infty}^{\infty} S(f)\, e^{j2\pi f t} df \cos(2\pi f_c t)$$

Using Euler's relation for the cosine,

$$(s(t)\cos(2\pi f_c t)) = \frac{1}{2}\int_{-\infty}^{\infty} S(f)\, e^{j2\pi(f+f_c)t} df + \frac{1}{2}\int_{-\infty}^{\infty} S(f)\, e^{j2\pi(f-f_c)t} df$$

$$(s(t)\cos(2\pi f_c t)) = \frac{1}{2}\int_{-\infty}^{\infty} S(f-f_c)\, e^{j2\pi f t} df + \frac{1}{2}\int_{-\infty}^{\infty} S(f+f_c)\, e^{j2\pi f t} df$$

$$(s(t)\cos(2\pi f_c t)) = \int_{-\infty}^{\infty} \frac{S(f-f_c) + S(f+f_c)}{2} e^{j2\pi f t} df$$

Exploiting the uniqueness property of the Fourier transform, we have

$$\mathcal{F}\left(s\left(t\right)\cos\left(2\pi f_c t\right)\right) = \frac{S\left(f - f_c\right) + S\left(f + f_c\right)}{2} \tag{4.37}$$

This component of the spectrum consists of the original signal's spectrum delayed and advanced *in frequency*. The spectrum of the amplitude modulated signal is shown in Figure 4.12.

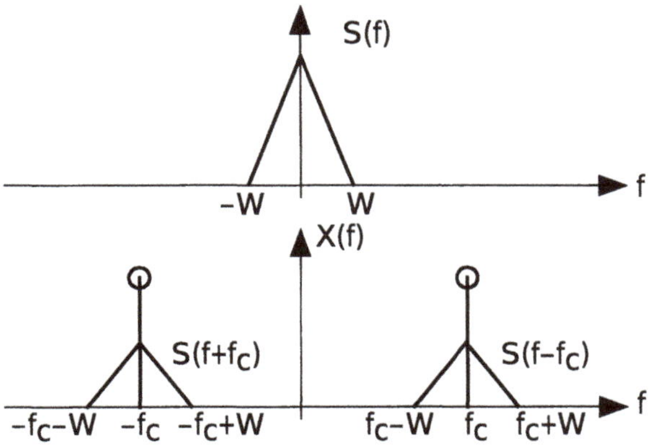

Figure 4.12: A signal which has a triangular shaped spectrum is shown in the top plot. Its highest frequency — the largest frequency containing power — is W Hz. Once amplitude modulated, the resulting spectrum has "lines" corresponding to the Fourier series components at $\pm f_c$ and the original triangular spectrum shifted to components at $\pm f_c$ and scaled by $\frac{1}{2}$.

Note how in this figure the signal $s(t)$ is defined in the frequency domain. To find its time domain representation, we simply use the inverse Fourier transform.

Exercise 4.14 *(Solution on p. 153.)*

What is the signal $s(t)$ that corresponds to the spectrum shown in the upper panel of Figure 4.12?

Exercise 4.15 *(Solution on p. 153.)*

What is the power in $x(t)$, the amplitude-modulated signal? Try the calculation in both the time and frequency domains.

In this example, we call the signal $s(t)$ a **baseband signal** because its power is contained at low frequencies. Signals such as speech and the Dow Jones averages are baseband signals. The baseband signal's **bandwidth** equals W, the highest frequency at which it has power. Since $x(t)$'s spectrum is confined to a frequency band not close to the origin (we assume ($f_c \gg W$)), we have a **bandpass signal**. The bandwidth of a bandpass signal is *not* its highest frequency, but the range of positive frequencies where the signal has power. Thus, in this example, the bandwidth is $2WHz$. Why a signal's bandwidth should depend on its spectral shape will become clear once we develop communications systems.

4.9 Linear Time Invariant Systems[14]

When we apply a periodic input to a linear, time-invariant system, the output is periodic and has Fourier series coefficients equal to the product of the system's frequency response and the input's Fourier coefficients (Filtering Periodic Signals (4.27)). The way we derived the spectrum of non-periodic signal from periodic ones makes it clear that the same kind of result works when the input is not periodic: *If $x(t)$ serves as*

[14]This content is available online at <http://cnx.org/content/m0048/2.18/>.

the input to a linear, time-invariant system having frequency response $H(f)$, the spectrum of the output is $X(f)H(f)$.

Example 4.6

Let's use this frequency-domain input-output relationship for linear, time-invariant systems to find a formula for the *RC*-circuit's response to a pulse input. We have expressions for the input's spectrum and the system's frequency response.

$$P(f) = e^{-(j\pi f\Delta)}\frac{\sin(\pi f\Delta)}{\pi f} \tag{4.38}$$

$$H(f) = \frac{1}{1+j2\pi fRC} \tag{4.39}$$

Thus, the output's Fourier transform equals

$$Y(f) = e^{-(j\pi f\Delta)}\frac{\sin(\pi f\Delta)}{\pi f}\frac{1}{1+j2\pi fRC} \tag{4.40}$$

You won't find this Fourier transform in our table, and the required integral is difficult to evaluate as the expression stands. This situation requires cleverness and an understanding of the Fourier transform's properties. In particular, recall Euler's relation for the sinusoidal term and note the fact that multiplication by a complex exponential in the frequency domain amounts to a time delay. Let's momentarily make the expression for $Y(f)$ more complicated.

$$\begin{aligned} e^{-(j\pi f\Delta)}\frac{\sin(\pi f\Delta)}{\pi f} &= e^{-(j\pi f\Delta)}\frac{e^{j\pi f\Delta}-e^{-(j\pi f\Delta)}}{j2\pi f} \\ &= \frac{1}{j2\pi f}\left(1-e^{-(j2\pi f\Delta)}\right) \end{aligned} \tag{4.41}$$

Consequently,

$$Y(f) = \frac{1}{j2\pi f}\left(1-e^{-(j\pi f\Delta)}\right)\frac{1}{1+j2\pi fRC} \tag{4.42}$$

The table of Fourier transform properties (Fourier Transform Properties, p. 129) suggests thinking about this expression as a *product* of terms.

- Multiplication by $\frac{1}{j2\pi f}$ means integration.
- Multiplication by the complex exponential $e^{-(j2\pi f\Delta)}$ means delay by Δ seconds in the time domain.
- The term $1-e^{-(j2\pi f\Delta)}$ means, in the time domain, subtract the time-delayed signal from its original.
- The inverse transform of the frequency response is $\frac{1}{RC}e^{-\left(\frac{t}{RC}\right)}u(t)$.

We can translate each of these frequency-domain products into time-domain operations *in any order we like* because the order in which multiplications occur doesn't affect the result. Let's start with the product of $\frac{1}{j2\pi f}$ (integration in the time domain) and the transfer function:

$$\frac{1}{j2\pi f}\frac{1}{1+j2\pi fRC} \leftrightarrow \left(1-e^{-\left(\frac{t}{RC}\right)}\right)u(t) \tag{4.43}$$

The middle term in the expression for $Y(f)$ consists of the difference of two terms: the constant 1 and the complex exponential $e^{-(j2\pi f\Delta)}$. Because of the Fourier transform's linearity, we simply subtract the results.

$$Y(f) \leftrightarrow \left(1-e^{-\left(\frac{t}{RC}\right)}\right)u(t) - \left(1-e^{-\left(\frac{t-\Delta}{RC}\right)}\right)u(t-\Delta) \tag{4.44}$$

Note that in delaying the signal how we carefully included the unit step. The second term in this result does not begin until $t = \Delta$. Thus, the waveforms shown in the Filtering Periodic Signals (Figure 4.10: Filtering a period signal) example mentioned above are exponentials. We say that the **time constant** of an exponentially decaying signal equals the time it takes to decrease by $\frac{1}{e}$ of its original value. Thus, the time-constant of the rising and falling portions of the output equal the product of the circuit's resistance and capacitance.

Exercise 4.16 ***(Solution on p. 153.)***

Derive the filter's output by considering the terms in (4.41) in the order given. Integrate last rather than first. You should get the same answer.

In this example, we used the table extensively to find the inverse Fourier transform, relying mostly on what multiplication by certain factors, like $\frac{1}{j2\pi f}$ and $e^{-(j2\pi f\Delta)}$, meant. We essentially treated multiplication by these factors as if they were transfer functions of some fictitious circuit. The transfer function $\frac{1}{j2\pi f}$ corresponded to a circuit that integrated, and $e^{-(j2\pi f\Delta)}$ to one that delayed. We even implicitly interpreted the circuit's transfer function as the input's spectrum! This approach to finding inverse transforms – breaking down a complicated expression into products and sums of simple components – is the engineer's way of breaking down the problem into several subproblems that are much easier to solve and then gluing the results together. Along the way we may make the system serve as the input, but in the rule $Y(f) = X(f)H(f)$, which term is the input and which is the transfer function is merely a notational matter (we labeled one factor with an X and the other with an H).

4.9.1 Transfer Functions

The notion of a transfer function applies well beyond linear circuits. Although we don't have all we need to demonstrate the result as yet, *all* linear, time-invariant systems have a frequency-domain input-output relation given by the product of the input's Fourier transform and the system's transfer function. Thus, linear circuits are a special case of linear, time-invariant systems. As we tackle more sophisticated problems in transmitting, manipulating, and receiving information, we will assume linear systems having certain properties (transfer functions) *without* worrying about what circuit has the desired property. At this point, you may be concerned that this approach is glib, and rightly so. Later we'll show that by involving software that we really don't need to be concerned about constructing a transfer function from circuit elements and op-amps.

4.9.2 Commutative Transfer Functions

Another interesting notion arises from the commutative property of multiplication (exploited in an example above (Example 4.6)): We can rather arbitrarily chose an order in which to apply each product. Consider a cascade of two linear, time-invariant systems. Because the Fourier transform of the first system's output is $X(f)H_1(f)$ and it serves as the second system's input, the cascade's output spectrum is $X(f)H_1(f)H_2(f)$. Because this product also equals $X(f)H_2(f)H_1(f)$, the *cascade having the linear systems in the opposite order yields the same result.* Furthermore, the cascade acts like a *single* linear system, having transfer function $H_1(f)H_2(f)$. This result applies to other configurations of linear, time-invariant systems as well; see this Frequency Domain Problem (Problem 4.12). Engineers exploit this property by determining what transfer function they want, then breaking it down into components arranged according to standard configurations. Using the fact that op-amp circuits can be connected in cascade with the transfer function equaling the product of its component's transfer function (see this analog signal processing problem (Problem 3.37)), we find a ready way of realizing designs. We now understand why op-amp implementations of transfer functions are so important.

4.10 Modeling the Speech Signal[15]

Vocal Tract

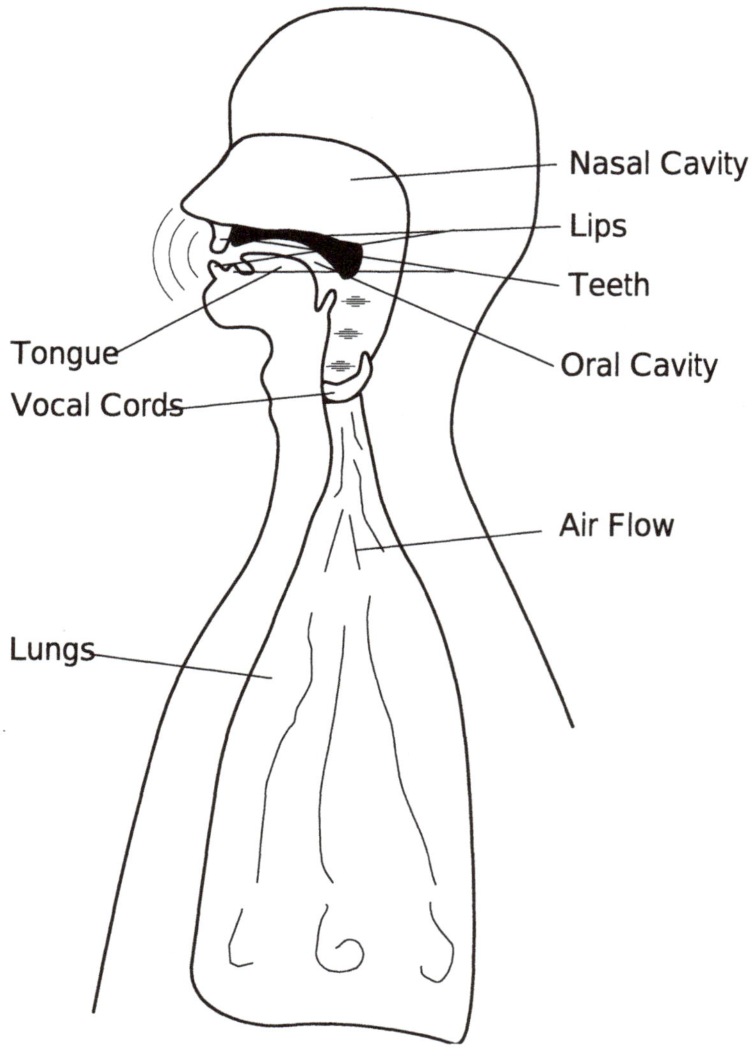

Figure 4.13: The vocal tract is shown in cross-section. Air pressure produced by the lungs forces air through the vocal cords that, when under tension, produce puffs of air that excite resonances in the vocal and nasal cavities. What are not shown are the brain and the musculature that control the entire speech production process.

[15]This content is available online at <http://cnx.org/content/m0049/2.25/>.

Model of the Vocal Tract

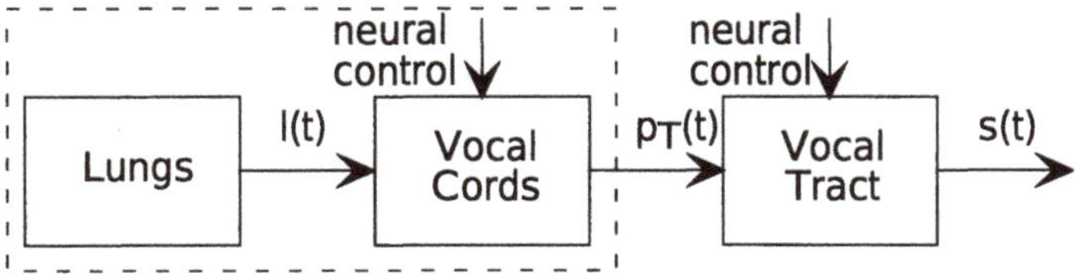

Figure 4.14: The systems model for the vocal tract. The signals $l(t)$, $p_T(t)$, and $s(t)$ are the air pressure provided by the lungs, the periodic pulse output provided by the vocal cords, and the speech output respectively. Control signals from the brain are shown as entering the systems from the top. Clearly, these come from the same source, but for modeling purposes we describe them separately since they control different aspects of the speech signal.

The information contained in the spoken word is conveyed by the speech signal. Because we shall analyze several speech transmission and processing schemes, we need to understand the speech signal's structure – what's special about the speech signal – and how we can describe and *model* speech production. This modeling effort consists of finding a system's description of how relatively unstructured signals, arising from simple sources, are given structure by passing them through an interconnection of systems to yield speech. For speech and for many other situations, system choice is governed by the physics underlying the actual production process. Because the fundamental equation of acoustics – the wave equation – applies here and is linear, we can use linear systems in our model with a fair amount of accuracy. The naturalness of linear system models for speech does not extend to other situations. In many cases, the underlying mathematics governed by the physics, biology, and/or chemistry of the problem are nonlinear, leaving linear systems models as approximations. Nonlinear models are far more difficult at the current state of knowledge to understand, and information engineers frequently prefer linear models because they provide a greater level of comfort, but not necessarily a sufficient level of accuracy.

Figure 4.13 (Vocal Tract) shows the actual speech production system and Figure 4.14 (Model of the Vocal Tract) shows the model speech production system. The characteristics of the model depends on whether you are saying a vowel or a consonant. We concentrate first on the vowel production mechanism. When the vocal cords are placed under tension by the surrounding musculature, air pressure from the lungs causes the vocal cords to vibrate. To visualize this effect, take a rubber band and hold it in front of your lips. If held open when you blow through it, the air passes through more or less freely; this situation corresponds to "breathing mode". If held tautly and close together, blowing through the opening causes the sides of the rubber band to vibrate. This effect works best with a wide rubber band. You can imagine what the airflow is like on the opposite side of the rubber band or the vocal cords. Your lung power is the simple source referred to earlier; it can be modeled as a constant supply of air pressure. The vocal cords respond to this input by vibrating, which means the output of this system is some periodic function.

Exercise 4.17 ***(Solution on p. 153.)***

Note that the vocal cord system takes a constant input and produces a periodic airflow that corresponds to its output signal. Is this system linear or nonlinear? Justify your answer.

Singers modify vocal cord tension to change the pitch to produce the desired musical note. Vocal cord tension is governed by a control input to the musculature; in system's models we represent control inputs as signals coming into the top or bottom of the system. Certainly in the case of speech and in many other cases as well, it is the control input that carries information, impressing it on the system's output. The change of signal structure resulting from varying the control input enables information to be conveyed by the signal, a process generically known as **modulation**. In singing, musicality is largely conveyed by pitch; in western speech, pitch is much less important. A sentence can be read in a monotone fashion without completely destroying the information expressed by the sentence. However, the difference between a statement and a question is frequently expressed by pitch changes. For example, note the sound differences between "Let's go to the park." and "Let's go to the park?";

For some consonants, the vocal cords vibrate just as in vowels. For example, the so-called nasal sounds "n" and "m" have this property. For others, the vocal cords do not produce a periodic output. Going back to mechanism, when consonants such as "f" are produced, the vocal cords are placed under much less tension, which results in turbulent flow. The resulting output airflow is quite erratic, so much so that we describe it as being **noise**. We define noise carefully later when we delve into communication problems.

The vocal cords' periodic output can be well described by the periodic pulse train $p_T(t)$ as shown in the periodic pulse signal (Figure 4.1), with T denoting the pitch period. The spectrum of this signal (4.9) contains harmonics of the frequency $\frac{1}{T}$, what is known as the **pitch frequency** or the **fundamental frequency** $F0$. The primary difference between adult male and female/prepubescent speech is pitch. Before puberty, pitch frequency for normal speech ranges between 150-400 Hz for both males and females. After puberty, the vocal cords of males undergo a physical change, which has the effect of lowering their pitch frequency to the range 80-160 Hz. If we could examine the vocal cord output, we could probably discern whether the speaker was male or female. This difference is also readily apparent in the speech signal itself.

To simplify our speech modeling effort, we shall assume that the pitch period is constant. With this simplification, we collapse the vocal-cord-lung system as a simple source that produces the periodic pulse signal (Figure 4.14 (Model of the Vocal Tract)). The sound pressure signal thus produced enters the mouth behind the tongue, creates acoustic disturbances, and exits primarily through the lips and to some extent through the nose. Speech specialists tend to name the mouth, tongue, teeth, lips, and nasal cavity the **vocal tract**. The physics governing the sound disturbances produced in the vocal tract and those of an organ pipe are quite similar. Whereas the organ pipe has the simple physical structure of a straight tube, the cross-section of the vocal tract "tube" varies along its length because of the positions of the tongue, teeth, and lips. It is these positions that are controlled by the brain to produce the vowel sounds. Spreading the lips, bringing the teeth together, and bringing the tongue toward the front portion of the roof of the mouth produces the sound "ee." Rounding the lips, spreading the teeth, and positioning the tongue toward the back of the oral cavity produces the sound "oh." These variations result in a linear, time-invariant system that has a frequency response typified by several peaks, as shown in Figure 4.15 (Speech Spectrum).

These peaks are known as **formants**. Thus, speech signal processors would say that the sound "oh" has a higher first formant frequency than the sound "ee," with $F2$ being much higher during "ee." $F2$ and $F3$ (the second and third formants) have more energy in "ee" than in "oh." Rather than serving as a filter, rejecting high or low frequencies, the vocal tract serves to *shape* the spectrum of the vocal cords. In the time domain, we have a periodic signal, the pitch, serving as the input to a linear system. We know that the output—the speech signal we utter and that is heard by others and ourselves—will also be periodic. Example time-domain speech signals are shown in Figure 4.15 (Speech Spectrum), where the periodicity is quite apparent.

Exercise 4.18 *(Solution on p. 153.)*

From the waveform plots shown in Figure 4.15 (Speech Spectrum), determine the pitch period and the pitch frequency.

Since speech signals are periodic, speech has a Fourier series representation given by a linear circuit's response to a periodic signal (4.27). Because the acoustics of the vocal tract are linear, we know that the spectrum of the output equals the product of the pitch signal's spectrum and the vocal tract's frequency response. We thus obtain the **fundamental model of speech production.**

$$S(f) = P_T(f) H_V(f) \tag{4.45}$$

Here, $H_V(f)$ is the transfer function of the vocal tract system. The Fourier series for the vocal cords' output, derived in this equation (p. 112), is

$$c_k = Ae^{-\left(\frac{j\pi k\Delta}{T}\right)}\frac{\sin\left(\frac{\pi k\Delta}{T}\right)}{\pi k} \tag{4.46}$$

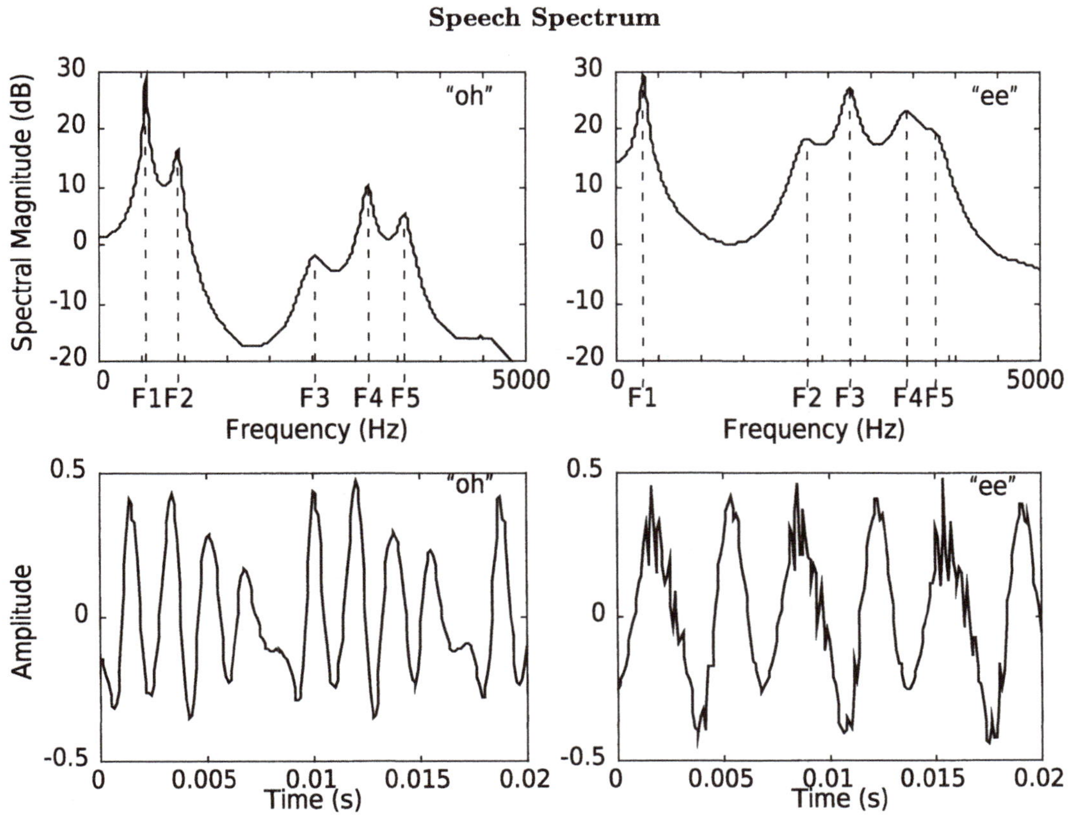

Figure 4.15: The ideal frequency response of the vocal tract as it produces the sounds "oh" and "ee" are shown on the top left and top right, respectively. The spectral peaks are known as formants, and are numbered consecutively from low to high frequency. The bottom plots show speech waveforms corresponding to these sounds.

and is plotted on the top in Figure 4.16 (voice spectrum). If we had, for example, a male speaker with about a 110 Hz pitch ($T \approx 9.1ms$) saying the vowel "oh", the spectrum of his speech *predicted by our model* is shown in Figure 4.16(b) (voice spectrum).

The model spectrum idealizes the measured spectrum, and captures all the important features. The measured spectrum certainly demonstrates what are known as **pitch lines**, and we realize from our model that they are due to the vocal cord's periodic excitation of the vocal tract. The vocal tract's shaping of the line spectrum is clearly evident, but difficult to discern exactly, especially at the higher frequencies. The model transfer function for the vocal tract makes the formants much more readily evident.

Exercise 4.19 ***(Solution on p. 153.)***

The Fourier series coefficients for speech are related to the vocal tract's transfer function only at the frequencies $\frac{k}{T}$, $k \in \{1, 2, \dots\}$; see previous result (4.9). Would male or female speech tend to have a more clearly identifiable formant structure when its spectrum is computed? Consider, for example, how the spectrum shown on the right in Figure 4.16 (voice spectrum) would change if the pitch were twice as high ($\approx 300Hz$).

voice spectrum

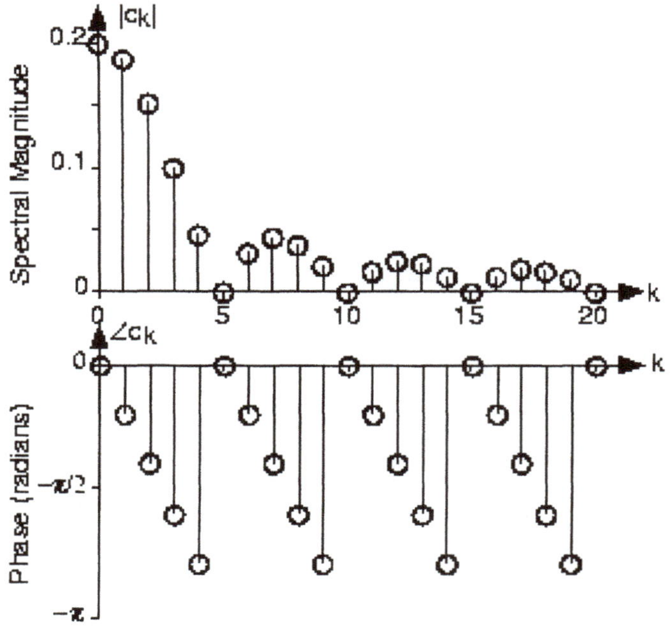

(a) pulse

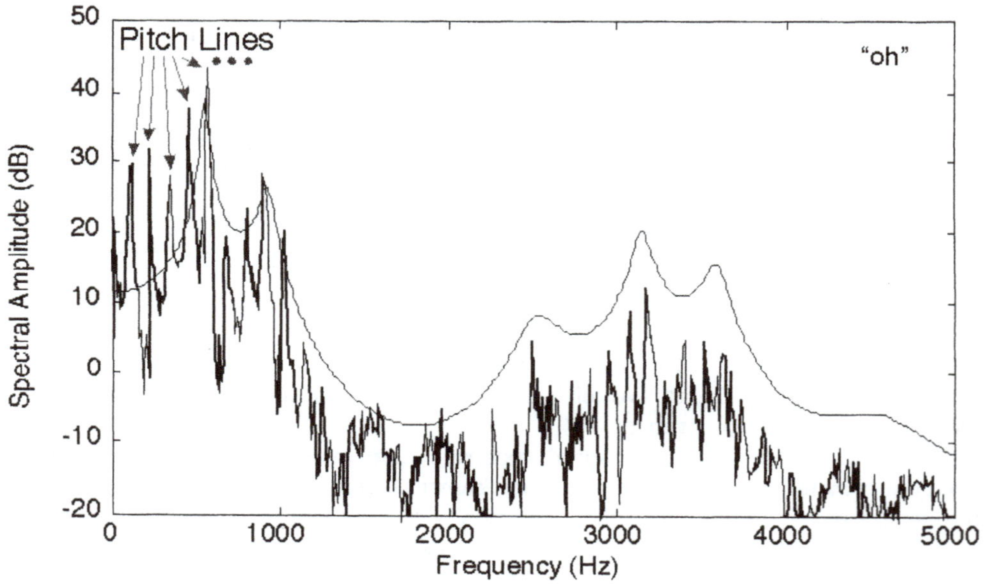

(b) voice spectrum

Figure 4.16: The vocal tract's transfer function, shown as the thin, smooth line, is superimposed on the spectrum of actual male speech corresponding to the sound "oh." The pitch lines corresponding to harmonics of the pitch frequency are indicated. (a) The vocal cords' output spectrum $P_T(f)$. (b) The vocal tract's transfer function, $H_V(f)$ and the speech spectrum.

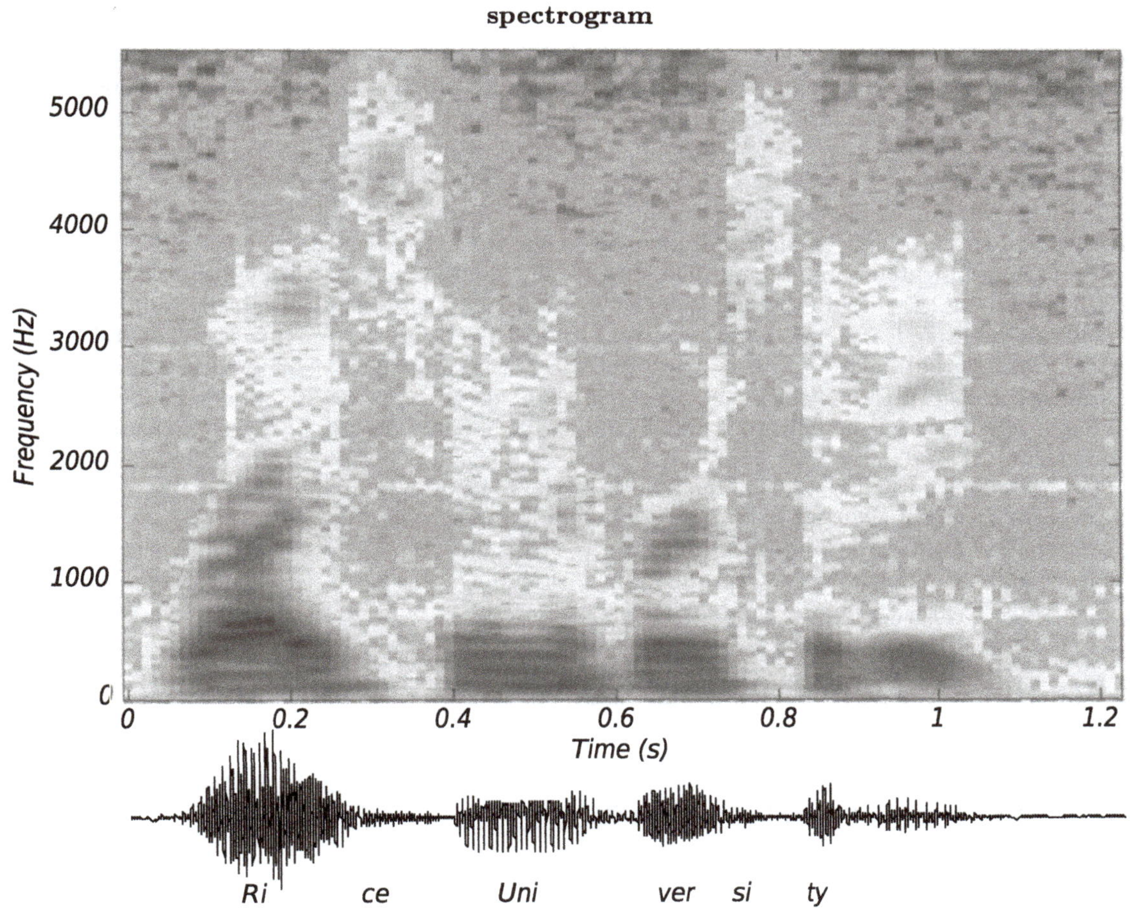

Figure 4.17: Displayed is the spectrogram of the author saying "Rice University." Blue indicates low energy portion of the spectrum, with red indicating the most energetic portions. Below the spectrogram is the time-domain speech signal, where the periodicities can be seen.

When we speak, pitch and the vocal tract's transfer function are not static; they change according to their control signals to produce speech. Engineers typically display how the speech spectrum changes over time with what is known as a spectrogram (Section 5.10) Figure 4.17 (spectrogram). Note how the line spectrum, which indicates how the pitch changes, is visible during the vowels, but not during the consonants (like the *ce* in "Rice").

The fundamental model for speech indicates how engineers use the physics underlying the signal generation process and exploit its structure to produce a systems model that suppresses the physics while emphasizing how the signal is "constructed." From everyday life, we know that speech contains a wealth of information. We want to determine how to transmit and receive it. Efficient and effective speech transmission requires us to know the signal's properties and its structure (as expressed by the fundamental model of speech production). We see from Figure 4.17 (spectrogram), for example, that speech contains significant energy from zero frequency up to around 5 kHz.

Effective speech transmission systems must be able to cope with signals having this bandwidth. It is interesting that one system that does *not* support this 5 kHz bandwidth is the telephone: Telephone systems act like a **bandpass filter** passing energy between about 200 Hz and 3.2 kHz. The most important consequence of this filtering is the removal of high frequency energy. In our sample utterance, the "ce" sound in "Rice"" contains most of its energy above 3.2 kHz; this filtering effect is why it is extremely difficult to distinguish the sounds "s" and "f" over the telephone. Try this yourself: Call a friend and determine if they can distinguish between the words "six" and "fix". If you say these words in isolation so that no context provides a hint about which word you are saying, your friend will not be able to tell them apart. Radio does support this bandwidth (see more about AM and FM radio systems (Section 6.11)).

Efficient speech transmission systems exploit the speech signal's special structure: What makes speech speech? You can conjure many signals that span the same frequencies as speech—car engine sounds, violin music, dog barks—but don't sound at all like speech. We shall learn later that transmission of *any* 5 kHz bandwidth signal requires about 80 kbps (thousands of bits per second) to transmit digitally. *Speech* signals can be transmitted using less than 1 kbps because of its special structure. To reduce the "digital bandwidth" so drastically means that engineers spent many years to develop signal processing and coding methods that could capture the special characteristics of speech without destroying how it sounds. If you used a speech transmission system to send a violin sound, it would arrive horribly distorted; speech transmitted the same way would sound fine.

Exploiting the special structure of speech requires going beyond the capabilities of analog signal processing systems. Many speech transmission systems work by finding the speaker's pitch and the formant frequencies. Fundamentally, we need to do more than filtering to determine the speech signal's structure; we need to manipulate signals in more ways than are possible with analog systems. Such flexibility is achievable (but not without some loss) with programmable *digital* systems.

4.11 Frequency Domain Problems[16]

Problem 4.1: Simple Fourier Series
Find the complex Fourier series representations of the following signals without explicitly calculating Fourier integrals. What is the signal's period in each case?

a) $s(t) = \sin(t)$
b) $s(t) = \sin^2(t)$
c) $s(t) = \cos(t) + 2\cos(2t)$
d) $s(t) = \cos(2t)\cos(t)$
e) $s(t) = \cos\left(10\pi t + \frac{\pi}{6}\right)(1 + \cos(2\pi t))$
f) $s(t)$ given by the depicted waveform (Figure 4.18).

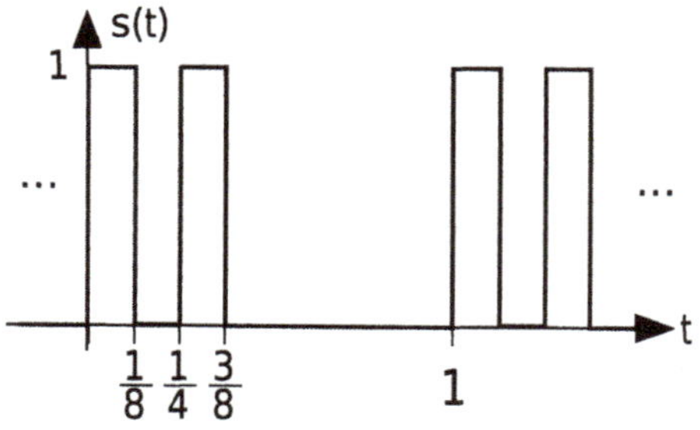

Figure 4.18

Problem 4.2: Fourier Series
Find the Fourier series representation for the following periodic signals (Figure 4.19). For the third signal,

[16]This content is available online at <http://cnx.org/content/m10350/2.32/>.

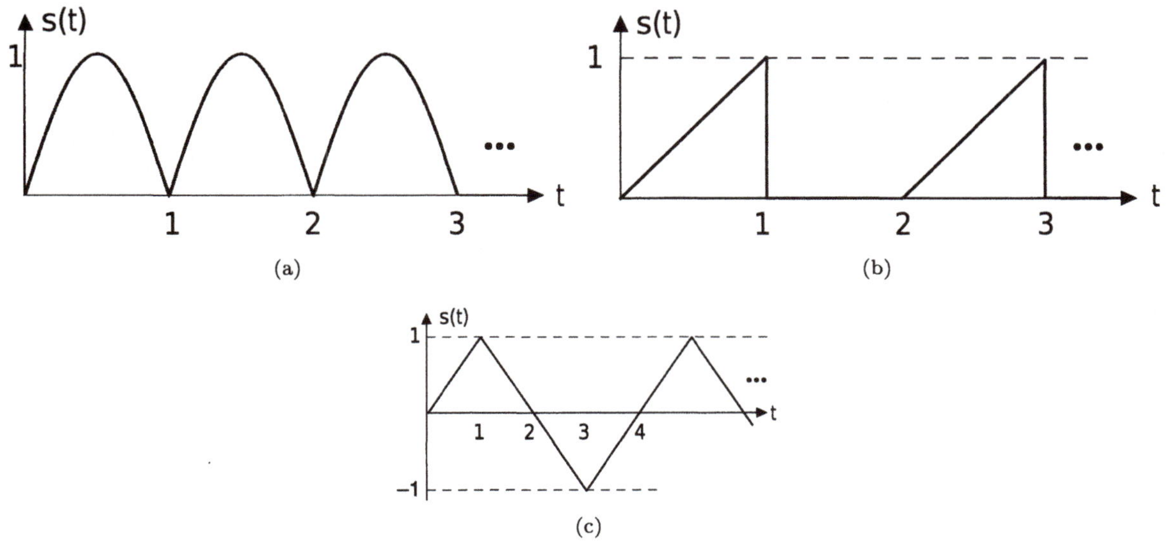

Figure 4.19

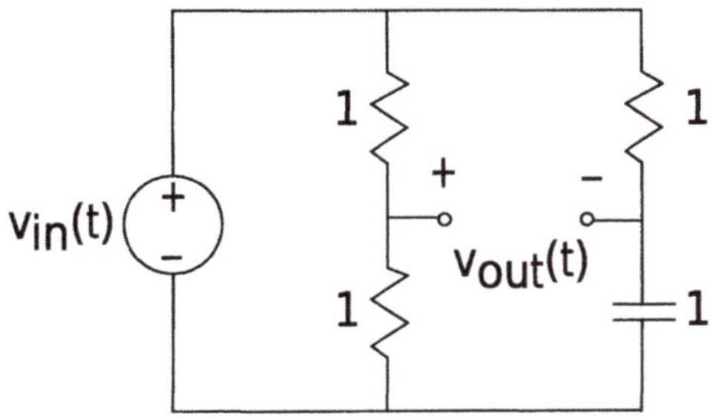

Figure 4.20

find the complex Fourier series for the triangle wave *without* performing the usual Fourier integrals. Hint: How is this signal related to one for which you already have the series?

Problem 4.3: Phase Distortion

We can learn about phase distortion by returning to circuits and investigate the following circuit (Figure 4.20).

a) Find this filter's transfer function.
b) Find the magnitude and phase of this transfer function. How would you characterize this circuit?
c) Let $v_{in}(t)$ be a square-wave of period T. What is the Fourier series for the output voltage?
d) Use Matlab to find the output's waveform for the cases $T = 0.01$ and $T = 2$. What value of T delineates the two kinds of results you found? The software in `fourier2.m` might be useful.
e) Instead of the depicted circuit, the square wave is passed through a system that delays its input, which

applies a linear phase shift to the signal's spectrum. Let the delay τ be $\frac{T}{4}$. Use the transfer function of a delay to compute using Matlab the Fourier series of the output. Show that the square wave is indeed delayed.

Problem 4.4: Approximating Periodic Signals
Often, we want to approximate a reference signal by a somewhat simpler signal. To assess the quality of an approximation, the most frequently used error measure is the mean-squared error. For a periodic signal $s(t)$,

$$\epsilon^2 = \frac{1}{T}\int_0^T (s(t) - \tilde{s}(t))^2 dt$$

where $s(t)$ is the reference signal and $\tilde{s}(t)$ its approximation. One convenient way of finding approximations for periodic signals is to truncate their Fourier series.

$$\tilde{s}(t) = \sum_{k=-K}^{K} \left(c_k e^{j\frac{2\pi k}{T}t}\right)$$

The point of this problem is to analyze whether this approach is the best (*i.e.*, always minimizes the mean-squared error).

a) Find a frequency-domain expression for the approximation error when we use the truncated Fourier series as the approximation.
b) Instead of truncating the series, let's generalize the nature of the approximation to including any set of $2K+1$ terms: We'll always include the c_0 and the negative indexed term corresponding to c_k. What selection of terms minimizes the mean-squared error? Find an expression for the mean-squared error resulting from your choice.
c) Find the Fourier series for the depicted signal (Figure 4.21). Use Matlab to find the truncated approximation and best approximation involving two terms. Plot the mean-squared error as a function of K for both approximations.

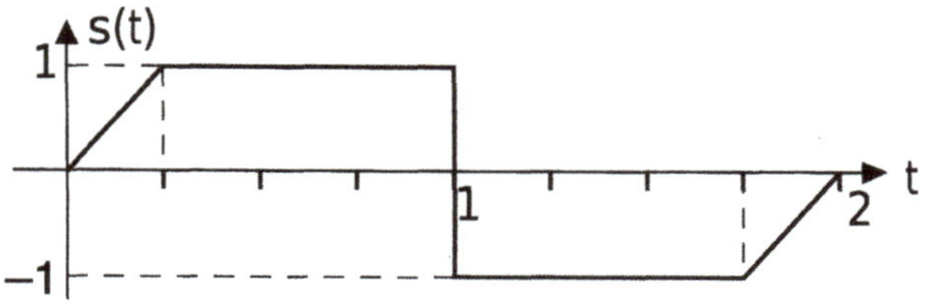

Figure 4.21

Problem 4.5: Long, Hot Days
The daily temperature is a consequence of several effects, one of them being the sun's heating. If this were the dominant effect, then daily temperatures would be proportional to the number of daylight hours. The plot (Figure 4.22) shows that the average daily high temperature does *not* behave that way.

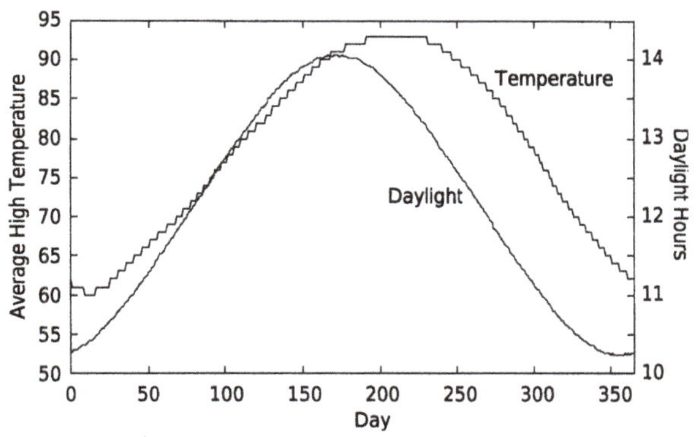

Figure 4.22

In this problem, we want to understand the temperature component of our environment using Fourier series and linear system theory. The file **temperature.mat** contains these data (daylight hours in the first row, corresponding average daily highs in the second) for Houston, Texas.

a) Let the length of day serve as the sole input to a system having an output equal to the average daily temperature. Examining the plots of input and output, would you say that the system is linear or not? How did you reach you conclusion?
b) Find the first five terms (c_0, ... , c_4) of the complex Fourier series for each signal.
c) What is the harmonic distortion in the two signals? Exclude c_0 from this calculation.
d) Because the harmonic distortion is small, let's concentrate only on the first harmonic. What is the phase shift between input and output signals?
e) Find the transfer function of the simplest possible linear model that would describe the data. Characterize and interpret the structure of this model. In particular, give a physical explanation for the phase shift.
f) Predict what the output would be if the model had no phase shift. Would days be hotter? If so, by how much?

Problem 4.6: Fourier Transform Pairs
Find the Fourier or inverse Fourier transform of the following.

a) $x(t) = e^{-(a|t|)}$,
b) $x(t) = te^{-(at)}u(t)$
c) $X(f) = \begin{cases} 1 \text{ if } |f| < W \\ 0 \text{ if } |f| > W \end{cases}$
d) $x(t) = e^{-(at)}\cos(2\pi f_0 t)\,u(t)$

Problem 4.7: Duality in Fourier Transforms
"Duality" means that the Fourier transform and the inverse Fourier transform are very similar. Consequently, the waveform $s(t)$ in the time domain and the spectrum $s(f)$ have a Fourier transform and an inverse Fourier transform, respectively, that are very similar.

a) Calculate the Fourier transform of the signal shown below (Figure 4.23(a)).
b) Calculate the inverse Fourier transform of the spectrum shown below (Figure 4.23(b)).
c) How are these answers related? What is the general relationship between the Fourier transform of $s(t)$ and the inverse transform of $s(f)$?

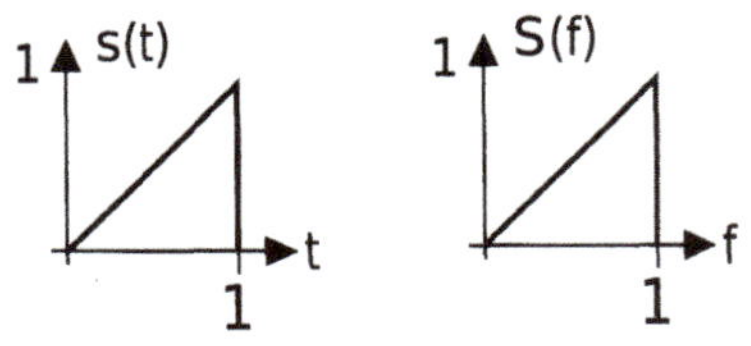

(a) (b)

Figure 4.23

Problem 4.8: Spectra of Pulse Sequences
Pulse sequences occur often in digital communication and in other fields as well. What are their spectral properties?

a) Calculate the Fourier transform of the single pulse shown below (Figure 4.24(a)).
b) Calculate the Fourier transform of the two-pulse sequence shown below (Figure 4.24(b)).
c) Calculate the Fourier transform for the *ten*-pulse sequence shown in below (Figure 4.24(c)). You should look for a general expression that holds for sequences of any length.
d) Using Matlab, plot the magnitudes of the three spectra. Describe how the spectra change as the number of repeated pulses increases.

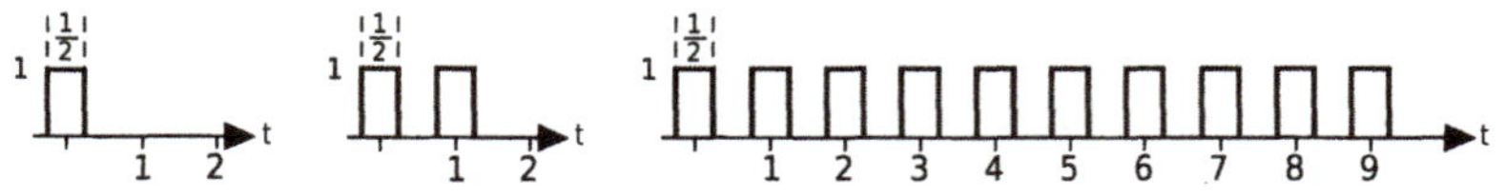

(a) (b) (c)

Figure 4.24

Problem 4.9: Lowpass Filtering a Square Wave
Let a square wave (period T) serve as the input to a first-order lowpass system constructed as a RC filter. We want to derive an expression for the time-domain response of the filter to this input.

a) First, consider the response of the filter to a simple pulse, having unit amplitude and width $\frac{T}{2}$. Derive an expression for the filter's output to this pulse.
b) Noting that the square wave is a superposition of a sequence of these pulses, what is the filter's response to the square wave?
c) The nature of this response should change as the relation between the square wave's period and the filter's cutoff frequency change. How long must the period be so that the response does *not* achieve a relatively constant value between transitions in the square wave? What is the relation of the filter's cutoff frequency to the square wave's spectrum in this case?

Problem 4.10: Mathematics with Circuits
Simple circuits can implement simple mathematical operations, such as integration and differentiation. We want to develop an active circuit (it contains an op-amp) having an output that is proportional to the

integral of its input. For example, you could use an integrator in a car to determine distance traveled from the speedometer.

a) What is the transfer function of an integrator?
b) Find an op-amp circuit so that its voltage output is proportional to the integral of its input for all signals.

Problem 4.11: Where is that sound coming from?
We determine where sound is coming from because we have two ears and a brain. Sound travels at a relatively slow speed and our brain uses the fact that sound will arrive at one ear before the other. As shown here (Figure 4.25), a sound coming from the right arrives at the left ear τ seconds after it arrives at the right ear.

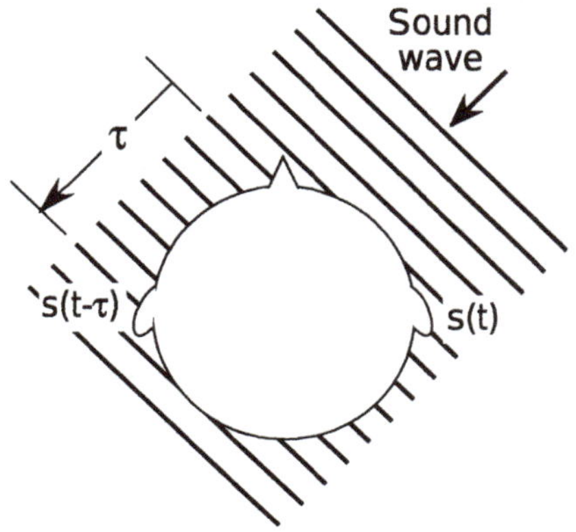

Figure 4.25

Once the brain finds this propagation delay, it can determine the sound direction. In an attempt to model what the brain might do, RU signal processors want to design an *optimal* system that delays each ear's signal by some amount then adds them together. Δ_l and Δ_r are the delays applied to the left and right signals respectively. The idea is to determine the delay values according to some criterion that is based on what is measured by the two ears.

a) What is the transfer function between the sound signal $s(t)$ and the processor output $y(t)$?
b) One way of determining the delay τ is to choose Δ_l and Δ_r to maximize the power in $y(t)$. How are these maximum-power processing delays related to τ?

Problem 4.12: Arrangements of Systems
Architecting a system of modular components means arranging them in various configurations to achieve some overall input-output relation. For each of the following (Figure 4.26), determine the overall transfer function between $x(t)$ and $y(t)$.

The overall transfer function for the cascade (first depicted system) is particularly interesting. What does it say about the effect of the ordering of linear, time-invariant systems in a cascade?

Problem 4.13: Filtering
Let the signal $s(t) = \frac{\sin(\pi t)}{\pi t}$ be the input to a linear, time-invariant filter having the transfer function shown below (Figure 4.27). Find the expression for $y(t)$, the filter's output.

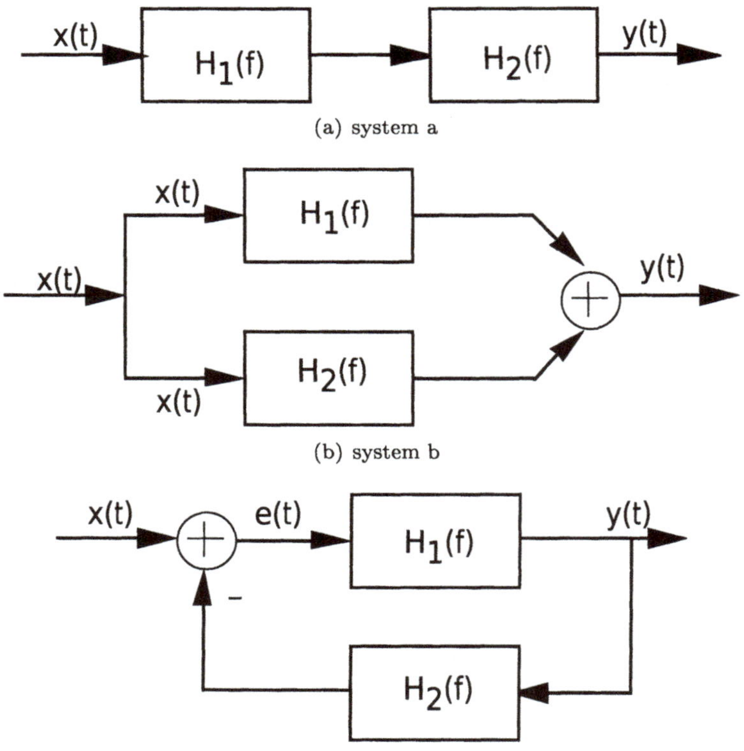

(a) system a

(b) system b

(c) system c

Figure 4.26

Problem 4.14: Circuits Filter!
A unit-amplitude pulse with duration of one second serves as the input to an RC-circuit having transfer function

$$H(f) = \frac{j2\pi f}{4 + j2\pi f}$$

a) How would you categorize this transfer function: lowpass, highpass, bandpass, other?
b) Find a circuit that corresponds to this transfer function.
c) Find an expression for the filter's output.

Problem 4.15: Reverberation
Reverberation corresponds to adding to a signal its delayed version.

a) Assuming τ represents the delay, what is the input-output relation for a reverberation system? Is the system linear and time-invariant? If so, find the transfer function; if not, what linearity or time-invariance criterion does reverberation violate.
b) A music group known as the ROwls is having trouble selling its recordings. The record company's engineer gets the idea of applying different delay to the low and high frequencies and adding the result to create a new musical effect. Thus, the ROwls' audio would be separated into two parts (one less

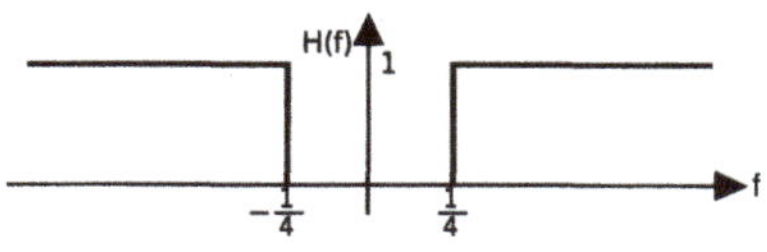

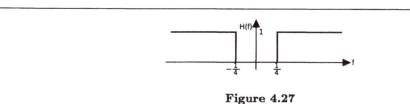

Figure 4.27

than the frequency f_0, the other greater than f_0), these would be delayed by τ_l and τ_h respectively, and the resulting signals added. Draw a block diagram for this new audio processing system, showing its various components.

c) How does the magnitude of the system's transfer function depend on the two delays?

Problem 4.16: Echoes in Telephone Systems
A frequently encountered problem in telephones is echo. Here, because of acoustic coupling between the ear piece and microphone in the handset, what you hear is also sent to the person talking. That person thus not only hears you, but also hears her own speech delayed (because of propagation delay over the telephone network) and attenuated (the acoustic coupling gain is less than one). Furthermore, the same problem applies to you as well: The acoustic coupling occurs in her handset as well as yours.

a) Develop a block diagram that describes this situation.
b) Find the transfer function between your voice and what the listener hears.
c) Each telephone contains a system for reducing echoes using electrical means. What simple system could null the echoes?

Problem 4.17: Demodulating an AM Signal
Let $m(t)$ denote the signal that has been amplitude modulated.

$$x(t) = A(1 + m(t))\sin(2\pi f_c t)$$

Radio stations try to restrict the amplitude of the signal $m(t)$ so that it is less than one in magnitude. The frequency f_c is very large compared to the frequency content of the signal. What we are concerned about here is not transmission, but reception.

a) The so-called coherent demodulator simply multiplies the signal $x(t)$ by a sinusoid having the same frequency as the carrier and lowpass filters the result. Analyze this receiver and show that it works. Assume the lowpass filter is ideal.
b) One issue in coherent reception is the phase of the sinusoid used by the receiver relative to that used by the transmitter. Assuming that the sinusoid of the receiver has a phase ϕ, how does the output depend on ϕ? What is the worst possible value for this phase?
c) The incoherent receiver is more commonly used because of the phase sensitivity problem inherent in coherent reception. Here, the receiver full-wave rectifies the received signal and lowpass filters the result (again ideally). Analyze this receiver. Does its output differ from that of the coherent receiver in a significant way?

Problem 4.18: Unusual Amplitude Modulation
We want to send a band-limited signal having the depicted spectrum (Figure 4.28(a)) with amplitude modulation in the usual way. I.B. Different suggests using the square-wave carrier shown below (Figure 4.28(b)). Well, it is different, but his friends wonder if any technique can demodulate it.

a) Find an expression for $X(f)$, the Fourier transform of the modulated signal.
b) Sketch the magnitude of $X(f)$, being careful to label important magnitudes and frequencies.

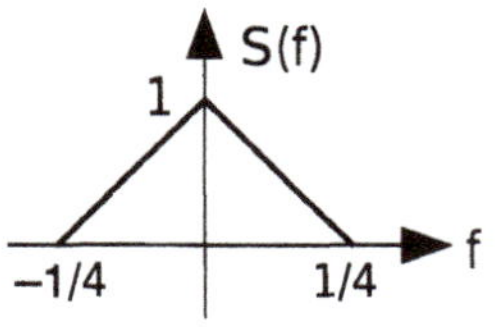

(a)

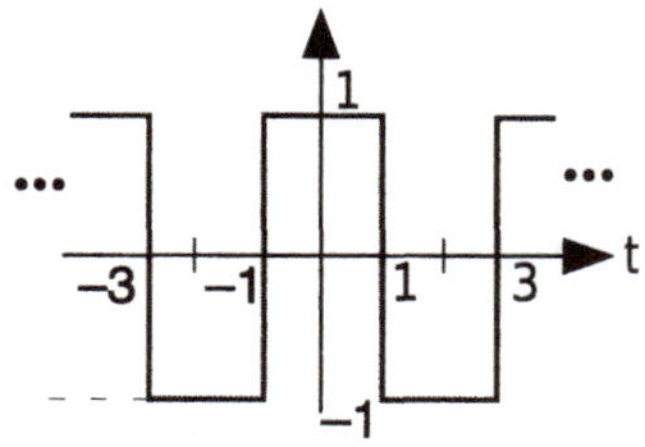

(b)

Figure 4.28

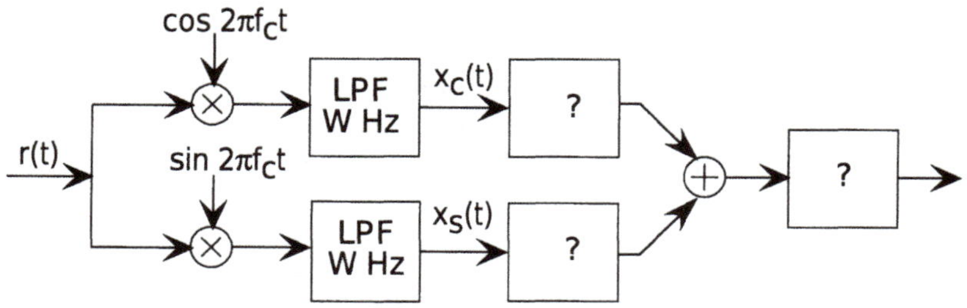

Figure 4.29

c) What demodulation technique obviously works?
d) I.B. challenges three of his friends to demodulate $x(t)$ some other way. One friend suggests modulating $x(t)$ with $\cos\left(\frac{\pi t}{2}\right)$, another wants to try modulating with $\cos(\pi t)$ and the third thinks $\cos\left(\frac{3\pi t}{2}\right)$ will work. Sketch the magnitude of the Fourier transform of the signal each student's approach produces. Which student comes closest to recovering the original signal? Why?

Problem 4.19: Sammy Falls Asleep...
While sitting in ELEC 241 class, he falls asleep during a critical time when an AM receiver is being described. The received signal has the form $r(t) = A(1 + m(t))\cos(2\pi f_c t + \phi)$ where the phase ϕ is unknown. The message signal is $m(t)$; it has a bandwidth of W Hz and a magnitude less than 1 ($|m(t)| < 1$). The phase ϕ is unknown. The instructor drew a diagram (Figure 4.29) for a receiver on the board; Sammy slept through the description of what the unknown systems where.

a) What are the signals $x_c(t)$ and $x_s(t)$?
b) What would you put in for the unknown systems that would guarantee that the final output contained the message regardless of the phase?

HINT: Think of a trigonometric identity that would prove useful.

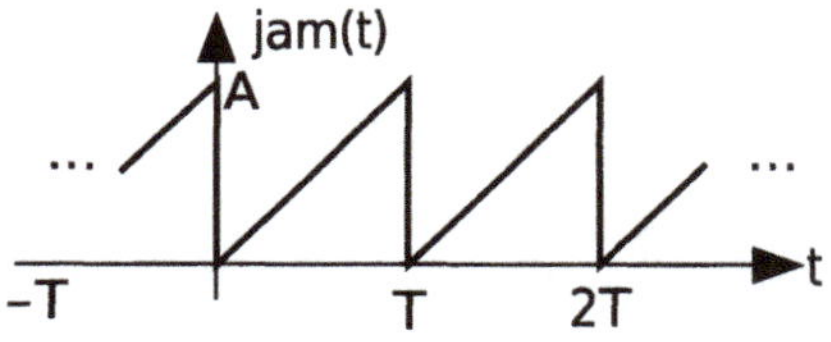

Figure 4.30

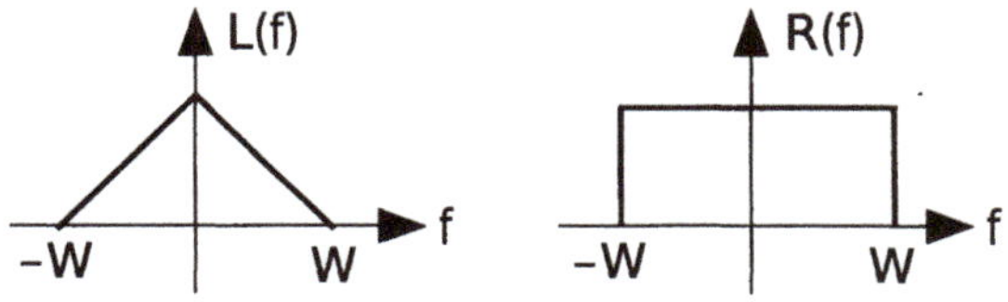

Figure 4.31

c) Sammy may have been asleep, but he can think of a far simpler receiver. What is it?

Problem 4.20: Jamming
Sid Richardson college decides to set up its own AM radio station KSRR. The resident electrical engineer decides that she can choose *any* carrier frequency and message bandwidth for the station. A rival college decides to **jam** its transmissions by transmitting a high-power signal that interferes with radios that try to receive KSRR. The jamming signal $jam\,(t)$ is what is known as a **sawtooth** wave (depicted in the following figure (Figure 4.30)) having a period known to KSRR's engineer.

a) Find the spectrum of the jamming signal.
b) Can KSRR entirely circumvent the attempt to jam it by carefully choosing its carrier frequency and transmission bandwidth? If so, find the station's carrier frequency and transmission bandwidth in terms of T, the period of the jamming signal; if not, show why not.

Problem 4.21: AM Stereo
A stereophonic signal consists of a "left" signal $l\,(t)$ and a "right" signal $r\,(t)$ that conveys sounds coming from an orchestra's left and right sides, respectively. To transmit these two signals simultaneously, the transmitter first forms the sum signal $s_+\,(t) = l\,(t) + r\,(t)$ and the difference signal $s_-\,(t) = l\,(t) - r\,(t)$. Then, the transmitter amplitude-modulates the difference signal with a sinusoid having frequency $2W$, where W is the bandwidth of the left and right signals. The sum signal and the modulated difference signal are added, the sum amplitude-modulated to the radio station's carrier frequency f_c, and transmitted. Assume the spectra of the left and right signals are as shown (Figure 4.31).

a) What is the expression for the transmitted signal? Sketch its spectrum.
b) Show the block diagram of a stereo AM receiver that can yield the left and right signals as separate outputs.
c) What signal would be produced by a conventional coherent AM receiver that expects to receive a standard AM signal conveying a message signal having bandwidth W?

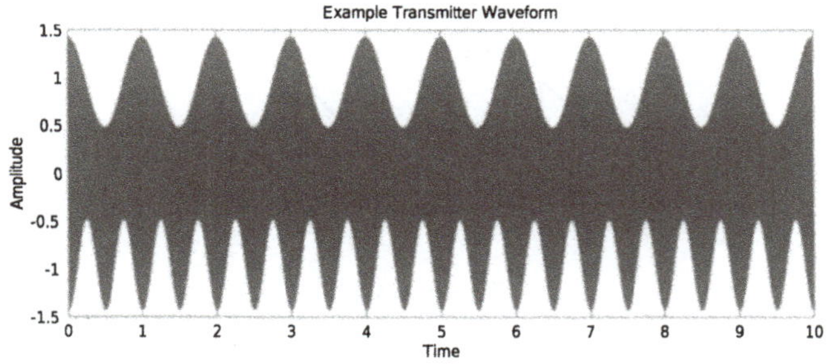

Figure 4.32

Problem 4.22: Novel AM Stereo Method
A clever engineer has submitted a patent for a new method for transmitting two signals *simultaneously* in the *same* transmission bandwidth as commercial AM radio. As shown (Figure 4.32), her approach is to modulate the positive portion of the carrier with one signal and the negative portion with a second. In detail the two message signals $m_1(t)$ and $m_2(t)$ are bandlimited to W Hz and have maximal amplitudes equal to 1. The carrier has a frequency f_c much greater than W. The transmitted signal $x(t)$ is given by

$$x(t) = \begin{cases} A(1 + am_1(t))\sin(2\pi f_c t) & \text{if } \sin(2\pi f_c t) \geq 0 \\ A(1 + am_2(t))\sin(2\pi f_c t) & \text{if } \sin(2\pi f_c t) < 0 \end{cases}$$

In all cases, $0 < a < 1$. The plot shows the transmitted signal when the messages are sinusoids: $m_1(t) = \sin(2\pi f_m t)$ and $m_2(t) = \sin(2\pi 2 f_m t)$ where $2f_m < W$. You, as the patent examiner, must determine whether the scheme meets its claims and is useful.

a) Provide a more concise expression for the transmitted signal $x(t)$ than given above.
b) What is the receiver for this scheme? It would yield both $m_1(t)$ and $m_2(t)$ from $x(t)$.
c) Find the spectrum of the positive portion of the transmitted signal.
d) Determine whether this scheme satisfies the design criteria, allowing you to grant the patent. Explain your reasoning.

Problem 4.23: A Radical Radio Idea
An ELEC 241 student has the bright idea of using a square wave instead of a sinusoid as an AM carrier. The transmitted signal would have the form

$$x(t) = A(1 + m(t))\, sq_T(t)$$

where the message signal $m(t)$ would be amplitude-limited: $|m(t)| < 1$

a) Assuming the message signal is lowpass and has a bandwidth of W Hz, what values for the square wave's period T are feasible. In other words, do some combinations of W and T prevent reception?
b) Assuming reception is possible, can *standard* radios receive this innovative AM transmission? If so, show how a coherent receiver could demodulate it; if not, show how the coherent receiver's output would be corrupted. Assume that the message bandwidth $W = 5kHz$.

Problem 4.24: Secret Communication
An amplitude-modulated secret message $m(t)$ has the following form.

$$r(t) = A(1 + m(t))\cos(2\pi(f_c + f_0)t)$$

The message signal has a bandwidth of W Hz and a magnitude less than 1 ($|m(t)| < 1$). The idea is to offset the carrier frequency by f_0 Hz from standard radio carrier frequencies. Thus, "off-the-shelf" coherent demodulators would assume the carrier frequency has f_c Hz. Here, $f_0 < W$.

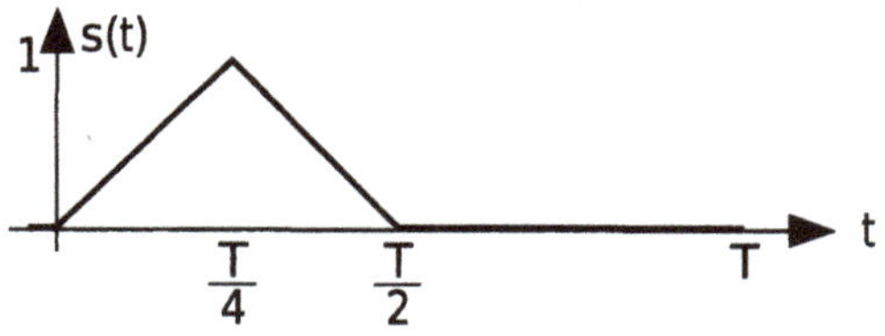

Figure 4.33

a) Sketch the spectrum of the demodulated signal produced by a coherent demodulator tuned to f_c Hz.
b) Will this demodulated signal be a "scrambled" version of the original? If so, how so; if not, why not?
c) Can you develop a receiver that can demodulate the message without knowing the offset frequency f_c?

Problem 4.25: Signal Scrambling
An excited inventor announces the discovery of a way of using analog technology to render music unlistenable without knowing the secret recovery method. The idea is to modulate the bandlimited message $m(t)$ by a special periodic signal $s(t)$ that is zero during half of its period, which renders the message unlistenable and superficially, at least, unrecoverable (Figure 4.33).

a) What is the Fourier series for the periodic signal?
b) What are the restrictions on the period T so that the message signal can be recovered from $m(t)\,s(t)$?
c) ELEC 241 students think they have "broken" the inventor's scheme and are going to announce it to the world. How would they recover the original message *without* having detailed knowledge of the modulating signal?

Solutions to Exercises in Chapter 4

Solution to Exercise 4.1 (p. 110)
Because of Euler's relation,

$$\sin(2\pi ft) = \frac{1}{2j}e^{+j2\pi ft} - \frac{1}{2j}e^{-(j2\pi ft)} \qquad (4.47)$$

Thus, $c_1 = \frac{1}{2j}$, $c_{-1} = -\left(\frac{1}{2j}\right)$, and the other coefficients are zero.

Solution to Exercise 4.2 (p. 113)
$c_0 = \frac{A\Delta}{T}$. This quantity clearly corresponds to the periodic pulse signal's average value.

Solution to Exercise 4.3 (p. 114)
Write the coefficients of the complex Fourier series in Cartesian form as $c_k = A_k + jB_k$ and substitute into the expression for the complex Fourier series.

$$\sum_{k=-\infty}^{\infty}\left(c_k e^{j\frac{2\pi kt}{T}}\right) = \sum_{k=-\infty}^{\infty}\left((A_k + jB_k)\, e^{j\frac{2\pi kt}{T}}\right)$$

Simplifying each term in the sum using Euler's formula,

$$\begin{aligned}(A_k + jB_k)\, e^{j\frac{2\pi kt}{T}} &= (A_k + jB_k)\left(\cos\left(\tfrac{2\pi kt}{T}\right) + j\sin\left(\tfrac{2\pi kt}{T}\right)\right)\\ &= A_k\cos\left(\tfrac{2\pi kt}{T}\right) - B_k\sin\left(\tfrac{2\pi kt}{T}\right) + j\left(A_k\sin\left(\tfrac{2\pi kt}{T}\right) + B_k\cos\left(\tfrac{2\pi kt}{T}\right)\right)\end{aligned}$$

We now combine terms that have the same frequency index *in magnitude*. Because the signal is real-valued, the coefficients of the complex Fourier series have conjugate symmetry: $c_{-k} = c_k^*$ or $A_{-k} = A_k$ and $B_{-k} = -B_k$. After we add the positive-indexed and negative-indexed terms, each term in the Fourier series becomes $2A_k\cos\left(\frac{2\pi kt}{T}\right) - 2B_k\sin\left(\frac{2\pi kt}{T}\right)$. To obtain the classic Fourier series (4.11), we must have $2A_k = a_k$ and $2B_k = -b_k$.

Solution to Exercise 4.4 (p. 115)
The average of a set of numbers is the sum divided by the number of terms. Viewing signal integration as the limit of a Riemann sum, the integral corresponds to the average.

Solution to Exercise 4.5 (p. 115)
We found that the complex Fourier series coefficients are given by $c_k = \frac{2}{j\pi k}$. The coefficients are pure imaginary, which means $a_k = 0$. The coefficients of the sine terms are given by $b_k = -(2\mathrm{Im}(c_k))$ so that

$$b_k = \begin{cases}\frac{4}{\pi k} \text{ if } k \text{ odd}\\ 0 \text{ if } k \text{ even}\end{cases}$$

Thus, the Fourier series for the square wave is

$$\mathrm{sq}(t) = \sum_{k\in\{1,3,\dots\}}\left(\frac{4}{\pi k}\sin\left(\frac{2\pi kt}{T}\right)\right) \qquad (4.48)$$

Solution to Exercise 4.6 (p. 117)
The rms value of a sinusoid equals its amplitude divided by $\sqrt{2}$. As a half-wave rectified sine wave is zero during half of the period, its rms value is $\frac{A}{2\sqrt{2}}$.

Solution to Exercise 4.7 (p. 117)
Total harmonic distortion equals $\frac{\sum_{k=2}^{\infty}\left({a_k}^2+{b_k}^2\right)}{{a_1}^2+{b_1}^2}$. Clearly, this quantity is most easily computed in the frequency domain. However, the numerator equals the square of the signal's rms value minus the power in the average and the power in the first harmonic.

Solution to Exercise 4.8 (p. 120)

Total harmonic distortion in the square wave is $1 - \frac{1}{2}\left(\frac{4}{\pi}\right)^2 = 20\%$.

Solution to Exercise 4.9 (p. 123)

N signals directly encoded require a bandwidth of $\frac{N}{T}$. Using a binary representation, we need $\frac{\log_2 N}{T}$. For $N = 128$, the binary-encoding scheme has a factor of $\frac{7}{128} = 0.05$ smaller bandwidth. Clearly, binary encoding is superior.

Solution to Exercise 4.10 (p. 123)

We can use N different amplitude values at only one frequency to represent the various letters.

Solution to Exercise 4.11 (p. 126)

Because the filter's gain at zero frequency equals one, the average output values equal the respective average input values.

Solution to Exercise 4.12 (p. 128)

$$\mathcal{F}(S(f)) = \int_{-\infty}^{\infty} S(f)\, e^{-(j2\pi f t)} df = \int_{-\infty}^{\infty} S(f)\, e^{+j2\pi f(-t)} df = s(-t)$$

Solution to Exercise 4.13 (p. 128)

$\mathcal{F}(\mathcal{F}(\mathcal{F}(\mathcal{F}(s(t))))) = s(t)$. We know that $\mathcal{F}(S(f)) = \int_{-\infty}^{\infty} S(f)\, e^{-(j2\pi f t)} df = \int_{-\infty}^{\infty} S(f)\, e^{+j2\pi f(-t)} df = s(-t)$. Therefore, two Fourier transforms applied to $s(t)$ yields $s(-t)$. We need two more to get us back where we started.

Solution to Exercise 4.14 (p. 131)

The signal is the inverse Fourier transform of the triangularly shaped spectrum, and equals $s(t) = W\left(\frac{\sin(\pi W t)}{\pi W t}\right)^2$

Solution to Exercise 4.15 (p. 131)

The result is most easily found in the spectrum's formula: the power in the signal-related part of $x(t)$ is half the power of the signal $s(t)$.

Solution to Exercise 4.16 (p. 133)

The inverse transform of the frequency response is $\frac{1}{RC} e^{-\left(\frac{t}{RC}\right)} u(t)$. Multiplying the frequency response by $1 - e^{-(j2\pi f \Delta)}$ means subtract from the original signal its time-delayed version. Delaying the frequency response's time-domain version by Δ results in $\frac{1}{RC} e^{\frac{-(t-\Delta)}{RC}} u(t-\Delta)$. Subtracting from the undelayed signal yields $\frac{1}{RC} e^{\frac{-t}{RC}} u(t) - \frac{1}{RC} e^{\frac{-(t-\Delta)}{RC}} u(t-\Delta)$. Now we integrate this sum. Because the integral of a sum equals the sum of the component integrals (integration is linear), we can consider each separately. Because integration and signal-delay are linear, the integral of a delayed signal equals the delayed version of the integral. The integral is provided in the example (4.44).

Solution to Exercise 4.17 (p. 135)

If the glottis were linear, a constant input (a zero-frequency sinusoid) should yield a constant output. The periodic output indicates nonlinear behavior.

Solution to Exercise 4.18 (p. 136)

In the bottom-left panel, the period is about 0.009 s, which equals a frequency of 111 Hz. The bottom-right panel has a period of about 0.0065 s, a frequency of 154 Hz.

Solution to Exercise 4.19 (p. 137)

Because males have a lower pitch frequency, the spacing between spectral lines is smaller. This closer spacing more accurately reveals the formant structure. Doubling the pitch frequency to 300 Hz for Figure 4.16 (voice spectrum) would amount to removing every other spectral line.

Chapter 5

Digital Signal Processing

5.1 Introduction to Digital Signal Processing[1]

Not only do we have analog signals — signals that are real- or complex-valued functions of a continuous variable such as time or space — we can define *digital* ones as well. Digital signals are *sequences*, functions defined only for the integers. We thus use the notation $s(n)$ to denote a discrete-time one-dimensional signal such as a digital music recording and $s(m, n)$ for a discrete-"time" two-dimensional signal like a photo taken with a digital camera. Sequences are fundamentally different than continuous-time signals. For example, continuity has no meaning for sequences.

Despite such fundamental differences, the theory underlying digital signal processing mirrors that for analog signals: Fourier transforms, linear filtering, and linear systems parallel what previous chapters described. These similarities make it easy to understand the definitions and why we need them, but the similarities should not be construed as "analog wannabes." We will discover that digital signal processing is *not* an approximation to analog processing. We must explicitly worry about the fidelity of converting analog signals into digital ones. The music stored on CDs, the speech sent over digital cellular telephones, and the video carried by digital television all evidence that analog signals can be accurately converted to digital ones and back again.

The key reason why digital signal processing systems have a technological advantage today is the *computer*: computations, like the Fourier transform, can be performed quickly enough to be calculated as the signal is produced, [2] and programmability means that the signal processing system can be easily changed. This flexibility has obvious appeal, and has been widely accepted in the marketplace. Programmability means that we can perform signal processing operations impossible with analog systems (circuits). We will also discover that digital systems enjoy an *algorithmic* advantage that contributes to rapid processing speeds: Computations can be restructured in non-obvious ways to speed the processing. This flexibility comes at a price, a consequence of how computers work. How do computers perform signal processing?

5.2 Introduction to Computer Organization[3]

5.2.1 Computer Architecture

To understand digital signal processing systems, we must understand a little about how computers compute. The modern definition of a *computer* is an electronic device that performs calculations on data, presenting

[1] This content is available online at <http://cnx.org/content/m10781/2.3/>.

[2] Taking a systems viewpoint for the moment, a system that produces its output as rapidly as the input arises is said to be a *real-time* system. All analog systems operate in real time; digital ones that depend on a computer to perform system computations may or may not work in real time. Clearly, we need real-time signal processing systems. Only recently have computers become fast enough to meet real-time requirements while performing non-trivial signal processing.

[3] This content is available online at <http://cnx.org/content/m10263/2.27/>.

the results to humans or other computers in a variety of (hopefully useful) ways.

Organization of a Simple Computer

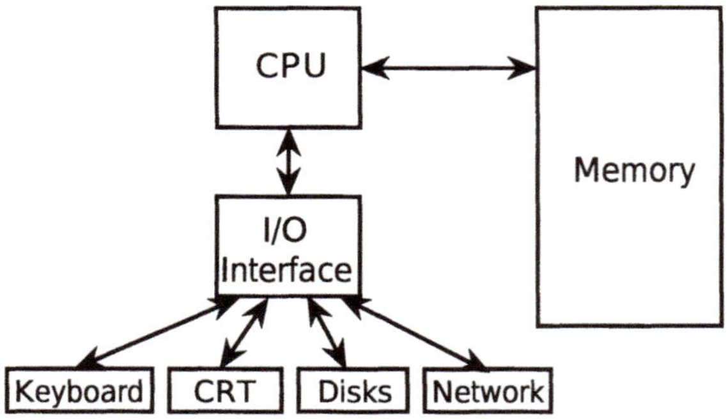

Figure 5.1: Generic computer hardware organization.

The generic computer contains *input* devices (keyboard, mouse, A/D (analog-to-digital) converter, etc.), a *computational unit*, and output devices (monitors, printers, D/A converters). The computational unit is the computer's heart, and usually consists of a *central processing unit* (CPU), a *memory*, and an input/output (I/O) interface. What I/O devices might be present on a given computer vary greatly.

- *A simple computer operates fundamentally in discrete time.* Computers are *clocked* devices, in which computational steps occur periodically according to ticks of a clock. This description belies clock speed: When you say "I have a 1 GHz computer," you mean that your computer takes 1 nanosecond to perform each step. That is incredibly fast! A "step" does not, unfortunately, necessarily mean a computation like an addition; computers break such computations down into several stages, which means that the clock speed need not express the computational speed. Computational speed is expressed in units of millions of instructions/second (Mips). Your 1 GHz computer (clock speed) may have a computational speed of 200 Mips.
- *Computers perform integer (discrete-valued) computations.* Computer calculations can be numeric (obeying the laws of arithmetic), logical (obeying the laws of an algebra), or symbolic (obeying any law you like).[4] Each computer instruction that performs an elementary numeric calculation — an addition, a multiplication, or a division — does so only for integers. The sum or product of two integers is also an integer, but the quotient of two integers is likely to not be an integer. How does a computer deal with numbers that have digits to the right of the decimal point? This problem is addressed by using the so-called *floating-point* representation of real numbers. At its heart, however, this representation relies on integer-valued computations.

5.2.2 Representing Numbers

Focusing on numbers, all numbers can represented by the *positional notation system.* [5] The b-ary positional representation system uses the position of digits ranging from 0 to b-1 to denote a number. The quantity b is

[4] An example of a symbolic computation is sorting a list of names.

[5] Alternative number representation systems exist. For example, we could use stick figure counting or Roman numerals. These were useful in ancient times, but very limiting when it comes to arithmetic calculations: ever tried to divide two Roman numerals?

known as the *base* of the number system. Mathematically, positional systems represent the positive integer n as

$$n = \sum_{k=0}^{\infty} \left(d_k b^k\right) \quad , \quad d_k \in \{0, \ldots, b-1\} \tag{5.1}$$

and we succinctly express n in base-b as $n_b = d_N d_{N-1} \ldots d_0$. The number 25 in base 10 equals $2 \times 10^1 + 5 \times 10^0$, so that the *digits* representing this number are $d_0 = 5$, $d_1 = 2$, and all other d_k equal zero. This same number in *binary* (base 2) equals 11001 ($1 \times 2^4 + 1 \times 2^3 + 0 \times 2^2 + 0 \times 2^1 + 1 \times 2^0$) and 19 in hexadecimal (base 16). Fractions between zero and one are represented the same way.

$$f = \sum_{k=-\infty}^{-1} \left(d_k b^k\right) \quad , \quad d_k \in \{0, \ldots, b-1\} \tag{5.2}$$

All numbers can be represented by their sign, integer and fractional parts. Complex numbers (Section 2.1) can be thought of as two real numbers that obey special rules to manipulate them.

Humans use base 10, commonly assumed to be due to us having ten fingers. Digital computers use the base 2 or *binary* number representation, each digit of which is known as a **bit** (*b*inary dig*it*).

Number representations on computers

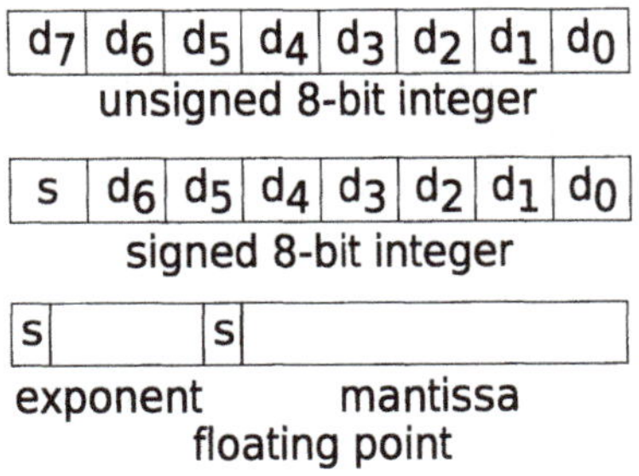

Figure 5.2: The various ways numbers are represented in binary are illustrated. The number of bytes for the exponent and mantissa components of floating point numbers varies.

Here, each bit is represented as a voltage that is either "high" or "low," thereby representing "1" or "0," respectively. To represent signed values, we tack on a special bit—the **sign bit**—to express the sign. The computer's memory consists of an ordered sequence of **bytes**, a collection of eight bits. A byte can therefore represent an unsigned number ranging from 0 to 255. If we take one of the bits and make it the sign bit, we can make the same byte to represent numbers ranging from -128 to 127. But a computer cannot represent *all* possible real numbers. The fault is not with the binary number system; rather having only a finite number of bytes is the problem. While a gigabyte of memory may seem to be a lot, it takes an infinite number of bits to represent π. Since we want to store many numbers in a computer's memory, we are restricted to those that have a *finite* binary representation. Large integers can be represented by an ordered sequence of bytes. Common lengths, usually expressed in terms of the number of bits, are 16, 32, and 64. Thus, an unsigned 32-bit number can represent integers ranging between 0 and $2^{32} - 1$ (4,294,967,295), a number almost big enough to enumerate every human in the world![6]

Exercise 5.1 *(Solution on p. 204.)*

For both 32-bit and 64-bit integer representations, what are the largest numbers that can be represented if a sign bit must also be included.

[6] You need one more bit to do that.

While this system represents integers well, how about numbers having nonzero digits to the right of the decimal point? In other words, how are numbers that have fractional parts represented? For such numbers, the binary representation system is used, but with a little more complexity. The *floating-point* system uses a number of bytes - typically 4 or 8 - to represent the number, but with one byte (sometimes two bytes) reserved to represent the *exponent* e of a power-of-two multiplier for the number - the *mantissa* m - expressed by the remaining bytes.

$$x = m2^{e} \tag{5.3}$$

The mantissa is usually taken to be a binary fraction having a magnitude in the range $\left[\frac{1}{2}, 1\right)$, which means that the binary representation is such that $d_{-1} = 1$. [7] The number zero is an exception to this rule, and it is the *only* floating point number having a zero fraction. The sign of the mantissa represents the sign of the number and the exponent can be a signed integer.

A computer's representation of integers is either perfect or only approximate, the latter situation occurring when the integer exceeds the range of numbers that a limited set of bytes can represent. Floating point representations have similar representation problems: *if* the number x can be multiplied/divided by enough powers of two to yield a fraction lying between 1/2 and 1 that has a *finite* binary-fraction representation, the number is represented exactly in floating point. Otherwise, we can only represent the number approximately, not catastrophically in error as with integers. For example, the number 2.5 equals 0.625×2^{2}, the fractional part of which has an exact binary representation. [8] However, the number 2.6 does *not* have an exact binary representation, and only be represented approximately in floating point. In *single precision floating point* numbers, which require 32 bits (one byte for the exponent and the remaining 24 bits for the mantissa), the number 2.6 will be represented as 2.600000079.... Note that this approximation has a much longer decimal expansion. This level of accuracy may not suffice in numerical calculations. *Double precision floating point numbers* consume 8 bytes, and *quadruple precision* 16 bytes. The more bits used in the mantissa, the greater the accuracy. This increasing accuracy means that more numbers can be represented exactly, but there are always some that cannot. Such inexact numbers have an infinite binary representation.[9] Realizing that real numbers can be only represented approximately is quite important, and underlies the entire field of *numerical analysis*, which seeks to predict the numerical accuracy of any computation.

Exercise 5.2 ***(Solution on p. 204.)***

What are the largest and smallest numbers that can be represented in 32-bit floating point? in 64-bit floating point that has sixteen bits allocated to the exponent? Note that both exponent and mantissa require a sign bit.

So long as the integers aren't too large, they can be represented exactly in a computer using the binary positional notation. Electronic circuits that make up the physical computer can add and subtract integers without error. (This statement isn't quite true; when does addition cause problems?)

---

[7] In some computers, this normalization is taken to an extreme: the leading binary digit is not explicitly expressed, providing an extra bit to represent the mantissa a little more accurately. This convention is known as the **hidden-ones notation**.

[8] See if you can find this representation.

[9] Note that there will *always* be numbers that have an infinite representation in any chosen positional system. The choice of base defines which do and which don't. If you were thinking that base 10 numbers would solve this inaccuracy, note that $1/3 = 0.333333....$ has an infinite representation in decimal (and binary for that matter), but has finite representation in base 3.

5.2.3 Computer Arithmetic and Logic

The binary addition and multiplication tables are

$$\left(\begin{array}{c} 0+0=0 \\ 0+1=1 \\ 1+1=10 \\ 1+0=1 \\ \\ 0\times 0=0 \\ 0\times 1=0 \\ 1\times 1=1 \\ 1\times 0=0 \end{array}\right) \quad (5.4)$$

Note that if carries are ignored,[10] subtraction of two single-digit binary numbers yields the same bit as addition. Computers use high and low voltage values to express a bit, and an array of such voltages express numbers akin to positional notation. Logic circuits perform arithmetic operations.

Exercise 5.3 ***(Solution on p. 204.)***

Add twenty-five and seven in base 2. Note the carries that might occur. Why is the result "nice"?

Also note that the logical operations of AND and OR are equivalent to binary addition (again if carries are ignored). The variables of logic indicate truth or falsehood. $A\bigcap B$, the AND of A and B, represents a statement that both A and B must be true for the statement to be true. You use this kind of statement to tell search engines that you want to restrict hits to cases where both of the events A and B occur. $A\bigcup B$, the OR of A and B, yields a value of truth if either is true. Note that if we represent truth by a "1" and falsehood by a "0," *binary multiplication corresponds to AND and addition (ignoring carries) to OR*. The Irish mathematician George Boole discovered this equivalence in the mid-nineteenth century. It laid the foundation for what we now call Boolean algebra, which expresses as equations logical statements. More importantly, any computer using base-2 representations and arithmetic can also easily evaluate logical statements. This fact makes an integer-based computational device much more powerful than might be apparent.

5.3 The Sampling Theorem[11]

5.3.1 Analog-to-Digital Conversion

Because of the way computers are organized, signal must be represented by a finite number of bytes. This restriction means that *both* the time axis and the amplitude axis must be **quantized**: They must each be a multiple of the integers. [12] Quite surprisingly, the Sampling Theorem allows us to quantize the time axis *without error* for some signals. The signals that can be sampled without introducing error are interesting, and as described in the next section, we can make a signal "samplable" by filtering. In contrast, no one has found a way of performing the amplitude quantization step without introducing an unrecoverable error. Thus, a signal's value can no longer be any real number. Signals processed by digital computers must be **discrete-valued**: their values must be proportional to the integers. Consequently, *analog-to-digital conversion introduces error.*

[10] A *carry* means that a computation performed at a given position affects other positions as well. Here, $1+1=10$ is an example of a computation that involves a carry.

[11] This content is available online at <http://cnx.org/content/m0050/2.18/>.

[12] We assume that we do not use floating-point A/D converters.

5.3.2 The Sampling Theorem

Digital transmission of information and digital signal processing all require signals to first be "acquired" by a computer. One of the most amazing and useful results in electrical engineering is that signals can be converted from a function of time into a sequence of numbers *without error*: We can convert the numbers back into the signal with (theoretically) *no* error. Harold Nyquist, a Bell Laboratories engineer, first derived this result, known as the Sampling Theorem, in the 1920s. It found no real application back then. Claude Shannon[13] , also at Bell Laboratories, revived the result once computers were made public after World War II.

The sampled version of the analog signal $s(t)$ is $s(nT_s)$, with T_s known as the **sampling interval.** Clearly, the value of the original signal at the sampling times is preserved; the issue is how the signal values *between* the samples can be reconstructed since they are lost in the sampling process. To characterize sampling, we approximate it as the product $x(t) = s(t) P_{T_s}(t)$, with $P_{T_s}(t)$ being the periodic pulse signal. The resulting signal, as shown in Figure 5.3 (Sampled Signal), has nonzero values only during the time intervals $\left(nT_s - \frac{\Delta}{2}, nT_s + \frac{\Delta}{2}\right)$, $n \in \{\dots, -1, 0, 1, \dots\}$.

Sampled Signal

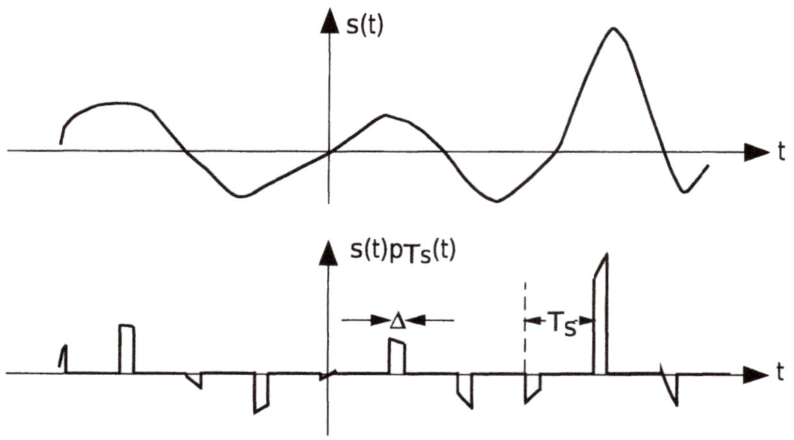

Figure 5.3: The waveform of an example signal is shown in the top plot and its sampled version in the bottom.

For our purposes here, we center the periodic pulse signal about the origin so that its Fourier series coefficients are real (the signal is even).

$$P_{T_s}(t) = \sum_{k=-\infty}^{\infty} \left(c_k e^{\frac{j2\pi kt}{T_s}} \right) \tag{5.5}$$

where

$$c_k = \frac{\sin\left(\frac{\pi k \Delta}{T_s}\right)}{\pi k} \tag{5.6}$$

If the properties of $s(t)$ and the periodic pulse signal are chosen properly, we can recover $s(t)$ from $x(t)$ by filtering.

[13]http://www.lucent.com/minds/infotheory/

To understand how signal values between the samples can be "filled" in, we need to calculate the sampled signal's spectrum. Using the Fourier series representation of the periodic sampling signal,

$$x(t) = \sum_{k=-\infty}^{\infty} \left(c_k e^{\frac{j2\pi kt}{T_s}} s(t) \right) \tag{5.7}$$

Considering each term in the sum separately, we need to know the spectrum of the product of the complex exponential and the signal. Evaluating this transform directly is quite easy.

$$\int_{-\infty}^{\infty} s(t)\, e^{\frac{j2\pi kt}{T_s}} e^{-(j2\pi ft)} dt = \int_{-\infty}^{\infty} s(t)\, e^{-\left(j2\pi\left(f-\frac{k}{T_s}\right)t\right)} dt = S\left(f - \frac{k}{T_s}\right) \tag{5.8}$$

Thus, the spectrum of the sampled signal consists of weighted (by the coefficients c_k) and delayed versions of the signal's spectrum (Figure 5.4 (aliasing)).

$$X(f) = \sum_{k=-\infty}^{\infty} \left(c_k S\left(f - \frac{k}{T_s}\right) \right) \tag{5.9}$$

In general, the terms in this sum overlap each other in the frequency domain, rendering recovery of the original signal impossible. This unpleasant phenomenon is known as **aliasing**.

aliasing

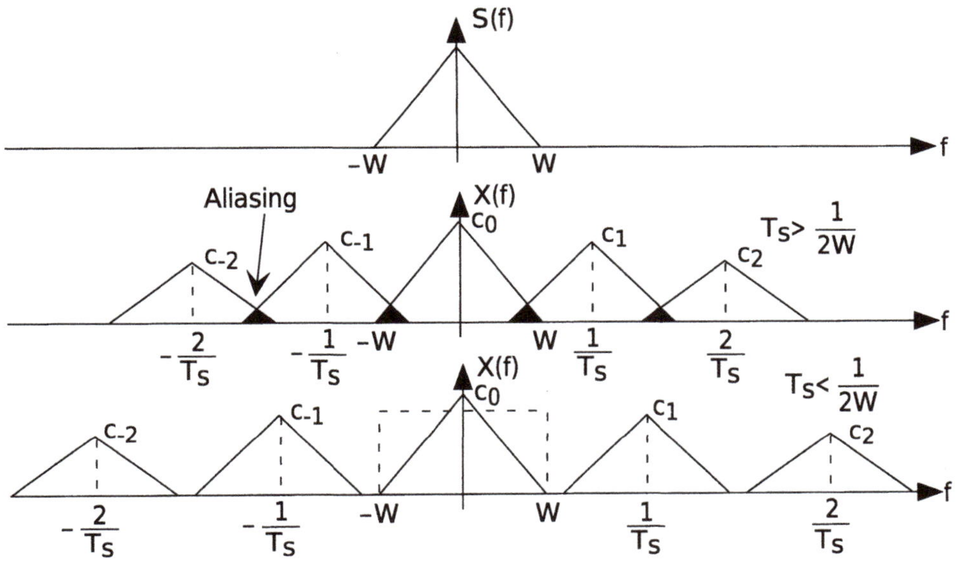

Figure 5.4: The spectrum of some bandlimited (to W Hz) signal is shown in the top plot. If the sampling interval T_s is chosen too large relative to the bandwidth W, aliasing will occur. In the bottom plot, the sampling interval is chosen sufficiently small to avoid aliasing. Note that if the signal were not bandlimited, the component spectra would *always* overlap.

If, however, we satisfy two conditions:

- The signal $s(t)$ is *bandlimited*—has power in a restricted frequency range—to W Hz, and
- the sampling interval T_s is small enough so that the individual components in the sum do not overlap—$T_s < 1/2W$,

aliasing will not occur. In this delightful case, we can recover the original signal by lowpass filtering $x(t)$ with a filter having a cutoff frequency equal to W Hz. These two conditions ensure the ability to recover a bandlimited signal from its sampled version: We thus have the **Sampling Theorem**.

Exercise 5.4 *(Solution on p. 204.)*

The Sampling Theorem (as stated) does not mention the pulse width Δ. What is the effect of this parameter on our ability to recover a signal from its samples (assuming the Sampling Theorem's two conditions are met)?

The frequency $\frac{1}{2T_s}$, known today as the **Nyquist frequency** and the **Shannon sampling frequency**, corresponds to the highest frequency at which a signal can contain energy and remain compatible with the Sampling Theorem. High-quality sampling systems ensure that no aliasing occurs by unceremoniously lowpass filtering the signal (cutoff frequency being slightly lower than the Nyquist frequency) before sampling. Such systems therefore vary the *anti-aliasing* filter's cutoff frequency as the sampling rate varies. Because such quality features cost money, many sound cards do *not* have anti-aliasing filters or, for that matter, post-sampling filters. They sample at high frequencies, 44.1 kHz for example, and hope the signal contains no frequencies above the Nyquist frequency (22.05 kHz in our example). If, however, the signal contains frequencies beyond the sound card's Nyquist frequency, the resulting aliasing can be impossible to remove.

Exercise 5.5 *(Solution on p. 204.)*

To gain a better appreciation of aliasing, sketch the spectrum of a sampled square wave. For simplicity consider only the spectral repetitions centered at $-\left(\frac{1}{T_s}\right)$, 0, $\frac{1}{T_s}$. Let the sampling interval T_s be 1; consider two values for the square wave's period: 3.5 and 4. Note in particular where the spectral lines go as the period decreases; some will move to the left and some to the right. What property characterizes the ones going the same direction?

If we satisfy the Sampling Theorem's conditions, the signal will change only slightly during each pulse. As we narrow the pulse, making Δ smaller and smaller, the nonzero values of the signal $s(t)\, p_{T_s}(t)$ will simply be $s(nT_s)$, the signal's **samples**. If indeed the Nyquist frequency equals the signal's highest frequency, at least two samples will occur within the period of the signal's highest frequency sinusoid. In these ways, the sampling signal captures the sampled signal's temporal variations in a way that leaves all the original signal's structure intact.

Exercise 5.6 *(Solution on p. 204.)*

What is the simplest bandlimited signal? Using this signal, convince yourself that less than two samples/period will not suffice to specify it. If the sampling rate $\frac{1}{T_s}$ is not high enough, what signal would your resulting undersampled signal become?

5.4 Amplitude Quantization[14]

The Sampling Theorem says that if we sample a bandlimited signal $s(t)$ fast enough, it can be recovered without error from its samples $s(nT_s)$, $n \in \{\dots, -1, 0, 1, \dots\}$. Sampling is only the first phase of acquiring data into a computer: Computational processing further requires that the samples be **quantized**: analog values are converted into digital (Section 1.2.2: Digital Signals) form. In short, we will have performed **analog-to-digital (A/D) conversion**.

[14]This content is available online at <http://cnx.org/content/m0051/2.21/>.

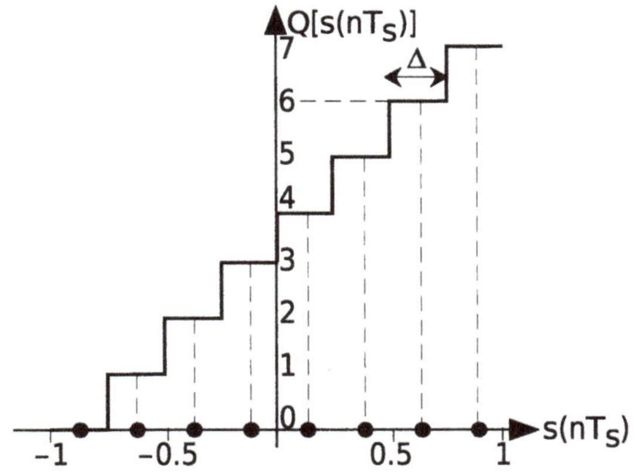

(a)

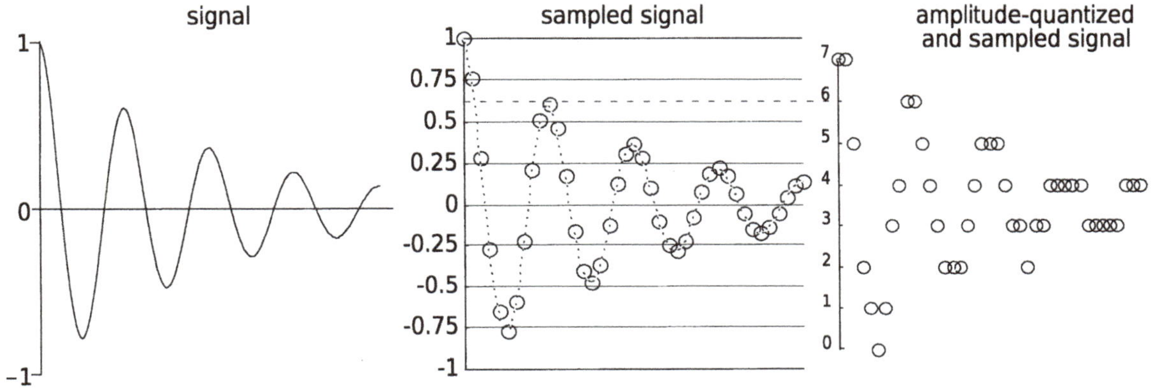

(b)

Figure 5.5: A three-bit A/D converter assigns voltage in the range $[-1, 1]$ to one of eight integers between 0 and 7. For example, all inputs having values lying between 0.5 and 0.75 are assigned the integer value six and, upon conversion back to an analog value, they all become 0.625. The width of a single quantization interval Δ equals $\frac{2}{2^B}$. The bottom panel shows a signal going through the analog-to-digital, where B is the number of bits used in the A/D conversion process (3 in the case depicted here). First it is sampled, then amplitude-quantized to three bits. Note how the sampled signal waveform becomes distorted after amplitude quantization. For example the two signal values between 0.5 and 0.75 become 0.625. This distortion is irreversible; it can be reduced (but not eliminated) by using more bits in the A/D converter.

A phenomenon reminiscent of the errors incurred in representing numbers on a computer prevents signal amplitudes from being converted with no error into a binary number representation. In analog-to-digital conversion, the signal is assumed to lie within a predefined range. Assuming we can scale the signal without affecting the information it expresses, we'll define this range to be $[-1, 1]$. Furthermore, the A/D converter assigns amplitude values in this range to a set of integers. A B-bit converter produces one of the integers $\{0, 1, \dots, 2^B - 1\}$ for each sampled input. Figure 5.5 shows how a three-bit A/D converter assigns input values to the integers. We define a **quantization interval** to be the range of values assigned to the same integer. Thus, for our example three-bit A/D converter, the quantization interval Δ is 0.25; in general, it is $\frac{2}{2^B}$.

Exercise 5.7 *(Solution on p. 204.)*

Recalling the plot of average daily highs in this frequency domain problem (Problem 4.5), why is this plot so jagged? Interpret this effect in terms of analog-to-digital conversion.

Because values lying anywhere within a quantization interval are assigned the same value for computer processing, *the original amplitude value cannot be recovered without error.* Typically, the D/A converter, the device that converts integers to amplitudes, assigns an amplitude equal to the value lying halfway in the quantization interval. The integer 6 would be assigned to the amplitude 0.625 in this scheme. The error introduced by converting a signal from analog to digital form by sampling and amplitude quantization then back again would be half the quantization interval for each amplitude value. Thus, the so-called *A/D* error equals half the width of a quantization interval: $\frac{1}{2^B}$. As we have fixed the input-amplitude range, the more bits available in the A/D converter, the smaller the quantization error.

To analyze the amplitude quantization error more deeply, we need to compute the **signal-to-noise** ratio, which equals the ratio of the signal power and the quantization error power. Assuming the signal is a sinusoid, the signal power is the square of the rms amplitude: $power(s) = \left(\frac{1}{\sqrt{2}}\right)^2 = \frac{1}{2}$. The illustration (Figure 5.6) details a single quantization interval.

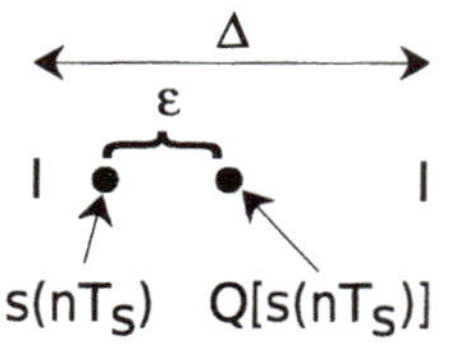

Figure 5.6: A single quantization interval is shown, along with a typical signal's value before amplitude quantization $s(nT_s)$ and after $Q(s(nT_s))$. ϵ denotes the error thus incurred.

Its width is Δ and the quantization error is denoted by ϵ. To find the power in the quantization error, we note that no matter into which quantization interval the signal's value falls, the error will have the same characteristics. To calculate the rms value, we must square the error and average it over the interval.

$$\begin{aligned} rms(\epsilon) &= \sqrt{\frac{1}{\Delta}\int_{-\left(\frac{\Delta}{2}\right)}^{\frac{\Delta}{2}} \epsilon^2 d\epsilon} \\ &= \left(\frac{\Delta^2}{12}\right)^{\frac{1}{2}} \end{aligned} \tag{5.10}$$

Since the quantization interval width for a B-bit converter equals $\frac{2}{2^B} = 2^{-(B-1)}$, we find that the signal-to-noise ratio for the analog-to-digital conversion process equals

$$SNR = \frac{\frac{1}{2}}{\frac{2^{-(2(B-1))}}{12}} = \frac{3}{2}2^{2B} = 6B + 10\log_{10}1.5dB \tag{5.11}$$

Thus, every bit increase in the A/D converter yields a 6 dB increase in the signal-to-noise ratio.

Exercise 5.8 *(Solution on p. 205.)*

This derivation assumed the signal's amplitude lay in the range $[-1, 1]$. What would the amplitude quantization signal-to-noise ratio be if it lay in the range $[-A, A]$?

Exercise 5.9 *(Solution on p. 205.)*

How many bits would be required in the A/D converter to ensure that the maximum amplitude quantization error was less than 60 db smaller than the signal's peak value?

Exercise 5.10 *(Solution on p. 205.)*

Music on a CD is stored to 16-bit accuracy. To what signal-to-noise ratio does this correspond?

Once we have acquired signals with an A/D converter, we can process them using digital hardware or software. It can be shown that if the computer processing is linear, the result of sampling, computer processing, and unsampling is equivalent to some analog linear system. Why go to all the bother if the same function can be accomplished using analog techniques? Knowing when digital processing excels and when it does not is an important issue.

5.5 Discrete-Time Signals and Systems[15]

Mathematically, analog signals are functions having as their independent variables continuous quantities, such as space and time. Discrete-time signals are functions defined on the integers; they are sequences. As with analog signals, we seek ways of decomposing discrete-time signals into simpler components. Because this approach leading to a better understanding of signal structure, we can exploit that structure to represent information (create ways of representing information with signals) and to extract information (retrieve the information thus represented). For symbolic-valued signals, the approach is different: We develop a common representation of all symbolic-valued signals so that we can embody the information they contain in a unified way. From an information representation perspective, the most important issue becomes, for both real-valued and symbolic-valued signals, efficiency: what is the most parsimonious and compact way to represent information so that it can be extracted later.

5.5.1 Real- and Complex-valued Signals

A discrete-time signal is represented symbolically as $s(n)$, where $n = \{\dots, -1, 0, 1, \dots\}$.

Cosine

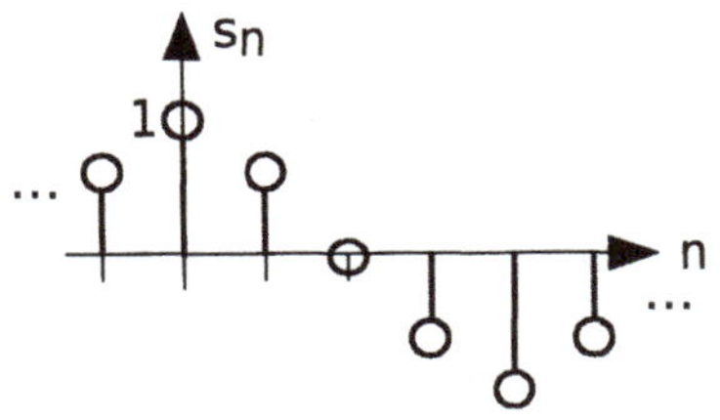

Figure 5.7: The discrete-time cosine signal is plotted as a stem plot. Can you find the formula for this signal?

We usually draw discrete-time signals as stem plots to emphasize the fact they are functions defined only on the integers. We can delay a discrete-time signal by an integer just as with analog ones. A signal delayed by m samples has the expression $s(n-m)$.

5.5.2 Complex Exponentials

The most important signal is, of course, the **complex exponential sequence.**

$$s(n) = e^{j2\pi fn} \tag{5.12}$$

[15]This content is available online at <http://cnx.org/content/m10342/2.13/>.

Note that the frequency variable f is dimensionless and that adding an integer to the frequency of the discrete-time complex exponential has no effect on the signal's value.

$$\begin{aligned} e^{j2\pi(f+m)n} &= e^{j2\pi fn}e^{j2\pi mn} \\ &= e^{j2\pi fn} \end{aligned} \quad (5.13)$$

This derivation follows because the complex exponential evaluated at an integer multiple of 2π equals one. Thus, the period of a discrete-time complex exponential equals one.

5.5.3 Sinusoids

Discrete-time sinusoids have the obvious form $s(n) = A\cos(2\pi fn + \phi)$. As opposed to analog complex exponentials and sinusoids that can have their frequencies be any real value, frequencies of their discrete-time counterparts yield unique waveforms *only* when f lies in the interval $\left(-\left(\frac{1}{2}\right), \frac{1}{2}\right]$. From the properties of the complex exponential, the sinusoid's period is always one; this choice of frequency interval will become evident later.

5.5.4 Unit Sample

The second-most important discrete-time signal is the **unit sample**, which is defined to be

$$\delta(n) = \begin{cases} 1 \text{ if } n = 0 \\ 0 \text{ otherwise} \end{cases} \quad (5.14)$$

Unit sample

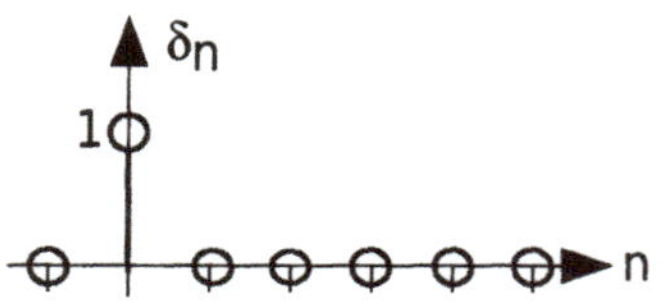

Figure 5.8: The unit sample.

Examination of a discrete-time signal's plot, like that of the cosine signal shown in Figure 5.7 (Cosine), reveals that all signals consist of a sequence of delayed and scaled unit samples. Because the value of a sequence at each integer m is denoted by $s(m)$ and the unit sample delayed to occur at m is written $\delta(n-m)$, we can decompose *any* signal as a sum of unit samples delayed to the appropriate location and scaled by the signal value.

$$s(n) = \sum_{m=-\infty}^{\infty} (s(m)\delta(n-m)) \quad (5.15)$$

This kind of decomposition is unique to discrete-time signals, and will prove useful subsequently.

5.5.5 Unit Step

The **unit sample** in discrete-time is well-defined at the origin, as opposed to the situation with analog signals.

$$u(n) = \begin{cases} 1 \text{ if } n \geq 0 \\ 0 \text{ if } n < 0 \end{cases} \quad (5.16)$$

5.5.6 Symbolic Signals

An interesting aspect of discrete-time signals is that their values do not need to be real numbers. We do have real-valued discrete-time signals like the sinusoid, but we also have signals that denote the sequence of characters typed on the keyboard. Such characters certainly aren't real numbers, and as a collection of possible signal values, they have little mathematical structure other than that they are members of a set. More formally, each element of the *symbolic-valued* signal $s(n)$ takes on one of the values $\{a_1, \dots, a_K\}$ which comprise the **alphabet** A. This technical terminology does not mean we restrict symbols to being members of the English or Greek alphabet. They could represent keyboard characters, bytes (8-bit quantities), integers that convey daily temperature. Whether controlled by software or not, discrete-time systems are ultimately constructed from digital circuits, which consist *entirely* of analog circuit elements. Furthermore, the transmission and reception of discrete-time signals, like e-mail, is accomplished with analog signals and systems. Understanding how discrete-time and analog signals and systems intertwine is perhaps the main goal of this course.

5.5.7 Discrete-Time Systems

Discrete-time systems can act on discrete-time signals in ways similar to those found in analog signals and systems. Because of the role of software in discrete-time systems, many more different systems can be envisioned and "constructed" with programs than can be with analog signals. In fact, a special class of analog signals can be converted into discrete-time signals, processed with software, and converted back into an analog signal, all without the incursion of error. For such signals, systems can be easily produced in software, with equivalent analog realizations difficult, if not impossible, to design.

5.6 Discrete-Time Fourier Transform (DTFT)[16]

The Fourier transform of the discrete-time signal $s(n)$ is defined to be

$$S\left(e^{j2\pi f}\right) = \sum_{n=-\infty}^{\infty} \left(s(n)\, e^{-(j2\pi f n)}\right) \tag{5.17}$$

Frequency here has no units. As should be expected, this definition is linear, with the transform of a sum of signals equaling the sum of their transforms. Real-valued signals have conjugate-symmetric spectra: $S\left(e^{-(j2\pi f)}\right) = S\left(e^{j2\pi f}\right)^*$.

Exercise 5.11 ***(Solution on p. 205.)***

A special property of the discrete-time Fourier transform is that it is periodic with period one: $S\left(e^{j2\pi(f+1)}\right) = S\left(e^{j2\pi f}\right)$. Derive this property from the definition of the DTFT.

Because of this periodicity, we need only plot the spectrum over one period to understand completely the spectrum's structure; typically, we plot the spectrum over the frequency range $\left[-\left(\frac{1}{2}\right), \frac{1}{2}\right]$. When the signal is real-valued, we can further simplify our plotting chores by showing the spectrum only over $\left[0, \frac{1}{2}\right]$; the spectrum at negative frequencies can be derived from positive-frequency spectral values.

When we obtain the discrete-time signal via sampling an analog signal, the Nyquist frequency (p. 162) corresponds to the discrete-time frequency $\frac{1}{2}$. To show this, note that a sinusoid having a frequency equal to the Nyquist frequency $\frac{1}{2T_s}$ has a sampled waveform that equals

$$\cos\left(2\pi \frac{1}{2Ts} nTs\right) = \cos(\pi n) = (-1)^n$$

[16]This content is available online at <http://cnx.org/content/m10247/2.28/>.

The exponential in the DTFT at frequency $\frac{1}{2}$ equals $e^{-\left(\frac{j2\pi n}{2}\right)} = e^{-(j\pi n)} = (-1)^n$, meaning that discrete-time frequency equals analog frequency multiplied by the sampling interval

$$f_D = f_A T_s \tag{5.18}$$

f_D and f_A represent discrete-time and analog frequency variables, respectively. The aliasing figure (Figure 5.4: aliasing) provides another way of deriving this result. As the duration of each pulse in the periodic sampling signal $p_{T_s}(t)$ narrows, the amplitudes of the signal's spectral repetitions, which are governed by the Fourier series coefficients (4.10) of $p_{T_s}(t)$, become increasingly equal. Examination of the periodic pulse signal (Figure 4.1) reveals that as Δ decreases, the value of c_0, the largest Fourier coefficient, decreases to zero: $|c_0| = \frac{A\Delta}{T_s}$. Thus, to maintain a mathematically viable Sampling Theorem, the amplitude A must increase as $\frac{1}{\Delta}$, becoming infinitely large as the pulse duration decreases. Practical systems use a small value of Δ, say $0.1 \cdot T_s$ and use amplifiers to rescale the signal. Thus, the sampled signal's spectrum becomes periodic with period $\frac{1}{T_s}$. Thus, the Nyquist frequency $\frac{1}{2T_s}$ corresponds to the frequency $\frac{1}{2}$.

Example 5.1

Let's compute the discrete-time Fourier transform of the exponentially decaying sequence $s(n) = a^n u(n)$, where $u(n)$ is the unit-step sequence. Simply plugging the signal's expression into the Fourier transform formula,

$$\begin{aligned} S\left(e^{j2\pi f}\right) &= \sum_{n=-\infty}^{\infty}\left(a^n u(n)\, e^{-(j2\pi f n)}\right) \\ &= \sum_{n=0}^{\infty}\left(\left(a e^{-(j2\pi f)}\right)^n\right) \end{aligned} \tag{5.19}$$

This sum is a special case of the **geometric series**.

$$\sum_{n=0}^{\infty}(\alpha^n) = \frac{1}{1-\alpha}\ , \quad |\alpha| < 1 \tag{5.20}$$

Thus, as long as $|a| < 1$, we have our Fourier transform.

$$S\left(e^{j2\pi f}\right) = \frac{1}{1 - a e^{-(j2\pi f)}} \tag{5.21}$$

Using Euler's relation, we can express the magnitude and phase of this spectrum.

$$|S\left(e^{j2\pi f}\right)| = \frac{1}{\sqrt{(1 - a\cos(2\pi f))^2 + a^2\sin^2(2\pi f)}} \tag{5.22}$$

$$\angle\left(S\left(e^{j2\pi f}\right)\right) = -\left(\tan^{-1}\left(\frac{a\sin(2\pi f)}{1 - a\cos(2\pi f)}\right)\right) \tag{5.23}$$

No matter what value of a we choose, the above formulae clearly demonstrate the periodic nature of the spectra of discrete-time signals. Figure 5.9 (Spectrum of exponential signal) shows indeed that the spectrum is a periodic function. We need only consider the spectrum between $-\left(\frac{1}{2}\right)$ and $\frac{1}{2}$ to unambiguously define it. When $a > 0$, we have a lowpass spectrum—the spectrum diminishes as frequency increases from 0 to $\frac{1}{2}$—with increasing a leading to a greater low frequency content; for $a < 0$, we have a highpass spectrum (Figure 5.10 (Spectra of exponential signals)).

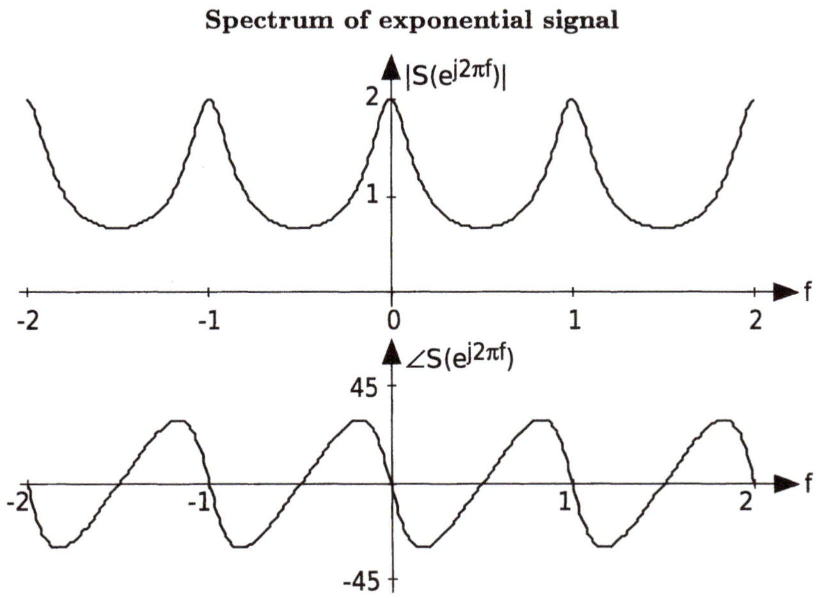

Figure 5.9: The spectrum of the exponential signal ($a = 0.5$) is shown over the frequency range [-2, 2], clearly demonstrating the periodicity of all discrete-time spectra. The angle has units of degrees.

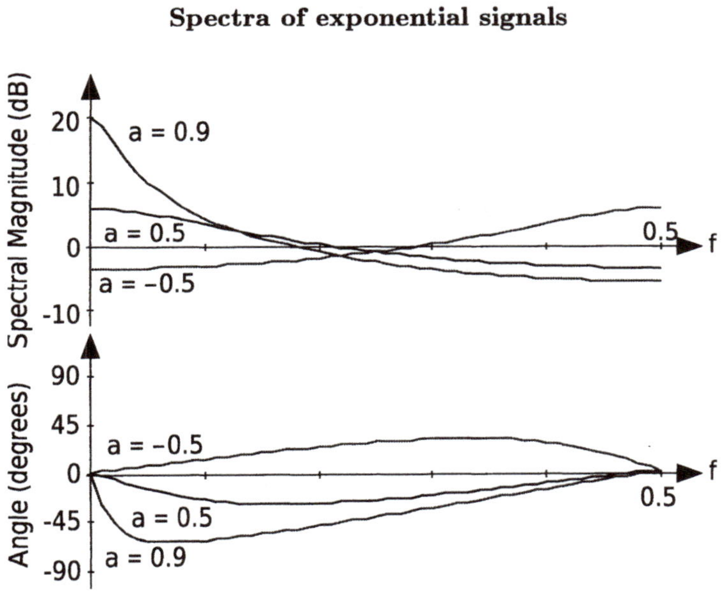

Figure 5.10: The spectra of several exponential signals are shown. What is the apparent relationship between the spectra for $a = 0.5$ and $a = -0.5$?

Example 5.2

Analogous to the analog pulse signal, let's find the spectrum of the length-N pulse sequence.

$$s(n) = \begin{cases} 1 \text{ if } 0 \leq n \leq N-1 \\ 0 \text{ otherwise} \end{cases} \tag{5.24}$$

The Fourier transform of this sequence has the form of a truncated geometric series.

$$S\left(e^{j2\pi f}\right) = \sum_{n=0}^{N-1} \left(e^{-(j2\pi fn)}\right) \tag{5.25}$$

For the so-called finite geometric series, we know that

$$\sum_{n=n_0}^{N+n_0-1} (\alpha^n) = \alpha^{n_0} \frac{1-\alpha^N}{1-\alpha} \tag{5.26}$$

for *all* values of α.

Exercise 5.12 ***(Solution on p. 205.)***

Derive this formula for the finite geometric series sum. The "trick" is to consider the difference between the series' sum and the sum of the series multiplied by α.

Applying this result yields (Figure 5.11 (Spectrum of length-ten pulse).)

$$\begin{aligned} S\left(e^{j2\pi f}\right) &= \frac{1-e^{-(j2\pi fN)}}{1-e^{-(j2\pi f)}} \\ &= e^{-(j\pi f(N-1))} \frac{\sin(\pi fN)}{\sin(\pi f)} \end{aligned} \tag{5.27}$$

The ratio of sine functions has the generic form of $\frac{\sin(Nx)}{\sin(x)}$, which is known as the **discrete-time sinc function** $dsinc(x)$. Thus, our transform can be concisely expressed as $S\left(e^{j2\pi f}\right) = e^{-(j\pi f(N-1))} dsinc(\pi f)$. The discrete-time pulse's spectrum contains many ripples, the number of which increase with N, the pulse's duration.

The inverse discrete-time Fourier transform is easily derived from the following relationship:

$$\int_{-\left(\frac{1}{2}\right)}^{\frac{1}{2}} e^{-(j2\pi fm)} e^{j2\pi fn} df = \begin{cases} 1 \text{ if } m = n \\ 0 \text{ if } m \neq n \end{cases} \tag{5.28}$$

Therefore, we find that

$$\begin{aligned} \int_{-\left(\frac{1}{2}\right)}^{\frac{1}{2}} S\left(e^{j2\pi f}\right) e^{j2\pi fn} df &= \int_{-\left(\frac{1}{2}\right)}^{\frac{1}{2}} \sum_m \left(s(m) e^{-(j2\pi fm)} e^{j2\pi fn}\right) df \\ &= \sum_m \left(s(m) \int_{-\left(\frac{1}{2}\right)}^{\frac{1}{2}} e^{(-(j2\pi f))(m-n)} df\right) \\ &= s(n) \end{aligned} \tag{5.29}$$

The Fourier transform pairs in discrete-time are

$$S\left(e^{j2\pi f}\right) = \sum_{n=-\infty}^{\infty} \left(s(n) e^{-(j2\pi fn)}\right) \tag{5.30}$$

Spectrum of length-ten pulse

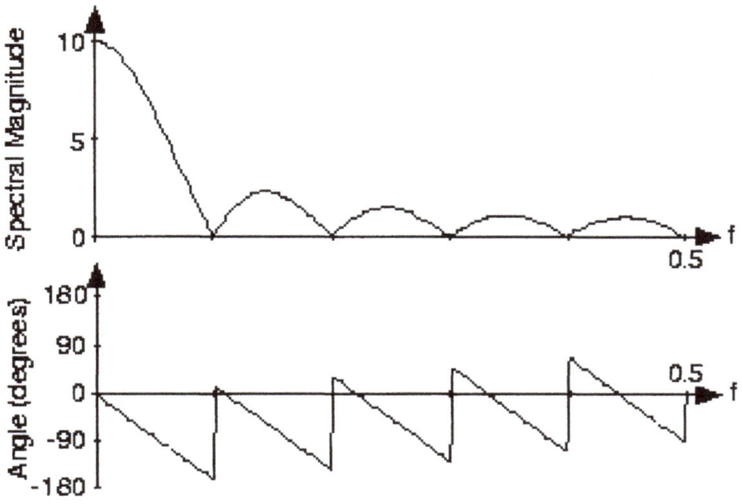

Figure 5.11: The spectrum of a length-ten pulse is shown. Can you explain the rather complicated appearance of the phase?

$$s(n) = \int_{-\left(\frac{1}{2}\right)}^{\frac{1}{2}} S\left(e^{j2\pi f}\right) e^{j2\pi fn} df \tag{5.31}$$

The properties of the discrete-time Fourier transform mirror those of the analog Fourier transform. The DTFT properties table [17] shows similarities and differences. One important common property is Parseval's Theorem.

$$\sum_{n=-\infty}^{\infty} \left(\left(|s(n)|\right)^2\right) = \int_{-\left(\frac{1}{2}\right)}^{\frac{1}{2}} \left(\left|S\left(e^{j2\pi f}\right)\right|\right)^2 df \tag{5.32}$$

To show this important property, we simply substitute the Fourier transform expression into the frequency-domain expression for power.

$$\begin{aligned} \int_{-\left(\frac{1}{2}\right)}^{\frac{1}{2}} \left(\left|S\left(e^{j2\pi f}\right)\right|\right)^2 df &= \int_{-\left(\frac{1}{2}\right)}^{\frac{1}{2}} \left(\sum_n \left(s(n) e^{-(j2\pi fn)}\right)\right) \sum_m \left(s(n)^* e^{j2\pi fm}\right) df \\ &= \sum_{(n,m)} \left(s(n)\, s(n)^* \int_{-\left(\frac{1}{2}\right)}^{\frac{1}{2}} e^{j2\pi f(m-n)} df\right) \end{aligned} \tag{5.33}$$

Using the orthogonality relation (5.28), the integral equals $\delta(m-n)$, where $\delta(n)$ is the unit sample (Figure 5.8: Unit sample). Thus, the double sum collapses into a single sum because nonzero values occur only when $n = m$, giving Parseval's Theorem as a result. We term $\sum_n \left(s^2(n)\right)$ the energy in the discrete-time signal $s(n)$ in spite of the fact that discrete-time signals don't consume (or produce for that matter) energy. This terminology is a carry-over from the analog world.

[17] "Discrete-Time Fourier Transform Properties" <http://cnx.org/content/m0506/latest/>

Exercise 5.13 *(Solution on p. 205.)*

Suppose we obtained our discrete-time signal from values of the product $s(t)\,p_{T_s}(t)$, where the duration of the component pulses in $p_{T_s}(t)$ is Δ. How is the discrete-time signal energy related to the total energy contained in $s(t)$? Assume the signal is bandlimited and that the sampling rate was chosen appropriate to the Sampling Theorem's conditions.

5.7 Discrete Fourier Transforms (DFT)[18]

The discrete-time Fourier transform (and the continuous-time transform as well) can be evaluated when we have an analytic expression for the signal. Suppose we just have a signal, such as the speech signal used in the previous chapter, for which there is no formula. How then would you compute the spectrum? For example, how did we compute a spectrogram such as the one shown in the speech signal example (Figure 4.17: spectrogram)? The Discrete Fourier Transform (DFT) allows the computation of spectra from discrete-time data. While in discrete-time we can *exactly* calculate spectra, for analog signals no similar exact spectrum computation exists. For analog-signal spectra, use must build special devices, which turn out in most cases to consist of A/D converters and discrete-time computations. Certainly discrete-time spectral analysis is more flexible than continuous-time spectral analysis.

The formula for the DTFT (5.17) is a sum, which conceptually can be easily computed save for two issues.

- *Signal duration.* The sum extends over the signal's duration, which must be finite to compute the signal's spectrum. It is exceedingly difficult to store an infinite-length signal in any case, so we'll assume that the signal extends over $[0, N-1]$.
- *Continuous frequency.* Subtler than the signal duration issue is the fact that the frequency variable is continuous: It may only need to span one period, like $\left[-\left(\frac{1}{2}\right), \frac{1}{2}\right]$ or $[0,1]$, but the DTFT formula as it stands requires evaluating the spectra at *all* frequencies within a period. Let's compute the spectrum at a few frequencies; the most obvious ones are the equally spaced ones $f = \frac{k}{K}$, $k \in \{0, \dots, K-1\}$.

We thus define the **discrete Fourier transform** (DFT) to be

$$S(k) = \sum_{n=0}^{N-1} \left(s(n)\, e^{-\left(\frac{j2\pi nk}{K}\right)} \right) \quad , \quad k \in \{0, \dots, K-1\} \tag{5.34}$$

Here, $S(k)$ is shorthand for $S\left(e^{j2\pi\frac{k}{K}}\right)$.

We can compute the spectrum at as many equally spaced frequencies as we like. Note that you can think about this computationally motivated choice as **sampling** the spectrum; more about this interpretation later. The issue now is how many frequencies are enough to capture how the spectrum changes with frequency. One way of answering this question is determining an inverse discrete Fourier transform formula: given $S(k)$, $k = \{0, \dots, K-1\}$ how do we find $s(n)$, $n = \{0, \dots, N-1\}$? Presumably, the formula will be of the form $s(n) = \sum_{k=0}^{K-1} \left(S(k)\, e^{\frac{j2\pi nk}{K}} \right)$. Substituting the DFT formula in this prototype inverse transform yields

$$s(n) = \sum_{k=0}^{K-1} \left(\sum_{m=0}^{N-1} \left(s(m)\, e^{-\left(j\frac{2\pi mk}{K}\right)} e^{j\frac{2\pi nk}{K}} \right) \right) \tag{5.35}$$

Note that the orthogonality relation we use so often has a different character now.

$$\sum_{k=0}^{K-1} \left(e^{-\left(j\frac{2\pi km}{K}\right)} e^{j\frac{2\pi kn}{K}} \right) = \begin{cases} K \text{ if } m = \{n, (n \pm K), (n \pm 2K), \dots\} \\ 0 \text{ otherwise} \end{cases} \tag{5.36}$$

[18]This content is available online at <http://cnx.org/content/m10249/2.26/>.

We obtain nonzero value whenever the two indices differ by multiples of K. We can express this result as $K\sum_l (\delta(m-n-lK))$. Thus, our formula becomes

$$s(n) = \sum_{m=0}^{N-1} \left(s(m) K \sum_{l=-\infty}^{\infty} (\delta(m-n-lK)) \right) \quad (5.37)$$

The integers n and m both range over $\{0, \dots, N-1\}$. To have an inverse transform, we need the sum to be a *single* unit sample for m, n in this range. If it did not, then $s(n)$ would equal a sum of values, and we would not have a valid transform: Once going into the frequency domain, we could not get back unambiguously! Clearly, the term $l=0$ always provides a unit sample (we'll take care of the factor of K soon). If we evaluate the spectrum at *fewer* frequencies than the signal's duration, the term corresponding to $m = n+K$ will also appear for some values of m, $n = \{0, \dots, N-1\}$. This situation means that our prototype transform equals $s(n) + s(n+K)$ for some values of n. The only way to eliminate this problem is to require $K \geq N$: We *must* have at least as many frequency samples as the signal's duration. In this way, we can return from the frequency domain we entered via the DFT.

Exercise 5.14 ***(Solution on p. 205.)***

When we have fewer frequency samples than the signal's duration, some discrete-time signal values equal the sum of the original signal values. Given the sampling interpretation of the spectrum, characterize this effect a different way.

Another way to understand this requirement is to use the theory of linear equations. If we write out the expression for the DFT as a set of linear equations,

$$s(0) + s(1) + \dots + s(N-1) = S(0) \quad (5.38)$$

$$s(0) + s(1) e^{(-j)\frac{2\pi}{K}} + \dots + s(N-1) e^{(-j)\frac{2\pi(N-1)}{K}} = S(1)$$

$$\vdots$$

$$s(0) + s(1) e^{(-j)\frac{2\pi(K-1)}{K}} + \dots + s(N-1) e^{(-j)\frac{2\pi(N-1)(K-1)}{K}} = S(K-1)$$

we have K equations in N unknowns if we want to find the signal from its sampled spectrum. This requirement is impossible to fulfill if $K < N$; we must have $K \geq N$. Our orthogonality relation essentially says that if we have a sufficient number of equations (frequency samples), the resulting set of equations can indeed be solved.

By convention, the number of DFT frequency values K is chosen to equal the signal's duration N. The discrete Fourier transform pair consists of

Discrete Fourier Transform Pair

$$S(k) = \sum_{n=0}^{N-1} \left(s(n) e^{-\left(j\frac{2\pi nk}{N}\right)} \right) \quad (5.39)$$

$$s(n) = \frac{1}{N} \sum_{k=0}^{N-1} \left(S(k) e^{j\frac{2\pi nk}{N}} \right) \quad (5.40)$$

Example 5.3

Use this demonstration to perform DFT analysis of a signal.

This is an unsupported media type. To view, please see http://cnx.org/content/m10249/latest/DFTanalysis.llb

---

Example 5.4
Use this demonstration to synthesize a signal from a DFT sequence.

This is an unsupported media type. To view, please see
http://cnx.org/content/m10249/latest/DFT_Component_Manipulation.llb

5.8 DFT: Computational Complexity[19]

We now have a way of computing the spectrum for an arbitrary signal: The Discrete Fourier Transform (DFT) (5.34) computes the spectrum at N equally spaced frequencies from a length- N sequence. An issue that never arises in analog "computation," like that performed by a circuit, is how much work it takes to perform the signal processing operation such as filtering. In computation, this consideration translates to the number of basic computational steps required to perform the needed processing. The number of steps, known as the **complexity**, becomes equivalent to how long the computation takes (how long must we wait for an answer). Complexity is not so much tied to specific computers or programming languages but to how many steps are required on any computer. Thus, a procedure's stated complexity says that the time taken will be **proportional** to some function of the amount of data used in the computation and the amount demanded.

For example, consider the formula for the discrete Fourier transform. For each frequency we chose, we must multiply each signal value by a complex number and add together the results. For a real-valued signal, each real-times-complex multiplication requires two real multiplications, meaning we have $2N$ multiplications to perform. To add the results together, we must keep the real and imaginary parts separate. Adding N numbers requires $N-1$ additions. Consequently, each frequency requires $2N+2(N-1)=4N-2$ basic computational steps. As we have N frequencies, the total number of computations is $N(4N-2)$.

In complexity calculations, we only worry about what happens as the data lengths increase, and take the dominant term—here the $4N^2$ term—as reflecting how much work is involved in making the computation. As multiplicative constants don't matter since we are making a "proportional to" evaluation, we find the DFT is an $O(N^2)$ computational procedure. This notation is read "order N-squared". Thus, if we double the length of the data, we would expect that the computation time to approximately quadruple.

Exercise 5.15 *(Solution on p. 205.)*
In making the complexity evaluation for the DFT, we assumed the data to be real. Three questions emerge. First of all, the spectra of such signals have conjugate symmetry, meaning that negative frequency components ($k=\left\{\frac{N}{2}+1,\dots,N+1\right\}$ in the DFT (5.34)) can be computed from the corresponding positive frequency components. Does this symmetry change the DFT's complexity? Secondly, suppose the data are complex-valued; what is the DFT's complexity now? Finally, a less important but interesting question is suppose we want K frequency values instead of N; now what is the complexity?

5.9 Fast Fourier Transform (FFT)[20]

One wonders if the DFT can be computed faster: Does another computational procedure – an **algorithm** – exist that can compute the same quantity, but more efficiently. We could seek methods that reduce the

[19]This content is available online at <http://cnx.org/content/m0503/2.11/>.
[20]This content is available online at <http://cnx.org/content/m10250/2.15/>.

constant of proportionality, but do not change the DFT's complexity $O\left(N^2\right)$. Here, we have something more dramatic in mind: Can the computations be restructured so that a *smaller* complexity results?

In 1965, IBM researcher Jim Cooley and Princeton faculty member John Tukey developed what is now known as the Fast Fourier Transform (FFT). It is an algorithm for computing that DFT that has order $O\left(NlogN\right)$ *for certain length inputs*. Now when the length of data doubles, the spectral computational time will not quadruple as with the DFT algorithm; instead, it approximately doubles. Later research showed that no algorithm for computing the DFT could have a smaller complexity than the FFT. Surprisingly, historical work has shown that Gauss[21] in the early nineteenth century developed the same algorithm, but did not publish it! After the FFT's rediscovery, not only was the computation of a signal's spectrum greatly speeded, but also the added feature of **algorithm** meant that computations had flexibility not available to analog implementations.

Exercise 5.16 *(Solution on p. 205.)*

Before developing the FFT, let's try to appreciate the algorithm's impact. Suppose a short-length transform takes 1 ms. We want to calculate a transform of a signal that is 10 times longer. Compare how much longer a straightforward implementation of the DFT would take in comparison to an FFT, both of which compute exactly the same quantity.

To derive the FFT, we assume that the signal's duration is a power of two: $N = 2^L$. Consider what happens to the even-numbered and odd-numbered elements of the sequence in the DFT calculation.

$$\begin{aligned} S(k) &= s(0) + s(2)\,e^{(-j)\frac{2\pi 2k}{N}} + \dots + s(N-2)\,e^{(-j)\frac{2\pi(N-2)k}{N}} \\ &\quad + s(1)\,e^{(-j)\frac{2\pi k}{N}} + s(3)\,e^{(-j)\frac{2\pi(2+1)k}{N}} + \dots + s(N-1)\,e^{(-j)\frac{2\pi(N-(2-1))k}{N}} \\ &= \left[s(0) + s(2)\,e^{(-j)\frac{2\pi k}{\frac{N}{2}}} + \dots + s(N-2)\,e^{(-j)\frac{2\pi\left(\frac{N}{2}-1\right)k}{\frac{N}{2}}}\right] \\ &\quad + \left[s(1) + s(3)\,e^{(-j)\frac{2}{\pi}} + \dots + s(N-1)\,e^{(-j)\frac{2\pi\left(\frac{N}{2}-1\right)k}{\frac{N}{2}}}\right] e^{\frac{-(j2\pi k)}{N}} \end{aligned} \tag{5.41}$$

Each term in square brackets has the **form** of a $\frac{N}{2}$-length DFT. The first one is a DFT of the even-numbered elements, and the second of the odd-numbered elements. The first DFT is combined with the second multiplied by the complex exponential $e^{-\left(\frac{j2\pi k}{N}\right)}$. The half-length transforms are each evaluated at frequency indices $k = 0, \dots, N-1$. Normally, the number of frequency indices in a DFT calculation range between zero and the transform length minus one. The **computational advantage** of the FFT comes from recognizing the periodic nature of the discrete Fourier transform. The FFT simply reuses the computations made in the half-length transforms and combines them through additions and the multiplication by $e^{-\left(\frac{j2\pi k}{N}\right)}$, which is not periodic over $\frac{N}{2}$, to rewrite the length-N DFT. Figure 5.12 (Length-8 DFT decomposition) illustrates this decomposition. As it stands, we now compute two length-$\frac{N}{2}$ transforms (complexity $2O\left(\frac{N^2}{4}\right)$), multiply one of them by the complex exponential (complexity $O(N)$), and add the results (complexity $O(N)$). At this point, the total complexity is still dominated by the half-length DFT calculations, but the proportionality coefficient has been reduced.

Now for the fun. Because $N = 2^L$, each of the half-length transforms can be reduced to two quarter-length transforms, each of these to two eighth-length ones, etc. This decomposition continues until we are left with length-2 transforms. This transform is quite simple, involving only additions. Thus, the first stage of the FFT has $\frac{N}{2}$ length-2 transforms (see the bottom part of Figure 5.12 (Length-8 DFT decomposition)). Pairs of these transforms are combined by adding one to the other multiplied by a complex exponential. Each pair requires 4 additions and 4 multiplications, giving a total number of computations equaling $8 \cdot \frac{N}{4} = \frac{N}{2}$. This number of computations does not change from stage to stage. Because the number of stages, the number of times the length can be divided by two, equals $\log_2 N$, the complexity of the FFT is $O(N \log_2 N)$.

[21] http://www-groups.dcs.st-and.ac.uk/~history/Mathematicians/Gauss.html

Length-8 DFT decomposition

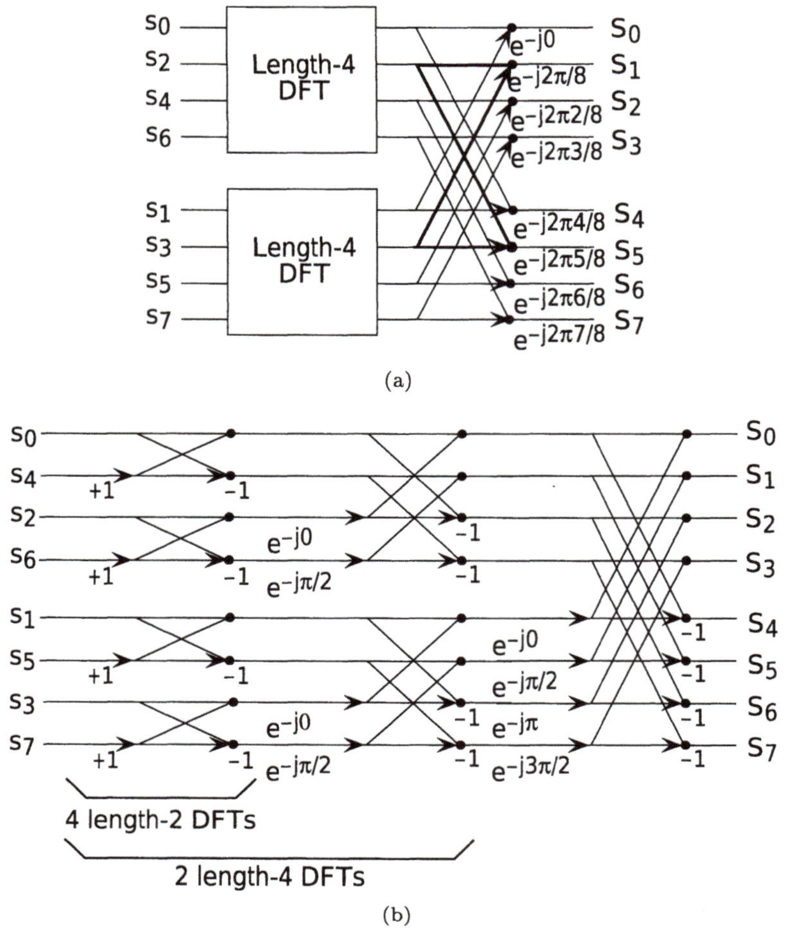

Figure 5.12: The initial decomposition of a length-8 DFT into the terms using even- and odd-indexed inputs marks the first phase of developing the FFT algorithm. When these half-length transforms are successively decomposed, we are left with the diagram shown in the bottom panel that depicts the length-8 FFT computation.

Doing an example will make computational savings more obvious. Let's look at the details of a length-8 DFT. As shown on Figure 5.13 (Butterfly), we first decompose the DFT into two length-4 DFTs, with the outputs added and subtracted together in pairs. Considering Figure 5.13 (Butterfly) as the frequency index goes from 0 through 7, we recycle values from the length-4 DFTs into the final calculation because of the periodicity of the DFT output. Examining how pairs of outputs are collected together, we create the basic computational element known as a **butterfly** (Figure 5.13 (Butterfly)).

Butterfly

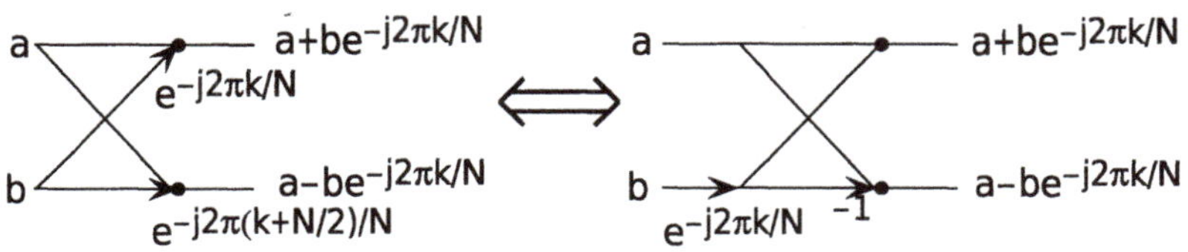

Figure 5.13: The basic computational element of the fast Fourier transform is the butterfly. It takes two complex numbers, represented by a and b, and forms the quantities shown. Each butterfly requires one complex multiplication and two complex additions.

By considering together the computations involving common output frequencies from the two half-length DFTs, we see that the two complex multiplies are related to each other, and we can reduce our computational work even further. By further decomposing the length-4 DFTs into two length-2 DFTs and combining their outputs, we arrive at the diagram summarizing the length-8 fast Fourier transform (Figure 5.12 (Length-8 DFT decomposition)). Although most of the complex multiplies are quite simple (multiplying by $e^{-(j\pi)}$ means negating real and imaginary parts), let's count those for purposes of evaluating the complexity as full complex multiplies. We have $\frac{N}{2}=4$ complex multiplies and $2N=16$ additions for each stage and $\log_2 N=3$ stages, making the number of basic computations $\frac{3N}{2}\log_2 N$ as predicted.

Exercise 5.17 *(Solution on p. 205.)*

Note that the ordering of the input sequence in the two parts of Figure 5.12 (Length-8 DFT decomposition) aren't quite the same. Why not? How is the ordering determined?

Other "fast" algorithms were discovered, all of which make use of how many common factors the transform length N has. In number theory, the number of prime factors a given integer has measures how *composite* it is. The numbers 16 and 81 are highly composite (equaling 2^4 and 3^4 respectively), the number 18 is less so ($2^1 \cdot 3^2$), and 17 not at all (it's prime). In over thirty years of Fourier transform algorithm development, the original Cooley-Tukey algorithm is far and away the most frequently used. It is so computationally efficient that power-of-two transform lengths are frequently used regardless of what the actual length of the data.

Exercise 5.18 *(Solution on p. 205.)*

Suppose the length of the signal were 500? How would you compute the spectrum of this signal using the Cooley-Tukey algorithm? What would the length N of the transform be?

5.10 Spectrograms[22]

We know how to acquire analog signals for digital processing (pre-filtering (Section 5.3), sampling (Section 5.3), and A/D conversion (Section 5.4)) and to compute spectra of discrete-time signals (using the FFT algorithm (Section 5.9)), let's put these various components together to learn how the spectrogram shown in Figure 5.14 (speech spectrogram), which is used to analyze speech (Section 4.10), is calculated. The speech was sampled at a rate of 11.025 kHz and passed through a 16-bit A/D converter.

POINT OF INTEREST: Music compact discs (CDs) encode their signals at a sampling rate of 44.1 kHz. We'll learn the rationale for this number later. The 11.025 kHz sampling rate for the speech is 1/4 of the CD sampling rate, and was the lowest available sampling rate commensurate with speech signal bandwidths available on my computer.

[22]This content is available online at <http://cnx.org/content/m0505/2.18/>.

Exercise 5.19 *(Solution on p. 205.)*

Looking at Figure 5.14 (speech spectrogram) the signal lasted a little over 1.2 seconds. How long was the sampled signal (in terms of samples)? What was the datarate during the sampling process in bps (bits per second)? Assuming the computer storage is organized in terms of bytes (8-bit quantities), how many bytes of computer memory does the speech consume?

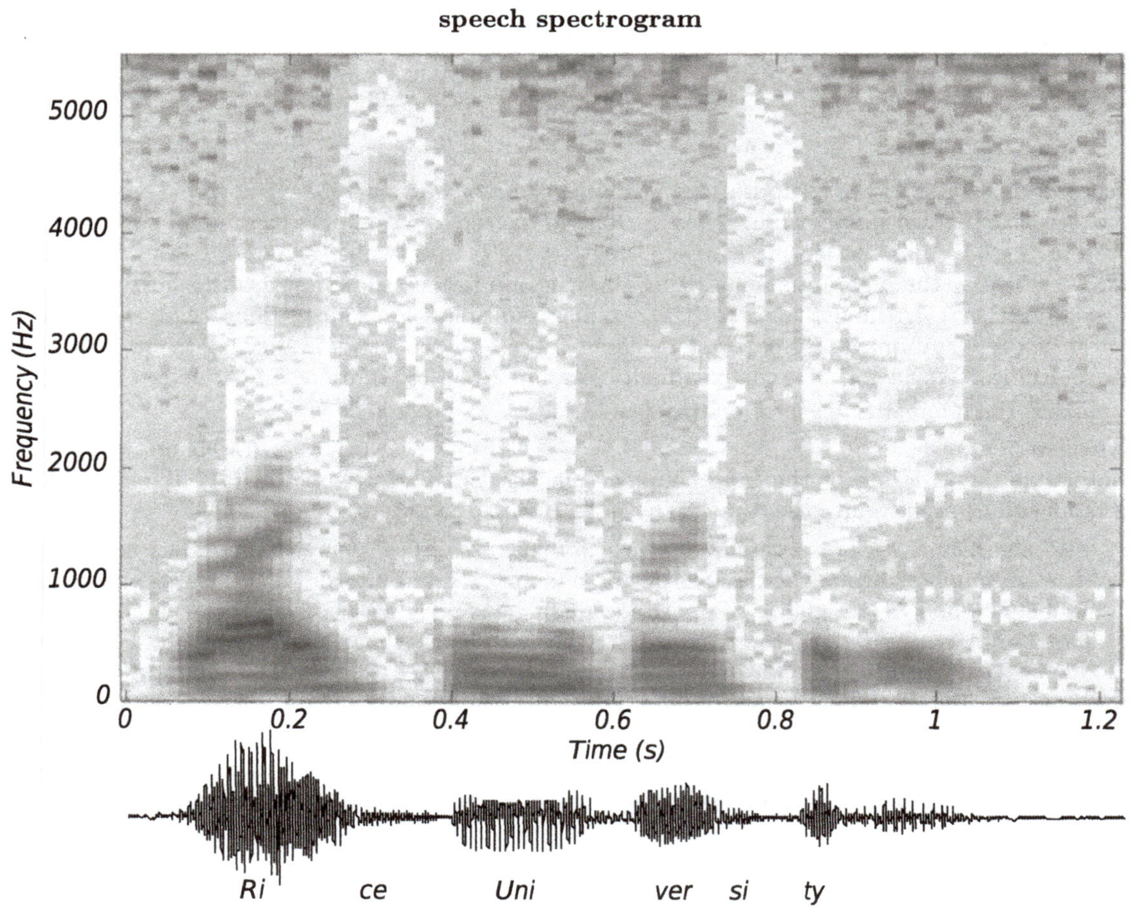

Figure 5.14

The resulting discrete-time signal, shown in the bottom of Figure 5.14 (speech spectrogram), clearly changes its character with time. To display these spectral changes, the long signal was sectioned into **frames**: comparatively short, contiguous groups of samples. Conceptually, a Fourier transform of each frame is calculated using the FFT. Each frame is not so long that significant signal variations are retained within a frame, but not so short that we lose the signal's spectral character. Roughly speaking, the speech signal's spectrum is evaluated over successive time segments and stacked side by side so that the x-axis corresponds to time and the y-axis frequency, with color indicating the spectral amplitude.

An important detail emerges when we examine each framed signal (Figure 5.15 (Spectrogram Hanning vs. Rectangular)).

Spectrogram Hanning vs. Rectangular

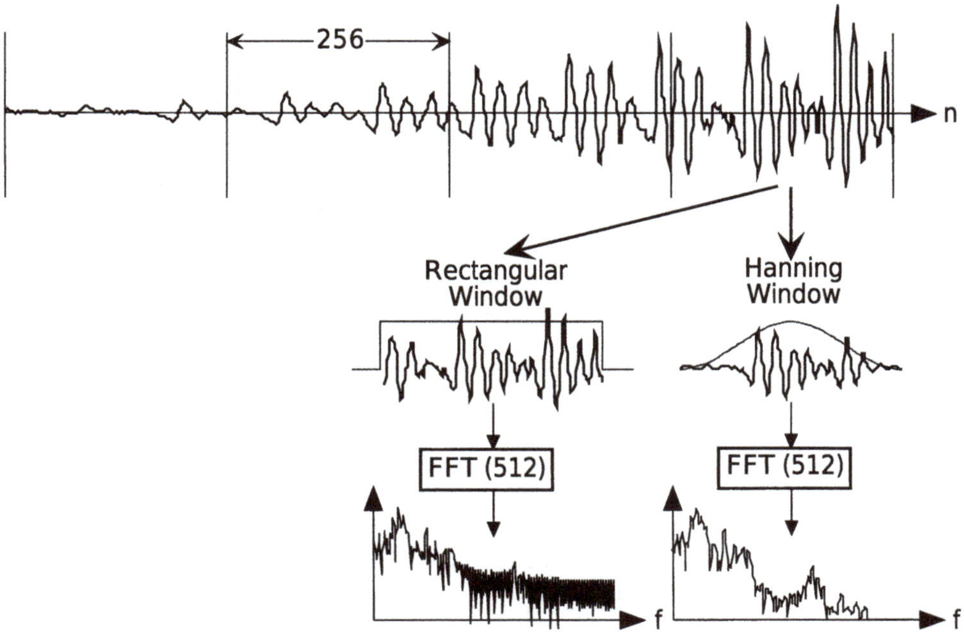

Figure 5.15: The top waveform is a segment 1024 samples long taken from the beginning of the "Rice University" phrase. Computing Figure 5.14 (speech spectrogram) involved creating frames, here demarked by the vertical lines, that were 256 samples long and finding the spectrum of each. If a rectangular window is applied (corresponding to extracting a frame from the signal), oscillations appear in the spectrum (middle of bottom row). Applying a Hanning window gracefully tapers the signal toward frame edges, thereby yielding a more accurate computation of the signal's spectrum at that moment of time.

At the frame's edges, the signal may change very abruptly, a feature not present in the original signal. A transform of such a segment reveals a curious oscillation in the spectrum, an artifact directly related to this sharp amplitude change. A better way to frame signals for spectrograms is to apply a **window**: Shape the signal values within a frame so that the signal decays gracefully as it nears the edges. This shaping is accomplished by multiplying the framed signal by the sequence $w(n)$. In sectioning the signal, we essentially applied a rectangular window: $w(n) = 1$, $0 \le n \le N-1$. A much more graceful window is the **Hanning window**; it has the cosine shape $w(n) = \frac{1}{2}\left(1 - \cos\left(\frac{2\pi n}{N}\right)\right)$. As shown in Figure 5.15 (Spectrogram Hanning vs. Rectangular), this shaping greatly reduces spurious oscillations in each frame's spectrum. Considering the spectrum of the Hanning windowed frame, we find that the oscillations resulting from applying the rectangular window obscured a formant (the one located at a little more than half the Nyquist frequency).

Exercise 5.20 ***(Solution on p. 206.)***

What might be the source of these oscillations? To gain some insight, what is the length- $2N$ discrete Fourier transform of a length-N pulse? The pulse emulates the rectangular window, and certainly has edges. Compare your answer with the length- $2N$ transform of a length- N Hanning window.

Hanning speech

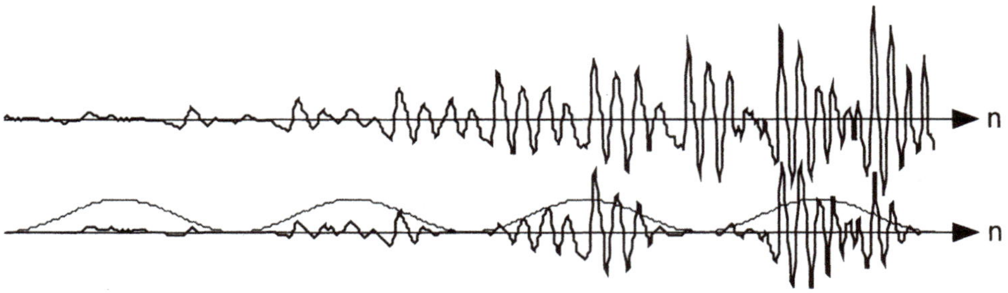

Figure 5.16: In comparison with the original speech segment shown in the upper plot, the non-overlapped Hanning windowed version shown below it is very ragged. Clearly, spectral information extracted from the bottom plot could well miss important features present in the original.

If you examine the windowed signal sections in sequence to examine windowing's affect on signal amplitude, we see that we have managed to amplitude-modulate the signal with the periodically repeated window (Figure 5.16 (Hanning speech)). To alleviate this problem, frames are overlapped (typically by half a frame duration). This solution requires more Fourier transform calculations than needed by rectangular windowing, but the spectra are much better behaved and spectral changes are much better captured.

The speech signal, such as shown in the speech spectrogram (Figure 5.14: speech spectrogram), is sectioned into overlapping, equal-length frames, with a Hanning window applied to each frame. The spectra of each of these is calculated, and displayed in spectrograms with frequency extending vertically, window time location running horizontally, and spectral magnitude color-coded. Figure 5.17 (Hanning windows) illustrates these computations.

Exercise 5.21 *(Solution on p. 206.)*

Why the specific values of 256 for N and 512 for K? Another issue is how was the length-512 transform of each length-256 windowed frame computed?

5.11 Discrete-Time Systems[23]

When we developed analog systems, interconnecting the circuit elements provided a natural starting place for constructing useful devices. In discrete-time signal processing, we are not limited by hardware considerations but by what can be constructed in software.

Exercise 5.22 *(Solution on p. 206.)*

One of the first analog systems we described was the amplifier (Section 2.6.2: Amplifiers). We found that implementing an amplifier was difficult in analog systems, requiring an op-amp at least. What is the discrete-time implementation of an amplifier? Is this especially hard or easy?

In fact, we will discover that frequency-domain implementation of systems, wherein we multiply the input signal's Fourier transform by a frequency response, is not only a viable alternative, but also a computationally efficient one. We begin with discussing the underlying mathematical structure of linear, shift-invariant systems, and devise how software filters can be constructed.

[23]This content is available online at <http://cnx.org/content/m0507/2.5/>.

Hanning windows

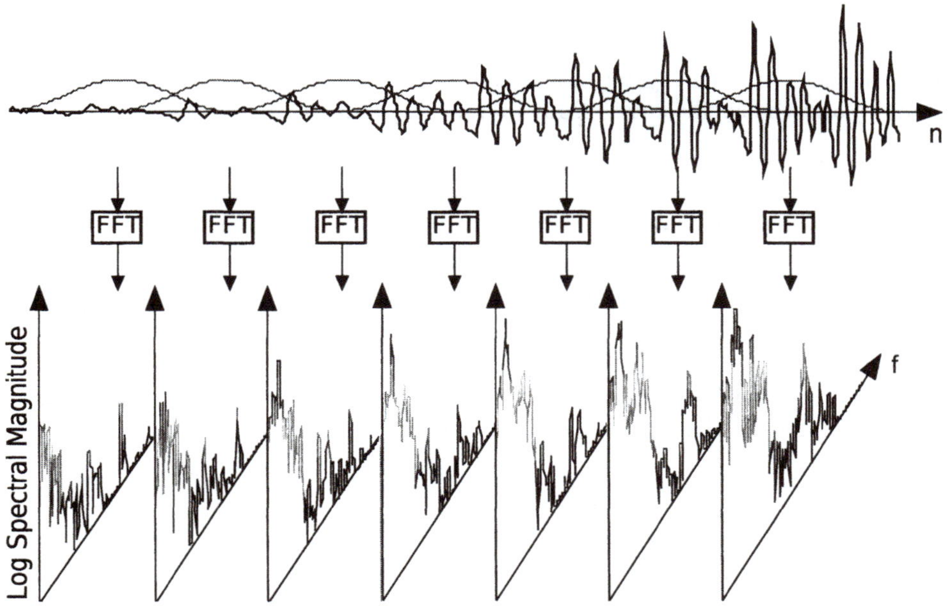

Figure 5.17: The original speech segment and the sequence of overlapping Hanning windows applied to it are shown in the upper portion. Frames were 256 samples long and a Hanning window was applied with a half-frame overlap. A length-512 FFT of each frame was computed, with the magnitude of the first 257 FFT values displayed vertically, with spectral amplitude values color-coded.

5.12 Discrete-Time Systems in the Time-Domain[24]

A discrete-time signal $s(n)$ is *delayed* by n_0 samples when we write $s(n-n_0)$, with $n_0>0$. Choosing n_0 to be negative advances the signal along the integers. As opposed to analog delays (Section 2.6.3: Delay), discrete-time delays can *only* be integer valued. In the frequency domain, delaying a signal corresponds to a linear phase shift of the signal's discrete-time Fourier transform: $\left(s(n-n_0)\leftrightarrow e^{-(j2\pi f n_0)}S\left(e^{j2\pi f}\right)\right)$.

Linear discrete-time systems have the superposition property.

$$S\left(a_1x_1(n)+a_2x_2(n)\right)=a_1S\left(x_1(n)\right)+a_2S\left(x_2(n)\right) \tag{5.42}$$

A discrete-time system is called **shift-invariant** (analogous to time-invariant analog systems (p. 28)) if delaying the input delays the corresponding output. If $S(x(n))=y(n)$, then a shift-invariant system has the property

$$S\left(x(n-n_0)\right)=y(n-n_0) \tag{5.43}$$

We use the term shift-invariant to emphasize that delays can only have integer values in discrete-time, while in analog signals, delays can be arbitrarily valued.

We want to concentrate on systems that are both linear and shift-invariant. It will be these that allow us the full power of frequency-domain analysis and implementations. Because we have no physical constraints in "constructing" such systems, we need only a mathematical specification. In analog systems, the differential equation specifies the input-output relationship in the time-domain. The corresponding discrete-time

[24]This content is available online at <http://cnx.org/content/m10251/2.22/>.

specification is the **difference equation**.

$$y(n) = a_1 y(n-1) + \cdots + a_p y(n-p) + b_0 x(n) + b_1 x(n-1) + \cdots + b_q x(n-q) \tag{5.44}$$

Here, the output signal $y(n)$ is related to its *past* values $y(n-l)$, $l = \{1, \dots, p\}$, and to the current and past values of the input signal $x(n)$. The system's characteristics are determined by the choices for the number of coefficients p and q and the coefficients' values $\{a_1, \dots, a_p\}$ and $\{b_0, b_1, \dots, b_q\}$.

> ASIDE: There is an asymmetry in the coefficients: where is a_0? This coefficient would multiply the $y(n)$ term in (5.44). We have essentially divided the equation by it, which does not change the input-output relationship. We have thus created the convention that a_0 is always one.

As opposed to differential equations, which only provide an *implicit* description of a system (we must somehow solve the differential equation), difference equations provide an *explicit* way of computing the output for any input. We simply express the difference equation by a program that calculates each output from the previous output values, and the current and previous inputs.

Difference equations are usually expressed in software with *for* loops. A MATLAB program that would compute the first 1000 values of the output has the form

```
for n=1:1000
    y(n) = sum(a.*y(n-1:-1:n-p)) + sum(b.*x(n:-1:n-q));
end
```

An important detail emerges when we consider making this program work; in fact, as written it has (at least) two bugs. What input and output values enter into the computation of $y(1)$? We need values for $y(0)$, $y(-1)$, ..., values we have not yet computed. To compute them, we would need more previous values of the output, which we have not yet computed. To compute these values, we would need even earlier values, ad infinitum. The way out of this predicament is to specify the system's **initial conditions**: we must provide the p output values that occurred before the input started. These values can be arbitrary, but the choice does impact how the system responds to a given input. *One* choice gives rise to a linear system: Make the initial conditions zero. The reason lies in the definition of a linear system (Section 2.6.6: Linear Systems): The only way that the output to a sum of signals can be the sum of the individual outputs occurs when the initial conditions in each case are zero.

Exercise 5.23 ***(Solution on p. 206.)***

The initial condition issue resolves making sense of the difference equation for inputs that start at some index. However, the program will not work because of a programming, not conceptual, error. What is it? How can it be "fixed?"

Example 5.5

Let's consider the simple system having $p = 1$ and $q = 0$.

$$y(n) = ay(n-1) + bx(n) \tag{5.45}$$

To compute the output at some index, this difference equation says we need to know what the previous output $y(n-1)$ and what the input signal is at that moment of time. In more detail, let's compute this system's output to a unit-sample input: $x(n) = \delta(n)$. Because the input is zero for negative indices, we start by trying to compute the output at $n = 0$.

$$y(0) = ay(-1) + b \tag{5.46}$$

What is the value of $y(-1)$? Because we have used an input that is zero for all negative indices, it is reasonable to assume that the output is also zero. Certainly, the difference equation would not describe a linear system (Section 2.6.6: Linear Systems) if the input that is zero for *all* time did not produce a zero output. With this assumption, $y(-1) = 0$, leaving $y(0) = b$. For $n > 0$, the input unit-sample is zero, which leaves us with the difference equation $y(n) = ay(n-1) \quad , \quad n > 0$. We can envision how the filter responds to this input by making a table.

$$y(n) = ay(n-1) + b\delta(n) \tag{5.47}$$

n	$x(n)$	$y(n)$
-1	0	0
0	1	b
1	0	ba
2	0	ba^2
:	0	:
n	0	ba^n

Figure 5.18

Coefficient values determine how the output behaves. The parameter b can be any value, and serves as a gain. The effect of the parameter a is more complicated (Figure 5.18). If it equals zero, the output simply equals the input times the gain b. For all non-zero values of a, the output lasts forever; such systems are said to be **IIR** (*I*nfinite *I*mpulse *R*esponse). The reason for this terminology is that the unit sample also known as the impulse (especially in analog situations), and the system's response to the "impulse" lasts forever. If a is positive and less than one, the output is a decaying exponential. When $a = 1$, the output is a unit step. If a is negative and greater than -1, the output oscillates while decaying exponentially. When $a = -1$, the output changes sign forever, alternating between b and $-b$. More dramatic effects when $|a| > 1$; whether positive or negative, the output signal becomes larger and larger, *growing* exponentially.

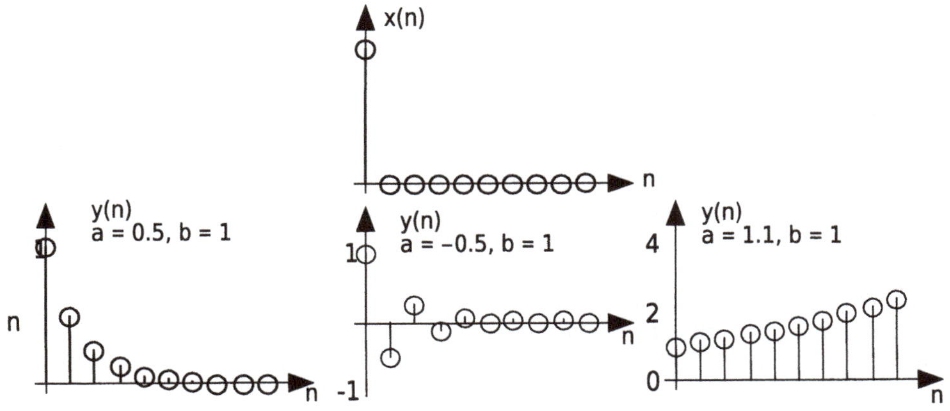

Figure 5.19: The input to the simple example system, a unit sample, is shown at the top, with the outputs for several system parameter values shown below.

Positive values of a are used in population models to describe how population size increases over time. Here, n might correspond to generation. The difference equation says that the number in the next generation is some multiple of the previous one. If this multiple is less than one, the population becomes extinct; if greater than one, the population flourishes. The same difference equation also describes the effect of compound interest on deposits. Here, n indexes the times at which compounding occurs (daily, monthly, etc.), a equals the compound interest rate plusone, and $b = 1$ (the bank provides no gain). In signal processing applications, we typically require that the output remain bounded for any input. For our example, that means that we restrict $|a| = 1$ and chose values for it and the gain according to the application.

Exercise 5.24 *(Solution on p. 206.)*

Note that the difference equation (5.44),

$$y(n) = a_1 y(n-1) + \cdots + a_p y(n-p) + b_0 x(n) + b_1 x(n-1) + \cdots + b_q x(n-q)$$

does not involve terms like $y(n+1)$ or $x(n+1)$ on the equation's right side. Can such terms also be included? Why or why not?

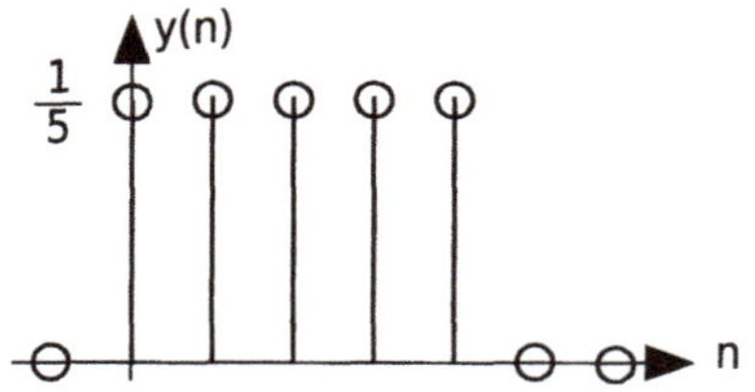

Figure 5.20: The plot shows the unit-sample response of a length-5 boxcar filter.

Example 5.6

A somewhat different system has no "a" coefficients. Consider the difference equation

$$y(n) = \frac{1}{q}(x(n) + \cdots + x(n-q+1)) \tag{5.48}$$

Because this system's output depends only on current and previous input values, we need not be concerned with initial conditions. When the input is a unit-sample, the output equals $\frac{1}{q}$ for $n = \{0, \dots, q-1\}$, then equals zero thereafter. Such systems are said to be **FIR** (*F*inite *I*mpulse *R*esponse) because their unit sample responses have finite duration. Plotting this response (Figure 5.20) shows that the unit-sample response is a pulse of width q and height $\frac{1}{q}$. This waveform is also known as a boxcar, hence the name **boxcar filter** given to this system. We'll derive its frequency response and develop its filtering interpretation in the next section. For now, note that the difference equation says that each output value equals the *average* of the input's current and previous values. Thus, the output equals the running average of input's previous q values. Such a system could be used to produce the average weekly temperature ($q = 7$) that could be updated daily.

This is an unsupported media type. To view, please see http://cnx.org/content/m10251/latest/DiscreteTimeSys.llb

5.13 Discrete-Time Systems in the Frequency Domain[25]

As with analog linear systems, we need to find the frequency response of discrete-time systems. We used impedances to derive directly from the circuit's structure the frequency response. The only structure we have so far for a discrete-time system is the difference equation. We proceed as when we used impedances: let the input be a complex exponential signal. When we have a linear, shift-invariant system, the output should also be a complex exponential of the same frequency, changed in amplitude and phase. These amplitude and phase changes comprise the frequency response we seek. The complex exponential input signal is $x(n) = Xe^{j2\pi fn}$. Note that this input occurs for *all* values of n. No need to worry about initial conditions here. Assume the output has a similar form: $y(n) = Ye^{j2\pi fn}$. Plugging these signals into the fundamental difference equation (5.44), we have

$$Ye^{j2\pi fn} = a_1Ye^{j2\pi f(n-1)} + \dots + a_pYe^{j2\pi f(n-p)} + b_0Xe^{j2\pi fn} + b_1Xe^{j2\pi f(n-1)} + \dots + b_qXe^{j2\pi f(n-q)} \quad (5.49)$$

The assumed output does indeed satisfy the difference equation if the output complex amplitude is related to the input amplitude by

$$Y = \frac{b_0 + b_1e^{-(j2\pi f)} + \dots + b_qe^{-(j2\pi qf)}}{1 - a_1e^{-(j2\pi f)} - \dots - a_pe^{-(j2\pi pf)}}X$$

This relationship corresponds to the system's frequency response or, by another name, its transfer function. We find that any discrete-time system defined by a difference equation has a transfer function given by

$$H\left(e^{j2\pi f}\right) = \frac{b_0 + b_1e^{-(j2\pi f)} + \dots + b_qe^{-(j2\pi qf)}}{1 - a_1e^{-(j2\pi f)} - \dots - a_pe^{-(j2\pi pf)}} \quad (5.50)$$

Furthermore, because *any* discrete-time signal can be expressed as a superposition of complex exponential signals and because linear discrete-time systems obey the Superposition Principle, the transfer function relates the discrete-time Fourier transform of the system's output to the input's Fourier transform.

$$Y\left(e^{j2\pi f}\right) = X\left(e^{j2\pi f}\right)H\left(e^{j2\pi f}\right) \quad (5.51)$$

Example 5.7

The frequency response of the simple IIR system (difference equation given in a previous example (Example 5.5)) is given by

$$H\left(e^{j2\pi f}\right) = \frac{b}{1 - ae^{-(j2\pi f)}} \quad (5.52)$$

This Fourier transform occurred in a previous example; the exponential signal spectrum (Figure 5.10: Spectra of exponential signals) portrays the magnitude and phase of this transfer function. When the filter coefficient a is positive, we have a lowpass filter; negative a results in a highpass filter. The larger the coefficient in magnitude, the more pronounced the lowpass or highpass filtering.

Example 5.8

The length-q boxcar filter (difference equation found in a previous example (Example 5.6)) has the frequency response

$$H\left(e^{j2\pi f}\right) = \frac{1}{q}\sum_{m=0}^{q-1}\left(e^{-(j2\pi fm)}\right) \quad (5.53)$$

This expression amounts to the Fourier transform of the boxcar signal (Figure 5.20). There we found that this frequency response has a magnitude equal to the absolute value of $dsinc(\pi f)$; see the length-10 filter's frequency response (Figure 5.11: Spectrum of length-ten pulse). We see that boxcar filters–length-q signal averagers–have a lowpass behavior, having a cutoff frequency of $\frac{1}{q}$.

[25]This content is available online at <http://cnx.org/content/m0510/2.14/>.

Exercise 5.25 *(Solution on p. 206.)*

Suppose we multiply the boxcar filter's coefficients by a sinusoid: $b_m = \frac{1}{q}\cos(2\pi f_0 m)$ Use Fourier transform properties to determine the transfer function. How would you characterize this system: Does it act like a filter? If so, what kind of filter and how do you control its characteristics with the filter's coefficients?

These examples illustrate the point that systems described (and implemented) by difference equations serve as filters for discrete-time signals. The filter's *order* is given by the number p of denominator coefficients in the transfer function (if the system is IIR) or by the number q of numerator coefficients if the filter is FIR. When a system's transfer function has both terms, the system is usually IIR, and its order equals p regardless of q. By selecting the coefficients and filter type, filters having virtually any frequency response desired can be designed. This design flexibility can't be found in analog systems. In the next section, we detail how analog signals can be filtered by computers, offering a much greater range of filtering possibilities than is possible with circuits.

5.14 Filtering in the Frequency Domain[26]

Because we are interested in actual computations rather than analytic calculations, we must consider the details of the discrete Fourier transform. To compute the length-N DFT, we assume that the signal has a duration less than or equal to N. Because frequency responses have an explicit frequency-domain specification (5.49) in terms of filter coefficients, we don't have a direct handle on which signal has a Fourier transform equaling a given frequency response. Finding this signal is quite easy. First of all, note that the discrete-time Fourier transform of a unit sample equals one for all frequencies. Because of the input and output of linear, shift-invariant systems are related to each other by $Y\left(e^{j2\pi f}\right) = H\left(e^{j2\pi f}\right) X\left(e^{j2\pi f}\right)$, *a unit-sample input, which has* $X\left(e^{j2\pi f}\right) = 1$, *results in the output's Fourier transform equaling the system's transfer function.*

Exercise 5.26 *(Solution on p. 206.)*

This statement is a very important result. Derive it yourself.

In the time-domain, the output for a unit-sample input is known as the system's **unit-sample response**, and is denoted by $h(n)$. Combining the frequency-domain and time-domain interpretations of a linear, shift-invariant system's unit-sample response, we have that $h(n)$ and the transfer function are Fourier transform pairs *in terms of the discrete-time Fourier transform.*

$$\left(h(n) \leftrightarrow H\left(e^{j2\pi f}\right)\right) \quad (5.54)$$

Returning to the issue of how to use the DFT to perform filtering, we can analytically specify the frequency response, and derive the corresponding length-N DFT by sampling the frequency response.

$$H(k) = H\left(e^{\frac{j2\pi k}{N}}\right) \quad , \quad k = \{0, \dots, N-1\} \quad (5.55)$$

Computing the inverse DFT yields a length-N signal *no matter what the actual duration of the unit-sample response might be.* If the unit-sample response has a duration less than or equal to N (it's a FIR filter), computing the inverse DFT of the sampled frequency response indeed yields the unit-sample response. If, however, the duration exceeds N, errors are encountered. The nature of these errors is easily explained by appealing to the Sampling Theorem. By sampling in the frequency domain, we have the potential for aliasing in the time domain (sampling in one domain, be it time or frequency, can result in aliasing in the other) unless we sample fast enough. Here, the duration of the unit-sample response determines the minimal sampling rate that prevents aliasing. For FIR systems — they by definition have finite-duration unit sample responses — the number of required DFT samples equals the unit-sample response's duration: $N \geq q$.

[26]This content is available online at <http://cnx.org/content/m10257/2.16/>.

Exercise 5.27 *(Solution on p. 206.)*

Derive the minimal DFT length for a length-q unit-sample response using the Sampling Theorem. Because sampling in the frequency domain causes repetitions of the unit-sample response in the time domain, sketch the time-domain result for various choices of the DFT length N.

Exercise 5.28 *(Solution on p. 206.)*

Express the unit-sample response of a FIR filter in terms of difference equation coefficients. Note that the corresponding question for IIR filters is far more difficult to answer: Consider the example (Example 5.5).

For IIR systems, we cannot use the DFT to find the system's unit-sample response: aliasing of the unit-sample response will *always* occur. Consequently, we can only implement an IIR filter accurately in the time domain with the system's difference equation. *Frequency-domain implementations are restricted to FIR filters.*

Another issue arises in frequency-domain filtering that is related to time-domain aliasing, this time when we consider the output. Assume we have an input signal having duration N_x that we pass through a FIR filter having a length-$q+1$ unit-sample response. What is the duration of the output signal? The difference equation for this filter is

$$y(n) = b_0 x(n) + \dots + b_q x(n-q) \tag{5.56}$$

This equation says that the output depends on current and past input values, with the input value q samples previous defining the extent of the filter's *memory* of past input values. For example, the output at index N_x depends on $x(N_x)$ (which equals zero), $x(N_x-1)$, through $x(N_x-q)$. Thus, the output returns to zero only after the last input value passes through the filter's memory. As the input signal's last value occurs at index N_x-1, the last nonzero output value occurs when $n-q=N_x-1$ or $n=q+N_x-1$. Thus, the output signal's duration equals $q+N_x$.

Exercise 5.29 *(Solution on p. 206.)*

In words, we express this result as "The output's duration equals the input's duration plus the filter's duration minus one.". Demonstrate the accuracy of this statement.

The main theme of this result is that a filter's output extends longer than either its input or its unit-sample response. Thus, to avoid aliasing when we use DFTs, the dominant factor is not the duration of input or of the unit-sample response, but of the output. Thus, the number of values at which we must evaluate the frequency response's DFT must be at least $q+N_x$ and we must compute the same length DFT of the input. To accommodate a shorter signal than DFT length, we simply **zero-pad** the input: Ensure that for indices extending beyond the signal's duration that the signal is zero. Frequency-domain filtering, diagrammed in Figure 5.21, is accomplished by storing the filter's frequency response as the DFT $H(k)$, computing the input's DFT $X(k)$, multiplying them to create the output's DFT $Y(k)=H(k)X(k)$, and computing the inverse DFT of the result to yield $y(n)$.

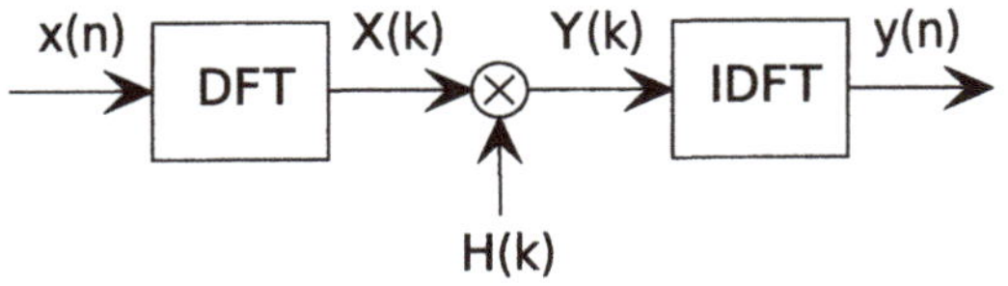

Figure 5.21: To filter a signal in the frequency domain, first compute the DFT of the input, multiply the result by the sampled frequency response, and finally compute the inverse DFT of the product. The DFT's length *must* be at least the sum of the input's and unit-sample response's duration minus one. We calculate these discrete Fourier transforms using the fast Fourier transform algorithm, of course.

Before detailing this procedure, let's clarify why so many new issues arose in trying to develop a frequency-domain implementation of linear filtering. The frequency-domain relationship between a filter's input and output is *always* true: $Y\left(e^{j2\pi f}\right) = H\left(e^{j2\pi f}\right) X\left(e^{j2\pi f}\right)$. This Fourier transforms in this result are discrete-time Fourier transforms; for example, $X\left(e^{j2\pi f}\right) = \sum_n \left(x\left(n\right) e^{-(j2\pi fn)}\right)$. Unfortunately, using this relationship to perform filtering is restricted to the situation when we have analytic formulas for the frequency response and the input signal. The reason why we had to "invent" the discrete Fourier transform (DFT) has the same origin: The spectrum resulting from the discrete-time Fourier transform depends on the *continuous* frequency variable f. That's fine for analytic calculation, but computationally we would have to make an uncountably infinite number of computations.

> NOTE: Did you know that two kinds of infinities can be meaningfully defined? A **countably infinite** quantity means that it can be associated with a limiting process associated with integers. An **uncountably infinite** quantity cannot be so associated. The number of rational numbers is countably infinite (the numerator and denominator correspond to locating the rational by row and column; the total number so-located can be counted, voila!); the number of irrational numbers is uncountably infinite. Guess which is "bigger?"

The DFT computes the Fourier transform at a finite set of frequencies — samples the true spectrum — which can lead to aliasing in the time-domain unless we sample sufficiently fast. The sampling interval here is $\frac{1}{K}$ for a length-K DFT: faster sampling to avoid aliasing thus requires a longer transform calculation. Since the longest signal among the input, unit-sample response and output is the output, it is that signal's duration that determines the transform length. We simply extend the other two signals with zeros (zero-pad) to compute their DFTs.

Example 5.9

Suppose we want to average daily stock prices taken over last year to yield a running weekly average (average over five trading sessions). The filter we want is a length-5 averager (as shown in the unit-sample response (Figure 5.20)), and the input's duration is 253 (365 calendar days minus weekend days and holidays). The output duration will be $253+5-1=257$, and this determines the transform length we need to use. Because we want to use the FFT, we are restricted to power-of-two transform lengths. We need to choose any FFT length that exceeds the required DFT length. As it turns out, 256 is a power of two ($2^8 = 256$), and this length just undershoots our required length. To use frequency domain techniques, we must use length-512 fast Fourier transforms.

Figure 5.22 shows the input and the filtered output. The MATLAB programs that compute the filtered output in the time and frequency domains are

```
Time Domain
h = [1 1 1 1 1]/5;
y = filter(h,1,[djia zeros(1,4)]);

Frequency Domain
h = [1 1 1 1 1]/5;
DJIA = fft(djia, 512);
H = fft(h, 512);
Y = H.*X;
y = ifft(Y);
```

> NOTE: The `filter` program has the feature that the length of its output equals the length of its input. To force it to produce a signal having the proper length, the program zero-pads the input appropriately.

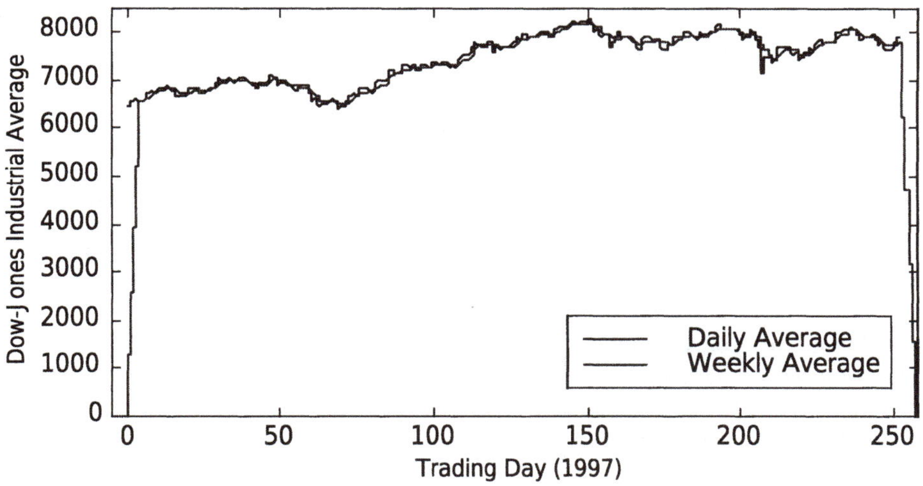

Figure 5.22: The blue line shows the Dow Jones Industrial Average from 1997, and the red one the length-5 boxcar-filtered result that provides a running weekly of this market index. Note the "edge" effects in the filtered output.

MATLAB's `fft` function automatically zero-pads its input if the specified transform length (its second argument) exceeds the signal's length. The frequency domain result will have a small imaginary component — largest value is 2.2×10^{-11} — because of the inherent finite precision nature of computer arithmetic. Because of the unfortunate misfit between signal lengths and favored FFT lengths, the number of arithmetic operations in the time-domain implementation is far less than those required by the frequency domain version: 514 versus 62,271. If the input signal had been one sample shorter, the frequency-domain computations would have been more than a factor of two less (28,696), but far more than in the time-domain implementation.

An interesting signal processing aspect of this example is demonstrated at the beginning and end of the output. The ramping up and down that occurs can be traced to assuming the input is zero before it begins and after it ends. The filter "sees" these initial and final values as the difference equation passes over the input. These artifacts can be handled in two ways: we can just ignore the edge effects or the data from previous and succeeding years' last and first week, respectively, can be placed at the ends.

5.15 Efficiency of Frequency-Domain Filtering[27]

To determine for what signal and filter durations a time- or frequency-domain implementation would be the most efficient, we need only count the computations required by each. For the time-domain, difference-equation approach, we need $N_x(2q+1)$. The frequency-domain approach requires three Fourier transforms, each requiring $\left(\frac{K}{2}\right)(\log_2 K)$ computations for a length-K FFT, and the multiplication of two spectra ($6K$ computations). The output-signal-duration-determined length must be at least $N_x + q$. Thus, we must compare

$$\left(N_x(2q+1) \leftrightarrow 6(N_x+q) + \frac{3}{2}(N_x+q)\log_2(N_x+q)\right)$$

[27]This content is available online at <http://cnx.org/content/m10279/2.14/>.

Exact analytic evaluation of this comparison is quite difficult (we have a transcendental equation to solve). Insight into this comparison is best obtained by dividing by N_x.

$$\left(2q+1 \leftrightarrow 6\left(1+\frac{q}{N_x}\right)+\frac{3}{2}\left(1+\frac{q}{N_x}\right)\log_2\left(N_x+q\right)\right)$$

With this manipulation, we are evaluating the number of computations per sample. For any given value of the filter's order q, the right side, the number of frequency-domain computations, will exceed the left if the signal's duration is long enough. However, for filter durations greater than about 10, as long as the input is at least 10 samples, the frequency-domain approach is faster *so long as the FFT's power-of-two constraint is advantageous.*

The frequency-domain approach is not yet viable; what will we do when the input signal is infinitely long? The difference equation scenario fits perfectly with the envisioned digital filtering structure (Figure 5.25), but so far we have required the input to have limited duration (so that we could calculate its Fourier transform). The solution to this problem is quite simple: Section the input into frames, filter each, and add the results together. To section a signal means expressing it as a linear combination of length-N_x non-overlapping "chunks." Because the filter is linear, filtering a sum of terms is equivalent to summing the results of filtering each term.

$$x(n)=\sum_{m=-\infty}^{\infty}\left(x\left(n-mN_x\right)\right)\Rightarrow y(n)=\sum_{m=-\infty}^{\infty}\left(y\left(n-mN_x\right)\right) \quad (5.57)$$

As illustrated in Figure 5.23, note that each filtered section has a duration longer than the input. Consequently, we must literally add the filtered sections together, not just butt them together.

Computational considerations reveal a substantial advantage for a frequency-domain implementation over a time-domain one. The number of computations for a time-domain implementation essentially remains constant whether we section the input or not. Thus, the number of computations for each output is $2q+1$. In the frequency-domain approach, computation counting changes because we need only compute the filter's frequency response $H(k)$ once, which amounts to a fixed overhead. We need only compute two DFTs and multiply them to filter a section. Letting N_x denote a section's length, the number of computations for a section amounts to $(N_x+q)\log_2(N_x+q)+6(N_x+q)$. In addition, we must add the filtered outputs together; the number of terms to add corresponds to the excess duration of the output compared with the input (q). The frequency-domain approach thus requires $\left(1+\frac{q}{N_x}\right)\log_2(N_x+q)+7\frac{q}{N_x}+6$ computations per output value. For even modest filter orders, the frequency-domain approach is much faster.

Exercise 5.30 *(Solution on p. 206.)*

Show that as the section length increases, the frequency domain approach becomes increasingly more efficient.

Note that the choice of section duration is arbitrary. Once the filter is chosen, we should section so that the required FFT length is precisely a power of two: Choose N_x so that $N_x+q=2^L$.

Implementing the digital filter shown in the A/D block diagram (Figure 5.25) with a frequency-domain implementation requires some additional signal management not required by time-domain implementations. Conceptually, a real-time, time-domain filter could accept each sample as it becomes available, calculate the difference equation, and produce the output value, all in less that the sampling interval T_s. Frequency-domain approaches don't operate on a sample-by-sample basis; instead, they operate on sections. They filter in real time by producing N_x outputs for the same number of inputs faster than N_xT_s. Because they generally take longer to produce an output section than the sampling interval duration, we must filter one section while accepting into memory the *next* section to be filtered. In programming, the operation of building up sections while computing on previous ones is known as **buffering**. Buffering can also be used in time-domain filters as well but isn't required.

Example 5.10

We want to lowpass filter a signal that contains a sinusoid and a significant amount of noise. The example shown in Figure 5.23 shows a portion of the noisy signal's waveform. If it weren't for the

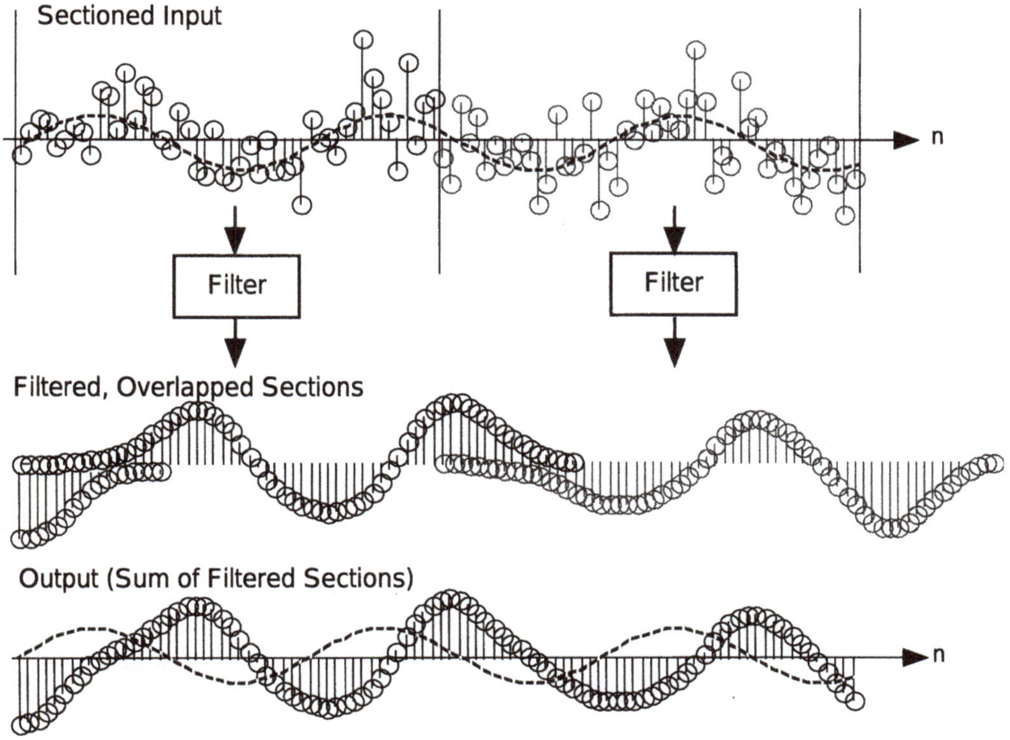

Figure 5.23: The noisy input signal is sectioned into length-48 frames, each of which is filtered using frequency-domain techniques. Each filtered section is added to other outputs that overlap to create the signal equivalent to having filtered the entire input. The sinusoidal component of the signal is shown as the red dashed line.

overlaid sinusoid, discerning the sine wave in the signal is virtually impossible. One of the primary applications of linear filters is **noise removal**: preserve the signal by matching filter's passband with the signal's spectrum and greatly reduce all other frequency components that may be present in the noisy signal.

A smart Rice engineer has selected a FIR filter having a unit-sample response corresponding a period-17 sinusoid: $h(n) = \frac{1}{17}\left(1 - \cos\left(\frac{2\pi n}{17}\right)\right)$, $n = \{0, \dots, 16\}$, which makes $q = 16$. Its frequency response (determined by computing the discrete Fourier transform) is shown in Figure 5.24. To apply, we can select the length of each section so that the frequency-domain filtering approach is maximally efficient: Choose the section length N_x so that $N_x + q$ is a power of two. To use a length-64 FFT, each section must be 48 samples long. Filtering with the difference equation would require 33 computations per output while the frequency domain requires a little over 16; this frequency-domain implementation is over twice as fast! Figure 5.23 shows how frequency-domain filtering works.

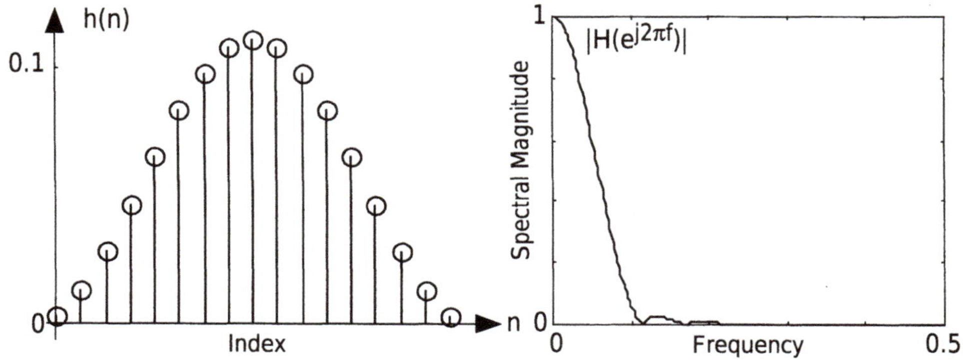

Figure 5.24: The figure shows the unit-sample response of a length-17 Hanning filter on the left and the frequency response on the right. This filter functions as a lowpass filter having a cutoff frequency of about 0.1.

We note that the noise has been dramatically reduced, with a sinusoid now clearly visible in the filtered output. Some residual noise remains because noise components within the filter's passband appear in the output as well as the signal.

Exercise 5.31 *(Solution on p. 207.)*

Note that when compared to the input signal's sinusoidal component, the output's sinusoidal component seems to be delayed. What is the source of this delay? Can it be removed?

5.16 Discrete-Time Filtering of Analog Signals[28]

Because of the Sampling Theorem (Section 5.3.2: The Sampling Theorem), we can process, in particular filter, analog signals "with a computer" by constructing the system shown in Figure 5.25. To use this system, we are assuming that the input signal has a lowpass spectrum and can be bandlimited without affecting important signal aspects. Bandpass signals can also be filtered digitally, but require a more complicated system. Highpass signals cannot be filtered digitally. Note that the input and output filters must be analog filters; trying to operate without them can lead to potentially very inaccurate digitization.

Another implicit assumption is that the digital filter can operate in *real time*: The computer and the filtering algorithm must be sufficiently fast so that outputs are computed faster than input values arrive. The sampling interval, which is determined by the analog signal's bandwidth, thus determines how long our program has to compute *each* output $y(n)$. The computational complexity for calculating each output with a difference equation (5.44) is $O(p+q)$. Frequency domain implementation of the filter is also possible. The idea begins by computing the Fourier transform of a length-N portion of the input $x(n)$, multiplying it by the filter's transfer function, and computing the inverse transform of the result. This approach seems overly complex and potentially inefficient. Detailing the complexity, however, we have $O(NlogN)$ for the two transforms (computed using the FFT algorithm) and $O(N)$ for the multiplication by the transfer function, which makes the total complexity $O(NlogN)$ *for N input values.* A frequency domain implementation thus requires $O(logN)$ computational complexity for each output value. The complexities of time-domain and frequency-domain implementations depend on different aspects of the filtering: The time-domain implementation depends on the combined orders of the filter while the frequency-domain implementation depends on the logarithm of the Fourier transform's length.

[28]This content is available online at <http://cnx.org/content/m0511/2.20/>.

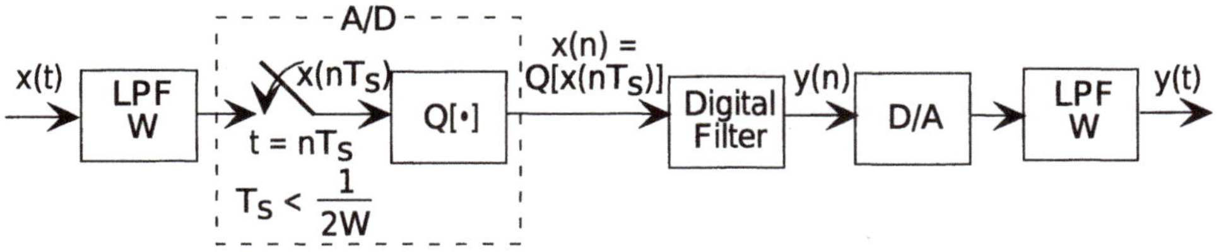

Figure 5.25: To process an analog signal digitally, the signal $x(t)$ must be filtered with an anti-aliasing filter (to ensure a bandlimited signal) before A/D conversion. This lowpass filter (LPF) has a cutoff frequency of W Hz, which determines allowable sampling intervals T_s . The greater the number of bits in the amplitude quantization portion $Q[\cdot]$ of the A/D converter, the greater the accuracy of the entire system. The resulting digital signal $x(n)$ can now be filtered in the time-domain with a difference equation or in the frequency domain with Fourier transforms. The resulting output $y(n)$ then drives a D/A converter and a second anti-aliasing filter (having the same bandwidth as the first one).

It could well be that in some problems the time-domain version is more efficient (more easily satisfies the real time requirement), while in others the frequency domain approach is faster. In the latter situations, it is the FFT algorithm for computing the Fourier transforms that enables the superiority of frequency-domain implementations. Because complexity considerations only express how algorithm running-time increases with system parameter choices, we need to detail both implementations to determine which will be more suitable for any given filtering problem. Filtering with a difference equation is straightforward, and the number of computations that must be made for each output value is $2(p+q)$.

Exercise 5.32 *(Solution on p. 207.)*

Derive this value for the number of computations for the general difference equation (5.44).

5.17 Digital Signal Processing Problems[29]

Problem 5.1: Sampling and Filtering

The signal $s(t)$ is bandlimited to 4 kHz. We want to sample it, but it has been subjected to various signal processing manipulations.

a) What sampling frequency (if any works) can be used to sample the result of passing $s(t)$ through an RC highpass filter with $R = 10k\Omega$ and $C = 8nF$?
b) What sampling frequency (if any works) can be used to sample the *derivative* of $s(t)$?
c) The signal $s(t)$ has been modulated by an 8 kHz sinusoid having an unknown phase: the resulting signal is $s(t)\sin(2\pi f_0 t + \phi)$, with $f_0 = 8kHz$ and $\phi = ?$ Can the modulated signal be sampled so that the *original* signal can be recovered from the modulated signal regardless of the phase value ϕ? If so, show how and find the smallest sampling rate that can be used; if not, show why not.

Problem 5.2: Non-Standard Sampling

Using the properties of the Fourier series can ease finding a signal's spectrum.

a) Suppose a signal $s(t)$ is periodic with period T. If c_k represents the signal's Fourier series coefficients, what are the Fourier series coefficients of $s\left(t - \frac{T}{2}\right)$?
b) Find the Fourier series of the signal $p(t)$ shown in Figure 5.26 (Pulse Signal).

[29]This content is available online at <http://cnx.org/content/m10351/2.33/>.

c) Suppose this signal is used to sample a signal bandlimited to $\frac{1}{T}Hz$. Find an expression for and sketch the spectrum of the sampled signal.

d) Does aliasing occur? If so, can a change in sampling rate prevent aliasing; if not, show how the signal can be recovered from these samples.

Pulse Signal

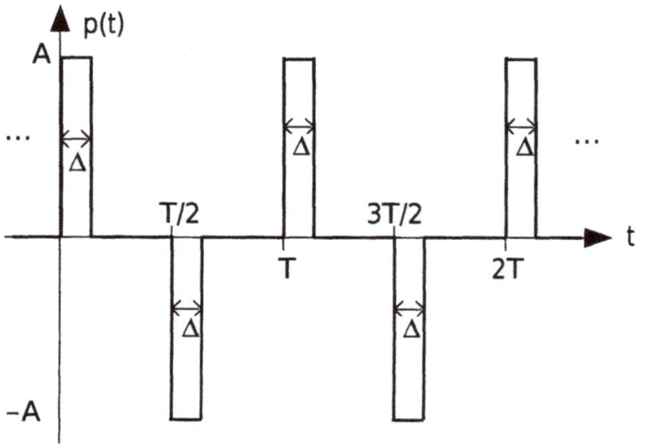

Figure 5.26

Problem 5.3: A Different Sampling Scheme

A signal processing engineer from Texas A&M claims to have developed an improved sampling scheme. He multiplies the bandlimited signal by the depicted periodic pulse signal to perform sampling (Figure 5.27).

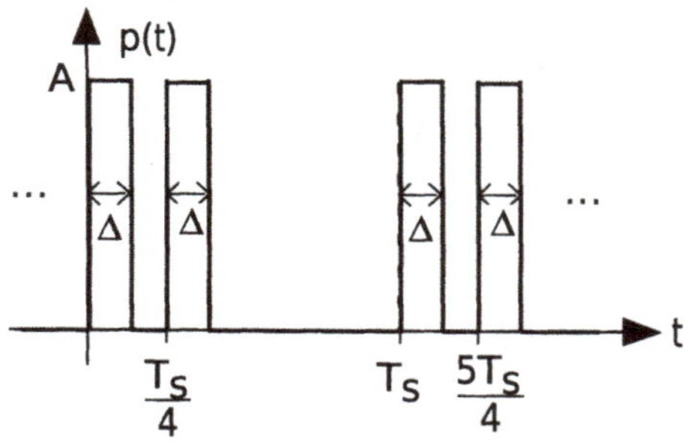

Figure 5.27

a) Find the Fourier spectrum of this signal.

b) Will this scheme work? If so, how should T_S be related to the signal's bandwidth? If not, why not?

Problem 5.4: Bandpass Sampling
The signal $s(t)$ has the indicated spectrum.

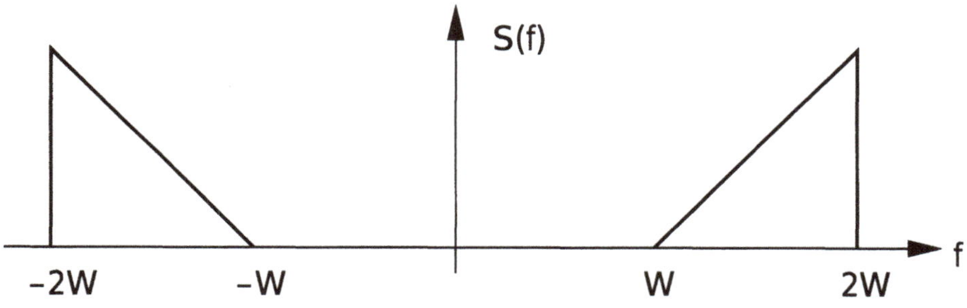

Figure 5.28

a) What is the minimum sampling rate for this signal suggested by the Sampling Theorem?
b) Because of the particular structure of this spectrum, one wonders whether a lower sampling rate could be used. Show that this is indeed the case, and find the system that reconstructs $s(t)$ from its samples.

Problem 5.5: Sampling Signals
If a signal is bandlimited to W Hz, we can sample it at any rate $\frac{1}{T_s} > 2W$ and recover the waveform exactly. This statement of the Sampling Theorem can be taken to mean that all information about the original signal can be extracted from the samples. While true in principle, you do have to be careful how you do so. In addition to the rms value of a signal, an important aspect of a signal is its peak value, which equals $max\{|s(t)|\}$.

a) Let $s(t)$ be a sinusoid having frequency W Hz. If we sample it at precisely the Nyquist rate, how accurately do the samples convey the sinusoid's amplitude? In other words, find the worst case example.
b) How fast would you need to sample for the amplitude estimate to be within 5% of the true value?
c) Another issue in sampling is the inherent amplitude quantization produced by A/D converters. Assume the maximum voltage allowed by the converter is V_{max} volts and that it quantizes amplitudes to b bits. We can express the quantized sample $Q(s(nT_s))$ as $s(nT_s) + \epsilon(t)$, where $\epsilon(t)$ represents the quantization error at the nth sample. Assuming the converter rounds, how large is maximum quantization error?
d) We can describe the quantization error as noise, with a power proportional to the square of the maximum error. What is the signal-to-noise ratio of the quantization error for a full-range sinusoid? Express your result in decibels.

Problem 5.6: Hardware Error
An A/D converter has a curious hardware problem: Every other sampling pulse is half its normal amplitude (Figure 5.29).

a) Find the Fourier series for this signal.
b) Can this signal be used to sample a bandlimited signal having highest frequency $W = \frac{1}{2T}$?

Problem 5.7: Simple D/A Converter
Commercial digital-to-analog converters don't work this way, but a simple circuit illustrates how they work. Let's assume we have a B-bit converter. Thus, we want to convert numbers having a B-bit representation

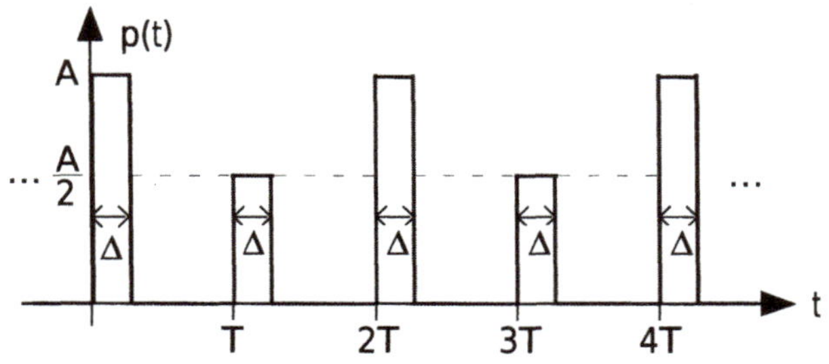

Figure 5.29

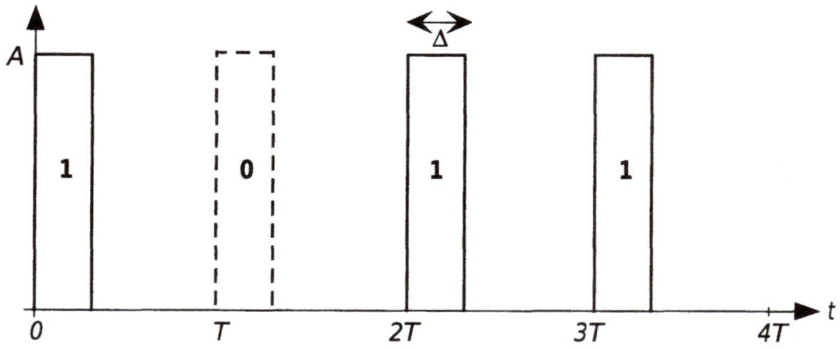

Figure 5.30

into a voltage proportional to that number. The first step taken by our simple converter is to represent the number by a sequence of B pulses occurring at multiples of a time interval T. The presence of a pulse indicates a "1" in the corresponding bit position, and pulse absence means a "0" occurred. For a 4-bit converter, the number 13 has the binary representation 1101 ($13_{10} = 1 \times 2^3 + 1 \times 2^2 + 0 \times 2^1 + 1 \times 2^0$) and would be represented by the depicted pulse sequence. Note that the pulse sequence is "backwards" from the binary representation. We'll see why that is.

This signal (Figure 5.30) serves as the input to a first-order RC lowpass filter. We want to design the filter and the parameters Δ and T so that the output voltage at time $4T$ (for a 4-bit converter) is proportional to the number. This combination of pulse creation and filtering constitutes our simple D/A converter. The requirements are

- The voltage at time $t = 4T$ should diminish by a factor of 2 the further the pulse occurs from this time. In other words, the voltage due to a pulse at $3T$ should be twice that of a pulse produced at $2T$, which in turn is twice that of a pulse at T, *etc.*
- The 4-bit D/A converter must support a 10 kHz sampling rate.

Show the circuit that works. How do the converter's parameters change with sampling rate and number of bits in the converter?

Problem 5.8: Discrete-Time Fourier Transforms
Find the Fourier transforms of the following sequences, where $s(n)$ is some sequence having Fourier transform $S\left(e^{j2\pi f}\right)$.

a) $(-1)^n s(n)$
b) $s(n)\cos(2\pi f_0 n)$
c) $x(n) = \begin{cases} s\left(\frac{n}{2}\right) \text{ if } n \text{ (even)} \\ 0 \text{ if } n \text{ (odd)} \end{cases}$
d) $ns(n)$

Problem 5.9: Spectra of Finite-Duration Signals
Find the indicated spectra for the following signals.

a) The discrete-time Fourier transform of $s(n) = \begin{cases} \cos^2\left(\frac{\pi}{4}n\right) \text{ if } n = \{-1, 0, 1\} \\ 0 \text{ if otherwise} \end{cases}$
b) The discrete-time Fourier transform of $s(n) = \begin{cases} n \text{ if } n = \{-2, -1, 0, 1, -2\} \\ 0 \text{ if otherwise} \end{cases}$
c) The discrete-time Fourier transform of $s(n) = \begin{cases} \sin\left(\frac{\pi}{4}n\right) \text{ if } n = \{0, \dots, 7\} \\ 0 \text{ if otherwise} \end{cases}$
d) The length-8 DFT of the previous signal.

Problem 5.10: Just Whistlin'
Sammy loves to whistle and decides to record and analyze his whistling in lab. He is a very good whistler; his whistle is a pure sinusoid that can be described by $s_a(t) = \sin(4000t)$. To analyze the spectrum, he samples his recorded whistle with a sampling interval of $T_S = 2.5 \times 10^{-4}$ to obtain $s(n) = s_a(nT_S)$. Sammy (wisely) decides to analyze a few samples at a time, so he grabs 30 consecutive, but arbitrarily chosen, samples. He calls this sequence $x(n)$ and realizes he can write it as

$$x(n) = \sin(4000nT_S + \theta), \quad n = \{0, \dots, 29\}$$

a) Did Sammy under- or over-sample his whistle?
b) What is the discrete-time Fourier transform of $x(n)$ and how does it depend on θ?
c) How does the 32-point DFT of $x(n)$ depend on θ?

Problem 5.11: Discrete-Time Filtering
We can find the input-output relation for a discrete-time filter much more easily than for analog filters. The key idea is that a sequence can be written as a weighted linear combination of unit samples.

a) Show that $x(n) = \sum_i (x(i)\delta(n-i))$ where $\delta(n)$ is the unit-sample.

$$\delta(n) = \begin{cases} 1 \text{ if } n = 0 \\ 0 \text{ otherwise} \end{cases}$$

b) If $h(n)$ denotes the *unit-sample response*—the output of a discrete-time linear, shift-invariant filter to a unit-sample input—find an expression for the output.
c) In particular, assume our filter is FIR, with the unit-sample response having duration $q + 1$. If the input has duration N, what is the duration of the filter's output to this signal?
d) Let the filter be a boxcar averager: $h(n) = \left(\frac{1}{q+1}\right)$ for $n = \{0, \dots, q\}$ and zero otherwise. Let the input be a pulse of unit height and duration N. Find the filter's output when $N = \frac{q+1}{2}$, q an odd integer.

Problem 5.12: A Digital Filter
A digital filter has the depicted (Figure 5.31) unit-sample reponse.

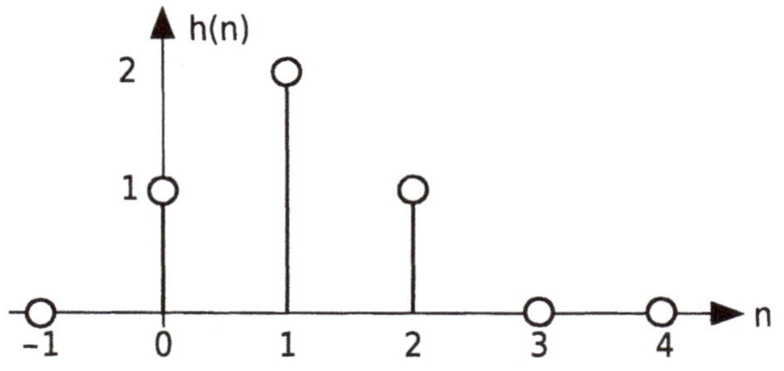

Figure 5.31

a) What is the difference equation that defines this filter's input-output relationship?
b) What is this filter's transfer function?
c) What is the filter's output when the input is $\sin\left(\frac{\pi n}{4}\right)$?

Problem 5.13: A Special Discrete-Time Filter
Consider a FIR filter governed by the difference equation

$$y(n) = \frac{1}{3}x(n+2) + \frac{2}{3}x(n+1) + x(n) + \frac{2}{3}x(n-1) + \frac{1}{3}x(n-2)$$

a) Find this filter's unit-sample response.
b) Find this filter's transfer function. Characterize this transfer function (*i.e.*, what classic filter category does it fall into).
c) Suppose we take a sequence and stretch it out by a factor of three.

$$x(n) = \begin{cases} s\left(\frac{n}{3}\right) \text{ if } n = 3m \;, \quad m = \{\dots, -1, 0, 1, \dots\} \\ 0 \text{ otherwise} \end{cases}$$

Sketch the sequence $x(n)$ for some example $s(n)$. What is the filter's output to this input? In particular, what is the output at the indices where the input $x(n)$ is intentionally zero? Now how would you characterize this system?

Problem 5.14: Simulating the Real World
Much of physics is governed by differntial equations, and we want to use signal processing methods to simulate physical problems. The idea is to replace the derivative with a discrete-time approximation and solve the resulting differential equation. For example, suppose we have the differential equation

$$\frac{d}{dt}y(t) + ay(t) = x(t)$$

and we approximate the derivative by

$$\frac{d}{dt}y(t)\mid_{t=nT} \approx \frac{y(nT) - y((n-1)T)}{T}$$

where T essentially amounts to a sampling interval.

a) What is the difference equation that must be solved to approximate the differential equation?
b) When $x(t) = u(t)$, the unit step, what will be the simulated output?
c) Assuming $x(t)$ is a sinusoid, how should the sampling interval T be chosen so that the approximation works well?

Problem 5.15: The DFT
Let's explore the DFT and its properties.

a) What is the length-K DFT of length-N boxcar sequence, where $N < K$?
b) Consider the special case where $K = 4$. Find the inverse DFT of the product of the DFTs of two length-3 boxcars.
c) If we could use DFTs to perform linear filtering, it should be true that the product of the input's DFT and the unit-sample response's DFT equals the output's DFT. So that you can use what you just calculated, let the input be a boxcar signal and the unit-sample response also be a boxcar. The result of part (b) would then be the filter's output *if* we could implement the filter with length-4 DFTs. Does the actual output of the boxcar-filter equal the result found in the previous part (list, item 2, p. 199)?
d) What would you need to change so that the product of the DFTs of the input and unit-sample response in this case equaled the DFT of the filtered output?

Problem 5.16: The Fast Fourier Transform
Just to determine how fast the FFT algorithm really is, we can take advantage of MATLAB's `fft` function. If `x` is a length-N vector, `fft(x)` computes the length-N transform using the most efficient algorithm it can. In other words, it does not automatically zero-pad the sequence and it will use the FFT algorithm if the length is a power of two. Let's count the number of arithmetic operations the `fft` program requires for lengths ranging from 2 to 1024.

a) For each length to be tested, generate a vector of random numbers, calculate the vector's transform, and determine how long it took. The program (Figure 5.32: Program) illustrates the computations.
b) Plot the vector of computation times. What lengths consume the most computations? What complexity do they seem to have? What lengths have the fewest computations?

Program

```
for n=2:1024,
  x = randn(1,n);
  t_start = cputime;
  fft(x);
  time(n) = cputime - t_start;
end
```

Figure 5.32

Problem 5.17: DSP Tricks
Sammy is faced with computing *lots* of discrete Fourier transforms. He will, or course, use the FFT algorithm, but he is behind schedule and needs to get his results as quickly as possible. He gets the idea of

computing *two* transforms at one time by computing the transform of $s(n) = s_1(n) + js_2(n)$, where $s_1(n)$ and $s_2(n)$ are two real-valued signals of which he needs to compute the spectra. The issue is whether he can retrieve the individual DFTs from the result or not.

a) What will be the DFT $S(k)$ of this complex-valued signal in terms of $S_1(k)$ and $S_2(k)$, the DFTs of the original signals?
b) Sammy's friend, an Aggie who knows some signal processing, says that retrieving the wanted DFTs is easy: "Just find the real and imaginary parts of $S(k)$." Show that this approach is too simplistic.
c) While his friend's idea is not correct, it does give him an idea. What approach will work? ***Hint***: Use the symmetry properties of the DFT.
d) How does the number of computations change with this approach? Will Sammy's idea ultimately lead to a faster computation of the required DFTs?

Problem 5.18: Discrete Cosine Transform (DCT)
The discrete cosine transform of a length-N sequence is defined to be

$$S_c(k) = \sum_{n=0}^{N-1} \left(s(n) \cos\left(\frac{2\pi nk}{2N}\right) \right)$$

Note that the number of frequency terms is $2N-1$: $k = \{0, \dots, 2N-1\}$.

a) Find the inverse DCT.
b) Does a Parseval's Theorem hold for the DCT?
c) You choose to transmit information about the signal $s(n)$ according to the DCT coefficients. You could only send one, which one would you send?

Problem 5.19: A Digital Filter
A digital filter is described by the following difference equation:

$$y(n) = ay(n-1) + ax(n) - x(n-1), a = \frac{1}{\sqrt{2}}$$

a) What is this filter's unit sample response?
b) What is this filter's transfer function?
c) What is this filter's output when the input is $\sin\left(\frac{\pi n}{4}\right)$?

Problem 5.20: Another Digital Filter
A digital filter is determined by the following difference equation.

$$y(n) = y(n-1) + x(n) - x(n-4)$$

a) Find this filter's unit sample response.
b) What is the filter's transfer function?
c) Find the filter's output when the input is the sinusoid $\sin\left(\frac{\pi n}{2}\right)$.

Problem 5.21: Yet Another Digital Filter
A filter has an input-output relationship given by the difference equation

$$y(n) = \frac{1}{4}x(n) + \frac{1}{2}x(n-1) + \frac{1}{4}x(n-2)$$

a) What is the filter's transfer function? How would you characterize it?
b) What is the filter's output when the input equals $\cos\left(\frac{\pi n}{2}\right)$?
c) What is the filter's output when the input is the depicted discrete-time square-wave (Figure 5.33)?

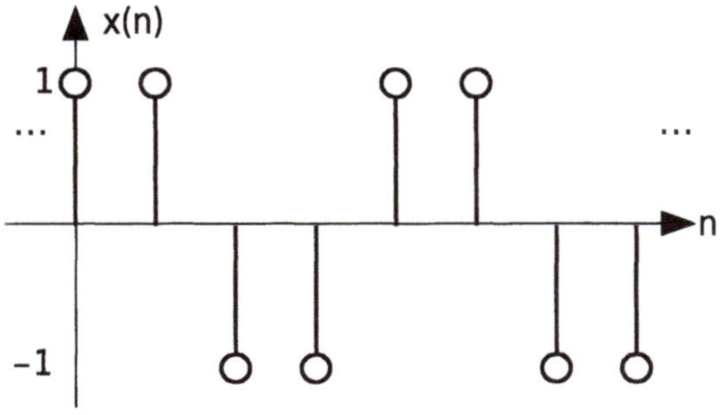

Figure 5.33

Problem 5.22: A Digital Filter in the Frequency Domain
We have a filter with the transfer function

$$H\left(e^{j2\pi f}\right) = e^{-(j2\pi f)}\cos(2\pi f)$$

operating on the input signal $x(n) = \delta(n) - \delta(n-2)$ that yields the output $y(n)$.

- What is the filter's unit-sample response?
- What is the discrete-Fourier transform of the output?
- What is the time-domain expression for the output?

Problem 5.23: Digital Filters
A discrete-time system is governed by the difference equation

$$y(n) = y(n-1) + \frac{x(n) + x(n-1)}{2}$$

a) Find the transfer function for this system.
b) What is this system's output when the input is $\sin\left(\frac{\pi n}{2}\right)$?
c) If the output is observed to be $y(n) = \delta(n) + \delta(n-1)$, then what is the input?

Problem 5.24: Digital Filtering
A digital filter has an input-output relationship expressed by the difference equation

$$y(n) = \frac{x(n) + x(n-1) + x(n-2) + x(n-3)}{4}$$

a) Plot the magnitude and phase of this filter's transfer function.
b) What is this filter's output when $x(n) = \cos\left(\frac{\pi n}{2}\right) + 2\sin\left(\frac{2\pi n}{3}\right)$?

Problem 5.25: Detective Work
The signal $x(n)$ equals $\delta(n) - \delta(n-1)$.

a) Find the length-8 DFT (discrete Fourier transform) of this signal.
b) You are told that when $x(n)$ served as the input to a linear FIR (finite impulse response) filter, the output was $y(n) = \delta(n) - \delta(n-1) + 2\delta(n-2)$. Is this statement true? If so, indicate why and find the system's unit sample response; if not, show why not.

Problem 5.26:
A discrete-time, shift invariant, linear system produces an output $y(n) = \{1, -1, 0, 0, \dots\}$ when its input $x(n)$ equals a unit sample.

a) Find the difference equation governing the system.
b) Find the output when $x(n) = \cos(2\pi f_0 n)$.
c) How would you describe this system's function?

Problem 5.27: Time Reversal has Uses
A discrete-time system has transfer function $H\left(e^{j2\pi f}\right)$. A signal $x(n)$ is passed through this system to yield the signal $w(n)$. The time-reversed signal $w(-n)$ is then passed through the system to yield the time-reversed output $y(-n)$. What is the transfer function between $x(n)$ and $y(n)$?

Problem 5.28: Removing "Hum"
The slang word "hum" represents power line waveforms that creep into signals because of poor circuit construction. Usually, the 60 Hz signal (and its harmonics) are added to the desired signal. What we seek are filters that can remove hum. In this problem, the signal and the accompanying hum have been sampled; we want to design a *digital* filter for hum removal.

a) Find filter coefficients for the length-3 FIR filter that can remove a sinusoid having *digital* frequency f_0 from its input.
b) Assuming the sampling rate is f_s to what analog frequency does f_0 correspond?
c) A more general approach is to design a filter having a frequency response *magnitude* proportional to the absolute value of a cosine: $|H\left(e^{j2\pi f}\right)| \propto |\cos(\pi f N)|$. In this way, not only can the fundamental but also its first few harmonics be removed. Select the parameter N and the sampling rate so that the frequencies at which the cosine equals zero correspond to 60 Hz and its odd harmonics through the fifth.
d) Find the difference equation that defines this filter.

Problem 5.29: Digital AM Receiver
Thinking that digital implementations are *always* better, our clever engineer wants to design a digital AM receiver. The receiver would bandpass the received signal, pass the result through an A/D converter, perform all the demodulation with digital signal processing systems, and end with a D/A converter to produce the analog message signal. Assume in this problem that the carrier frequency is always a large *even* multiple of the message signal's bandwidth W.

a) What is the smallest sampling rate that would be needed?
b) Show the block diagram of the least complex digital AM receiver.
c) Assuming the channel adds white noise and that a b-bit A/D converter is used, what is the output's signal-to-noise ratio?

Problem 5.30: DFTs
A problem on Samantha's homework asks for the *8-point* DFT of the discrete-time signal $\delta(n-1) + \delta(n-7)$.

a) What answer should Samantha obtain?
b) As a check, her group partner Sammy says that he computed the inverse DFT of her answer and got $\delta(n+1) + \delta(n-1)$. Does Sammy's result mean that Samantha's answer is wrong?
c) The homework problem says to lowpass-filter the sequence by multiplying its DFT by

$$H(k) = \begin{cases} 1 \text{ if } k = \{0, 1, 7\} \\ 0 \text{ otherwise} \end{cases}$$

and then computing the inverse DFT. Will this filtering algorithm work? If so, find the filtered output; if not, why not?

Problem 5.31: Stock Market Data Processing
Because a trading week lasts five days, stock markets frequently compute running averages each day over the previous five trading days to smooth price fluctuations. The technical stock analyst at the Buy-Lo–Sell-Hi brokerage firm has heard that FFT filtering techniques work better than any others (in terms of producing more accurate averages).

a) What is the difference equation governing the five-day averager for daily stock prices?
b) Design an efficient FFT-based filtering algorithm for the broker. How much data should be processed at once to produce an efficient algorithm? What length transform should be used?
c) Is the analyst's information correct that FFT techniques produce more accurate averages than any others? Why or why not?

Problem 5.32: Digital Filtering of Analog Signals
RU Electronics wants to develop a filter that would be used in analog applications, but that is implemented digitally. The filter is to operate on signals that have a 10 kHz bandwidth, and will serve as a lowpass filter.

a) What is the block diagram for your filter implementation? Explicitly denote which components are analog, which are digital (a computer performs the task), and which interface between analog and digital worlds.
b) What sampling rate must be used and how many bits must be used in the A/D converter for the acquired signal's signal-to-noise ratio to be at least 60 dB? For this calculation, assume the signal is a sinusoid.
c) If the filter is a length-128 FIR filter (the duration of the filter's unit-sample response equals 128), should it be implemented in the time or frequency domain?
d) Assuming $H\left(e^{j2\pi f}\right)$ is the transfer function of the digital filter, what is the transfer function of your system?

Problem 5.33: Signal Compression
Because of the slowness of the Internet, lossy signal compression becomes important if you want signals to be received quickly. An enterprising 241 student has proposed a scheme based on frequency-domain processing. First of all, he would section the signal into length-N blocks, and compute its N-point DFT. He then would discard (zero the spectrum) at *half* of the frequencies, quantize them to b-bits, and send these over the network. The receiver would assemble the transmitted spectrum and compute the inverse DFT, thus reconstituting an N-point block.

a) At what frequencies should the spectrum be zeroed to minimize the error in this lossy compression scheme?
b) The nominal way to represent a signal digitally is to use simple b-bit quantization of the time-domain waveform. How long should a section be in the proposed scheme so that the required number of bits/sample is smaller than that nominally required?
c) Assuming that effective compression can be achieved, would the proposed scheme yield satisfactory results?

Solutions to Exercises in Chapter 5

Solution to Exercise 5.1 (p. 157)
For b-bit signed integers, the largest number is $2^{b-1} - 1$. For $b = 32$, we have 2,147,483,647 and for $b = 64$, we have 9,223,372,036,854,775,807 or about 9.2×10^{18}.

Solution to Exercise 5.2 (p. 158)
In floating point, the number of bits in the exponent determines the largest and smallest representable numbers. For 32-bit floating point, the largest (smallest) numbers are $2^{\pm 127} = 1.7 \times 10^{38}$ (5.9×10^{-39}). For 64-bit floating point, the largest number is about 10^{9863}.

Solution to Exercise 5.3 (p. 159)
$25 = 11011_2$ and $7 = 111_2$. We find that $11001_2 + 111_2 = 100000_2 = 32$.

Solution to Exercise 5.4 (p. 162)
The only effect of pulse duration is to unequally weight the spectral repetitions. Because we are only concerned with the repetition centered about the origin, the pulse duration has no significant effect on recovering a signal from its samples.

Solution to Exercise 5.5 (p. 162)

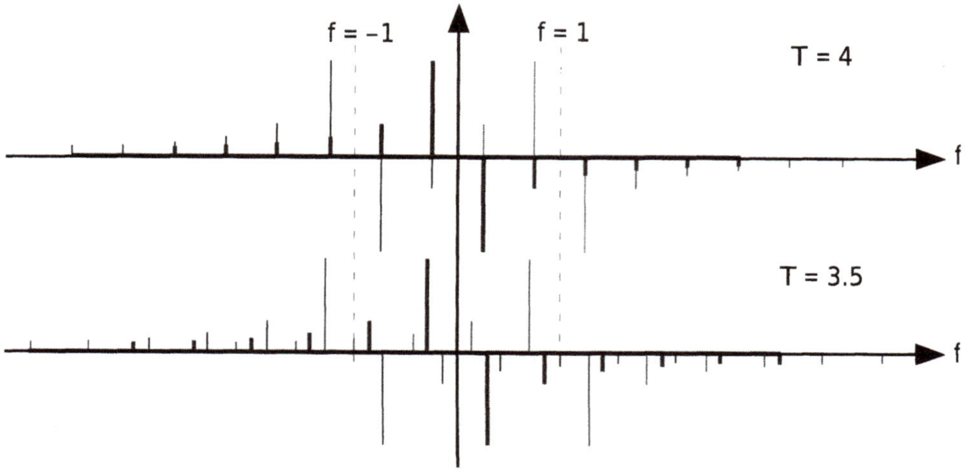

Figure 5.34

The square wave's spectrum is shown by the bolder set of lines centered about the origin. The dashed lines correspond to the frequencies about which the spectral repetitions (due to sampling with $T_s = 1$) occur. As the square wave's period decreases, the negative frequency lines move to the left and the positive frequency ones to the right.

Solution to Exercise 5.6 (p. 162)
The simplest bandlimited signal is the sine wave. At the Nyquist frequency, exactly two samples/period would occur. Reducing the sampling rate would result in fewer samples/period, and these samples would appear to have arisen from a lower frequency sinusoid.

Solution to Exercise 5.7 (p. 164)
The plotted temperatures were quantized to the nearest degree. Thus, the high temperature's amplitude was quantized as a form of A/D conversion.

Solution to Exercise 5.8 (p. 164)
The signal-to-noise ratio does not depend on the signal amplitude. With an A/D range of $[-A, A]$, the quantization interval $\Delta = \frac{2A}{2^B}$ and the signal's rms value (again assuming it is a sinusoid) is $\frac{A}{\sqrt{2}}$.

Solution to Exercise 5.9 (p. 164)
Solving $2^{-B} = .001$ results in $B = 10$ bits.

Solution to Exercise 5.10 (p. 164)
A 16-bit A/D converter yields a SNR of $6 \times 16 + 10\log_{10} 1.5 = 97.8$ dB.

Solution to Exercise 5.11 (p. 167)

$$\begin{aligned} S\left(e^{j2\pi(f+1)}\right) &= \sum_{n=-\infty}^{\infty}\left(s(n)\,e^{-(j2\pi(f+1)n)}\right) \\ &= \sum_{n=-\infty}^{\infty}\left(e^{-(j2\pi n)}s(n)\,e^{-(j2\pi fn)}\right) \\ &= \sum_{n=-\infty}^{\infty}\left(s(n)\,e^{-(j2\pi fn)}\right) \\ &= S\left(e^{j2\pi f}\right) \end{aligned} \tag{5.58}$$

Solution to Exercise 5.12 (p. 170)

$$\alpha \sum_{n=n_0}^{N+n_0-1}(\alpha^n) - \sum_{n=n_0}^{N+n_0-1}(\alpha^n) = \alpha^{N+n_0} - \alpha^{n_0}$$

which, after manipulation, yields the geometric sum formula.

Solution to Exercise 5.13 (p. 171)
If the sampling frequency exceeds the Nyquist frequency, the spectrum of the samples equals the analog spectrum, but over the normalized analog frequency fT. Thus, the energy in the sampled signal equals the original signal's energy multiplied by T.

Solution to Exercise 5.14 (p. 173)
This situation amounts to aliasing in the time-domain.

Solution to Exercise 5.15 (p. 174)
When the signal is real-valued, we may only need half the spectral values, but the complexity remains unchanged. If the data are complex-valued, which demands retaining all frequency values, the complexity is again the same. When only K frequencies are needed, the complexity is $O(KN)$.

Solution to Exercise 5.16 (p. 175)
If a DFT required 1ms to compute, and signal having ten times the duration would require 100ms to compute. Using the FFT, a 1ms computing time would increase by a factor of about $\log_2 10 = 3.3$, a factor of 30 less than the DFT would have needed.

Solution to Exercise 5.17 (p. 177)
The upper panel has not used the FFT algorithm to compute the length-4 DFTs while the lower one has. The ordering is determined by the algorithm.

Solution to Exercise 5.18 (p. 177)
The transform can have *any* greater than or equal to the actual duration of the signal. We simply "pad" the signal with zero-valued samples until a computationally advantageous signal length results. Recall that the FFT is an *algorithm* to compute the DFT (Section 5.7). Extending the length of the signal this way merely means we are sampling the frequency axis more finely than required. To use the Cooley-Tukey algorithm, the length of the resulting zero-padded signal can be 512, 1024, etc. samples long.

Solution to Exercise 5.19 (p. 178)
Number of samples equals $1.2 \times 11025 = 13230$. The datarate is $11025 \times 16 = 176.4$ kbps. The storage required would be 26460 bytes.

Solution to Exercise 5.20 (p. 179)
The oscillations are due to the boxcar window's Fourier transform, which equals the sinc function.

Solution to Exercise 5.21 (p. 180)
These numbers are powers-of-two, and the FFT algorithm can be exploited with these lengths. To compute a longer transform than the input signal's duration, we simply zero-pad the signal.

Solution to Exercise 5.22 (p. 180)
In discrete-time signal processing, an amplifier amounts to a multiplication, a very easy operation to perform.

Solution to Exercise 5.23 (p. 182)
The indices can be negative, and this condition is not allowed in MATLAB. To fix it, we must start the signals later in the array.

Solution to Exercise 5.24 (p. 184)
Such terms would require the system to know what future input or output values would be before the current value was computed. Thus, such terms can cause difficulties.

Solution to Exercise 5.25 (p. 186)
It now acts like a bandpass filter with a center frequency of f_0 and a bandwidth equal to *twice* of the original lowpass filter.

Solution to Exercise 5.26 (p. 186)
The DTFT of the unit sample equals a constant (equaling 1). Thus, the Fourier transform of the output equals the transfer function.

Solution to Exercise 5.27 (p. 186)
In sampling a discrete-time signal's Fourier transform L times equally over $[0, 2\pi)$ to form the DFT, the corresponding signal equals the periodic repetition of the original signal.

$$\left(S(k) \leftrightarrow \sum_{i=-\infty}^{\infty} (s(n - iL))\right) \tag{5.59}$$

To avoid aliasing (in the time domain), the transform length must equal or exceed the signal's duration.

Solution to Exercise 5.28 (p. 187)
The difference equation for an FIR filter has the form

$$y(n) = \sum_{m=0}^{q} (b_m x(n - m)) \tag{5.60}$$

The unit-sample response equals

$$h(n) = \sum_{m=0}^{q} (b_m \delta(n - m)) \tag{5.61}$$

which corresponds to the representation described in a problem (Example 5.6) of a length-q boxcar filter.

Solution to Exercise 5.29 (p. 187)
The unit-sample response's duration is $q + 1$ and the signal's N_x. Thus the statement is correct.

Solution to Exercise 5.30 (p. 190)
Let N denote the input's total duration. The time-domain implementation requires a total of $N(2q + 1)$ computations, or $2q + 1$ computations per input value. In the frequency domain, we split the input into $\frac{N}{N_x}$ sections, each of which requires $\left(1 + \frac{q}{N_x}\right) \log_2(N_x + q) + 7\frac{q}{N_x} + 6$ per input in the section. Because we divide *again* by N_x to find the number of computations per input value in the entire input, this quantity *decreases* as N_x increases. For the time-domain implementation, it stays constant.

Solution to Exercise 5.31 **(p. 192)**

The delay is *not* computational delay here–the plot shows the first output value is aligned with the filter's first input–although in real systems this is an important consideration. Rather, the delay is due to the filter's phase shift: A phase-shifted sinusoid is equivalent to a time-delayed one: $\cos(2\pi fn - \phi) = \cos\left(2\pi f\left(n - \frac{\phi}{2\pi f}\right)\right)$. All filters have phase shifts. This delay could be removed if the filter introduced no phase shift. Such filters do not exist in analog form, but digital ones can be programmed, but not in real time. Doing so would require the output to emerge before the input arrives!

Solution to Exercise 5.32 **(p. 193)**

We have $p+q+1$ multiplications and $p+q-1$ additions. Thus, the total number of arithmetic operations equals $2(p+q)$.

Chapter 6

Information Communication

6.1 Information Communication[1]

As far as a communications engineer is concerned, signals express information. Because systems manipulate signals, they also affect the information content. Information comes neatly packaged in both analog and digital forms. Speech, for example, is clearly an analog signal, and computer files consist of a sequence of bytes, a form of "discrete-time" signal despite the fact that the index sequences byte position, not time sample. *Communication systems* endeavor not to manipulate information, but to transmit it from one place to another, so-called **point-to-point communication**, from one place to many others, **broadcast communication**, or from many to many, like a telephone conference call or a chat room. Communication systems can be fundamentally analog, like radio, or digital, like computer networks.

This chapter develops a common theory that underlies how such systems work. We describe and analyze several such systems, some old like AM radio, some new like computer networks. The question as to which is better, analog or digital communication, has been answered, because of Claude Shannon's[2] fundamental work on a theory of information published in 1948, the development of cheap, high-performance computers, and the creation of high-bandwidth communication systems. *The answer is to use a digital communication strategy.* In most cases, you should convert all information-bearing signals into discrete-time, amplitude-quantized signals. Fundamentally digital signals, like computer files (which are a special case of symbolic signals), are in the proper form. Because of the Sampling Theorem, we know how to convert analog signals into digital ones. Shannon showed that once in this form, *a properly engineered system can communicate digital information with no error despite the fact that the communication channel thrusts noise onto all transmissions.* This startling result has no counterpart in analog systems; AM radio will remain noisy. The convergence of these theoretical and engineering results on communications systems has had important consequences in other arenas. The audio compact disc (CD) and the digital videodisk (DVD) are now considered digital communications systems, with communication design considerations used throughout.

Go back to the fundamental model of communication (Figure 1.4: Fundamental model of communication). Communications design begins with two fundamental considerations.

1. What is the nature of the information source, and to what extent can the receiver tolerate errors in the received information?
2. What are the channel's characteristics and how do they affect the transmitted signal?

In short, what are we going to send and how are we going to send it? Interestingly, digital as well as analog transmission are accomplished using analog signals, like voltages in Ethernet (an example of **wireline** communications) and electromagnetic radiation (**wireless**) in cellular telephone.

[1]This content is available online at <http://cnx.org/content/m0513/2.8/>.
[2]http://www.lucent.com/minds/infotheory/

6.2 Types of Communication Channels[3]

Electrical communications channels are either **wireline** or **wireless** channels. Wireline channels physically connect transmitter to receiver with a "wire" which could be a twisted pair, coaxial cable or optic fiber. Consequently, wireline channels are more private and much less prone to interference. Simple wireline channels connect a single transmitter to a single receiver: a **point-to-point** connection as with the telephone. Listening in on a conversation requires that the wire be tapped and the voltage measured. Some wireline channels operate in **broadcast** modes: one or more transmitter is connected to several receivers. One simple example of this situation is cable television. Computer networks can be found that operate in point-to-point or in broadcast modes. Wireless channels are much more public, with a transmitter's antenna radiating a signal that can be received by any antenna sufficiently close enough. In contrast to wireline channels where the receiver takes in only the transmitter's signal, the receiver's antenna will react to electromagnetic radiation coming from any source. This feature has two faces: The smiley face says that a receiver can take in transmissions from any source, letting receiver electronics select wanted signals and disregarding others, thereby allowing portable transmission and reception, while the frowny face says that interference and noise are much more prevalent than in wireline situations. A noisier channel subject to interference compromises the flexibility of wireless communication.

> POINT OF INTEREST: You will hear the term **tetherless networking** applied to completely wireless computer networks.

Maxwell's equations neatly summarize the physics of all electromagnetic phenomena, including circuits, radio, and optic fiber transmission.

$$\begin{aligned} \nabla \times \mathbf{E} &= -\left(\frac{\partial}{\partial t}(\mu \mathbf{H})\right) \\ \mathrm{div}(\epsilon \mathbf{E}) &= \rho \\ \nabla \times \mathbf{H} &= \sigma \mathbf{E} + \frac{\partial}{\partial t}(\epsilon \mathbf{E}) \\ div(\mu \mathbf{H}) &= 0 \end{aligned} \tag{6.1}$$

where $\mathbf{E}$ is the electric field, $\mathbf{H}$ the magnetic field, ϵ dielectric permittivity, μ magnetic permeability, σ electrical conductivity, and ρ is the charge density. Kirchoff's Laws represent special cases of these equations for circuits. We are not going to solve Maxwell's equations here; do bear in mind that a fundamental understanding of communications channels ultimately depends on fluency with Maxwell's equations. Perhaps the most important aspect of them is that they are **linear** with respect to the electrical and magnetic fields. Thus, the fields (and therefore the voltages and currents) resulting from two or more sources will *add*.

> POINT OF INTEREST: Nonlinear electromagnetic media do exist. The equations as written here are simpler versions that apply to free-space propagation and conduction in metals. Nonlinear media are becoming increasingly important in optic fiber communications, which are also governed by Maxwell's equations.

6.3 Wireline Channels[4]

Wireline channels were the first used for electrical communications in the mid-nineteenth century for the telegraph. Here, the channel is one of several wires connecting transmitter to receiver. The transmitter simply creates a voltage related to the message signal and applies it to the wire(s). We must have a circuit—a closed path—that supports current flow. In the case of single-wire communications, the earth is used as the

[3] This content is available online at <http://cnx.org/content/m0099/2.13/>.
[4] This content is available online at <http://cnx.org/content/m0100/2.27/>.

current's return path. In fact, the term **ground** for the reference node in circuits originated in single-wire telegraphs. You can imagine that the earth's electrical characteristics are highly variable, and they are. Single-wire metallic channels cannot support high-quality signal transmission having a bandwidth beyond a few hundred Hertz over any appreciable distance.

Coaxial Cable Cross-section

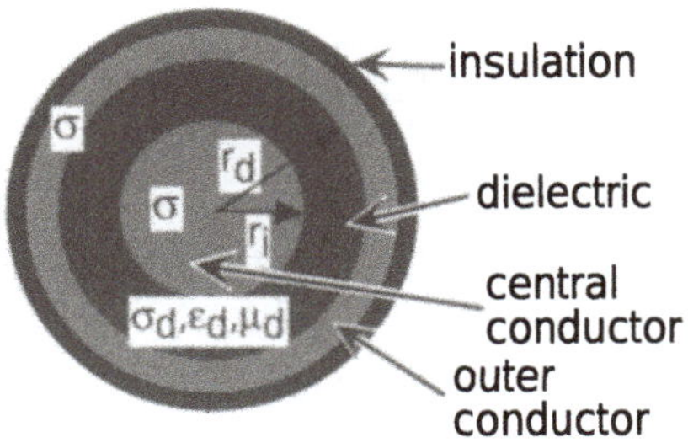

Figure 6.1: Coaxial cable consists of one conductor wrapped around the central conductor. This type of cable supports broader bandwidth signals than twisted pair, and finds use in cable television and Ethernet.

Consequently, most wireline channels today essentially consist of pairs of conducting wires Figure 6.1 (Coaxial Cable Cross-section), and the transmitter applies a message-related voltage across the pair. How these pairs of wires are physically configured greatly affects their transmission characteristics. One example is **twisted pair**, wherein the wires are wrapped about each other. Telephone cables are one example of a twisted pair channel. Another is **coaxial cable**, where a concentric conductor surrounds a central wire with a dielectric material in between. Coaxial cable, fondly called "co-ax" by engineers, is what Ethernet uses as its channel. In either case, wireline channels form a dedicated circuit between transmitter and receiver. As we shall find subsequently, several transmissions can share the circuit by amplitude modulation techniques; commercial cable TV is an example. These information-carrying circuits are designed so that interference from nearby electromagnetic sources is minimized. Thus, by the time signals arrive at the receiver, they are relatively interference- and noise-free.

Both twisted pair and co-ax are examples of **transmission lines**, which all have the circuit model shown in Figure 6.2 (Circuit Model for a Transmission Line) for an infinitesimally small length. This circuit model arises from solving Maxwell's equations for the particular transmission line geometry.

Circuit Model for a Transmission Line

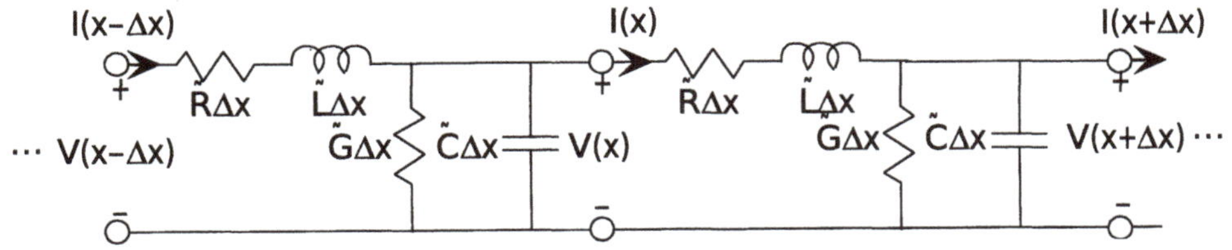

Figure 6.2: The so-called distributed parameter model for two-wire cables has the depicted circuit model structure. Element values depend on geometry and the properties of materials used to construct the transmission line.

The series resistance comes from the conductor used in the wires and from the conductor's geometry. The inductance and the capacitance derive from transmission line geometry, and the parallel conductance from the medium between the wire pair. Note that all the circuit elements have values expressed by the product of a constant times a length; this notation represents that element values here have per-unit-length units. For example, the series resistance $\tilde{R}$ has units of ohms/meter. For coaxial cable, the element values depend on the inner conductor's radius r_i, the outer radius of the dielectric r_d, the conductivity of the conductors σ, and the conductivity σ_d, dielectric constant ϵ_d, and magnetic permittivity μ_d of the dielectric as

$$\tilde{R} = \frac{1}{2\pi\delta\sigma}\left(\frac{1}{r_d} + \frac{1}{r_i}\right) \tag{6.2}$$

$$\tilde{C} = \frac{2\pi\epsilon_d}{\ln\left(\frac{r_d}{r_i}\right)}$$

$$\tilde{G} = \frac{2\pi\sigma_d}{\ln\left(\frac{r_d}{r_i}\right)}$$

$$\tilde{L} = \frac{\mu_d}{2\pi}\ln\left(\frac{r_d}{r_i}\right)$$

For twisted pair, having a separation d between the conductors that have conductivity σ and common radius r and that are immersed in a medium having dielectric and magnetic properties, the element values are then

$$\tilde{R} = \frac{1}{\pi r\delta\sigma} \tag{6.3}$$

$$\tilde{C} = \frac{\pi\epsilon}{\operatorname{arccosh}\left(\frac{d}{2r}\right)}$$

$$\tilde{G} = \frac{\pi\sigma}{\operatorname{arccosh}\left(\frac{d}{2r}\right)}$$

$$\tilde{L} = \frac{\mu}{\pi}\left(\frac{\delta}{2r} + \operatorname{arccosh}\left(\frac{d}{2r}\right)\right)$$

The voltage between the two conductors and the current flowing through them will depend on distance x along the transmission line as well as time. We express this dependence as $v(x,t)$ and $i(x,t)$. When we place a sinusoidal source at one end of the transmission line, these voltages and currents will also be sinusoidal because the transmission line model consists of linear circuit elements. As is customary in analyzing linear circuits, we express voltages and currents as the real part of complex exponential signals, and write circuit variables as a complex amplitude—here dependent on distance—times a complex exponential: $v(x,t) = \operatorname{Re}\left(V(x)e^{j2\pi ft}\right)$ and $i(x,t) = \operatorname{Re}\left(I(x)e^{j2\pi ft}\right)$. Using the transmission line circuit model, we find from KCL, KVL, and v-i relations the equations governing the complex amplitudes.

KCL at Center Node

$$I(x) = I(x - \Delta x) - V(x)\left(\tilde{G} + j2\pi f\tilde{C}\right)\Delta x \tag{6.4}$$

V-I relation for RL series

$$V(x) - V(x + \Delta x) = I(x)\left(\tilde{R} + j2\pi f\tilde{L}\right)\Delta x \tag{6.5}$$

Rearranging and taking the limit $\Delta x \to 0$ yields the so-called **transmission line equations.**

$$\frac{d}{dx}I(x) = \left(-\left(\tilde{G} + j2\pi f\tilde{C}\right)\right)V(x) \tag{6.6}$$

$$\frac{d}{dx}V(x) = \left(-\left(\widetilde{R} + j2\pi f\,\widetilde{L}\right)\right)I(x)$$

By combining these equations, we can obtain a single equation that governs how the voltage's or the current's complex amplitude changes with position along the transmission line. Taking the derivative of the second equation and plugging the first equation into the result yields the equation governing the voltage.

$$\frac{d^2}{dx^2}V(x) = \left(\widetilde{G} + j2\pi f\,\widetilde{C}\right)\left(\widetilde{R} + j2\pi f\,\widetilde{L}\right)V(x) \tag{6.7}$$

This equation's solution is

$$V(x) = V_+ e^{-(\gamma x)} + V_- e^{\gamma x} \tag{6.8}$$

Calculating its second derivative and comparing the result with our equation for the voltage can check this solution.

$$\begin{aligned}\frac{d^2}{dx^2}V(x) &= \gamma^2\left(V_+ e^{-(\gamma x)} + V_- e^{\gamma x}\right)\\ &= \gamma^2 V(x)\end{aligned} \tag{6.9}$$

Our solution works so long as the quantity γ satisfies

$$\begin{aligned}\gamma &= \pm\sqrt{\left(\widetilde{G} + j2\pi f\,\widetilde{C}\right)\left(\widetilde{R} + j2\pi f\,\widetilde{L}\right)}\\ &= \pm\left(a(f) + jb(f)\right)\end{aligned} \tag{6.10}$$

Thus, γ depends on frequency, and we express it in terms of real and imaginary parts as indicated. The quantities V_+ and V_- are constants determined by the source and physical considerations. For example, let the spatial origin be the middle of the transmission line model Figure 6.2 (Circuit Model for a Transmission Line). Because the circuit model contains simple circuit elements, physically possible solutions for voltage amplitude cannot increase with distance along the transmission line. Expressing γ in terms of its real and imaginary parts in our solution shows that such increases are a (mathematical) possibility. $V(x) = V_+ e^{(-(a+jb))x} + V_- e^{(a+jb)x}$ The voltage cannot increase without limit; because $a(f)$ is always positive, we must segregate the solution for negative and positive x. The first term will increase exponentially for $x < 0$ unless $V_+ = 0$ in this region; a similar result applies to V_- for $x > 0$. These physical constraints give us a cleaner solution.

$$V(x) = \begin{cases} V_+ e^{(-(a+jb))x} \text{ if } x > 0 \\ V_- e^{(a+jb)x} \text{ if } x < 0 \end{cases} \tag{6.11}$$

This solution suggests that voltages (and currents too) will decrease *exponentially* along a transmission line. The **space constant**, also known as the **attenuation constant**, is the distance over which the voltage decreases by a factor of $\frac{1}{e}$. It equals the reciprocal of $a(f)$, which depends on frequency, and is expressed by manufacturers in units of dB/m.

The presence of the imaginary part of γ, $b(f)$, also provides insight into how transmission lines work. Because the solution for $x > 0$ is proportional to $e^{-(jbx)}$, we know that the voltage's complex amplitude will *vary sinusoidally in space*. The complete solution for the voltage has the form

$$v(x,t) = \text{Re}\left(V_+ e^{-(ax)} e^{j(2\pi f t - bx)}\right) \tag{6.12}$$

The complex exponential portion has the form of a **propagating wave**. If we could take a snapshot of the voltage (take its picture at $t = t_1$), we would see a sinusoidally varying waveform along the transmission line. One period of this variation, known as the **wavelength**, equals $\lambda = \frac{2\pi}{b}$. If we were to take a second picture at some later time $t = t_2$, we would also see a sinusoidal voltage. Because

$$2\pi f t_2 - bx = 2\pi f(t_1 + t_2 - t_1) - bx = 2\pi f t_1 - b\left(x - \frac{2\pi f}{b}(t_2 - t_1)\right)$$

the second waveform appears to be the first one, but delayed—shifted to the right—in space. Thus, the voltage appeared to move to the right with a speed equal to $\frac{2\pi f}{b}$ (assuming $b > 0$). We denote this **propagation speed** by c, and it equals

$$c = \left| \frac{2\pi f}{\mathrm{Im}\left(\sqrt{\left(\tilde{G} + j2\pi f\,\tilde{C}\right)\left(\tilde{R} + j2\pi f\,\tilde{L}\right)}\right)} \right| \quad (6.13)$$

In the high-frequency region where $\left(j2\pi f\,\tilde{L} \gg \tilde{R}\right)$ and $\left(j2\pi f\,\tilde{C} \gg \tilde{G}\right)$, the quantity under the radical simplifies to $-4\pi^2 f^2\,\tilde{L}\tilde{C}$, and we find the propagation speed to be

$$\lim_{f\to\infty} c = \frac{1}{\sqrt{\tilde{L}\tilde{C}}} \quad (6.14)$$

For typical coaxial cable, this propagation speed is a fraction (one-third to two-thirds) of the speed of light.

Exercise 6.1 ***(Solution on p. 272.)***

Find the propagation speed in terms of physical parameters for both the coaxial cable and twisted pair examples.

By using the second of the transmission line equation (6.6), we can solve for the current's complex amplitude. Considering the spatial region $x > 0$, for example, we find that

$$\frac{d}{dx}V(x) = -(\gamma V(x)) = \left(-\left(\tilde{R} + j2\pi f\,\tilde{L}\right)\right) I(x)$$

which means that the ratio of voltage and current complex amplitudes does not depend on distance.

$$\begin{aligned} \frac{V(x)}{I(x)} &= \sqrt{\frac{\tilde{R} + j2\pi f\,\tilde{L}}{\tilde{G} + j2\pi f\,\tilde{C}}} \\ &= Z_0 \end{aligned} \quad (6.15)$$

The quantity Z_0 is known as the transmission line's **characteristic impedance**. Note that when the signal frequency is sufficiently high, the characteristic impedance is real, which means the transmission line appears resistive in this high-frequency regime.

$$\lim_{f\to\infty} Z_0 = \sqrt{\frac{\tilde{L}}{\tilde{C}}} \quad (6.16)$$

Typical values for characteristic impedance are 50 and 75 Ω.

A related transmission line is the optic fiber. Here, the electromagnetic field is light, and it propagates down a cylinder of glass. In this situation, we don't have two conductors—in fact we have none—and the energy is propagating in what corresponds to the dielectric material of the coaxial cable. Optic fiber communication has exactly the same properties as other transmission lines: Signal strength decays exponentially according to the fiber's space constant and propagates at some speed less than light would in free space. From the encompassing view of Maxwell's equations, the only difference is the electromagnetic signal's frequency. Because no electric conductors are present and the fiber is protected by an opaque "insulator," optic fiber transmission is interference-free.

Exercise 6.2 ***(Solution on p. 272.)***

From tables of physical constants, find the frequency of a sinusoid in the middle of the visible light range. Compare this frequency with that of a mid-frequency cable television signal.

To summarize, we use transmission lines for high-frequency wireline signal communication. In wireline communication, we have a direct, physical connection—a circuit—between transmitter and receiver. When we select the transmission line characteristics and the transmission frequency so that we operate in the high-frequency regime, signals are not filtered as they propagate along the transmission line: The characteristic impedance is real-valued—the tranmission line's equivalent impedance is a resistor—and all the signal's components at various frequencies propagate at the same speed. Transmitted signal amplitude does decay exponentially along the transmission line. Note that in the high-frequency regime that the space constant is approximately zero, which means the attenuation is quite small.

Exercise 6.3 ***(Solution on p. 272.)***

What is the limiting value of the space constant in the high frequency regime?

6.4 Wireless Channels[5]

Wireless channels exploit the prediction made by Maxwell's equation that electromagnetic fields propagate in free space like light. When a voltage is applied to an antenna, it creates an electromagnetic field that propagates in all directions (although antenna geometry affects how much power flows in any given direction) that induces electric currents in the receiver's antenna. Antenna geometry determines how energetic a field a voltage of a given frequency creates. In general terms, the dominant factor is the relation of the antenna's size to the field's wavelength. The fundamental equation relating frequency and wavelength for a propagating wave is

$$\lambda f = c$$

Thus, wavelength and frequency are inversely related: High frequency corresponds to small wavelengths. For example, a 1 MHz electromagnetic field has a wavelength of 300 m. Antennas having a size or distance from the ground comparable to the wavelength radiate fields most efficiently. Consequently, the lower the frequency the bigger the antenna must be. Because most information signals are baseband signals, having spectral energy at low frequencies, they must be modulated to higher frequencies to be transmitted over wireless channels.

For most antenna-based wireless systems, how the signal diminishes as the receiver moves further from the transmitter derives by considering how radiated power changes with distance from the transmitting antenna. An antenna radiates a given amount of power into free space, and ideally this power propagates without loss in all directions. Considering a sphere centered at the transmitter, the total power, which is found by integrating the radiated power over the surface of the sphere, must be constant regardless of the sphere's radius. This requirement results from the conservation of energy. Thus, if $p(d)$ represents the power integrated with respect to direction at a distance d from the antenna, the total power will be $p(d)\, 4\pi d^2$. For this quantity to be a constant, we must have

$$\left(p(d) \propto \frac{1}{d^2}\right)$$

which means that the received signal amplitude A_R must be proportional to the transmitter's amplitude A_T and inversely related to distance from the transmitter.

$$A_R = \frac{kA_T}{d} \tag{6.17}$$

for some value of the constant k. Thus, the further from the transmitter the receiver is located, the weaker the received signal. Whereas the attenuation found in wireline channels can be controlled by physical parameters and choice of transmission frequency, the inverse-distance attenuation found in wireless channels persists across all frequencies.

[5]This content is available online at <http://cnx.org/content/m0101/2.14/>.

Exercise 6.4 *(Solution on p. 272.)*

Why don't signals attenuate according to the inverse-square law in a conductor? What is the difference between the wireline and wireless cases?

The speed of propagation is governed by the dielectric constant μ_0 and magnetic permeability ϵ_0 of free space.

$$\begin{aligned} c &= \frac{1}{\sqrt{\mu_0 \epsilon_0}} \\ &= 3 \times 10^8 \text{m/s} \end{aligned} \quad (6.18)$$

Known familiarly as the speed of light, it sets an upper limit on how fast signals can propagate from one place to another. Because signals travel at a finite speed, a receiver senses a transmitted signal only after a time delay directly related to the propagation speed:

$$\Delta t = \frac{d}{c}$$

At the speed of light, a signal travels across the United States in 16 ms, a reasonably small time delay. If a lossless (zero space constant) coaxial cable connected the East and West coasts, this delay would be two to three times longer because of the slower propagation speed.

6.5 Line-of-Sight Transmission[6]

Long-distance transmission over either kind of channel encounters attenuation problems. Losses in wireline channels are explored in the Circuit Models module (Section 6.3), where repeaters can extend the distance between transmitter and receiver beyond what passive losses the wireline channel imposes. In wireless channels, not only does radiation loss occur (p. 215), but also one antenna may not "see" another because of the earth's curvature.

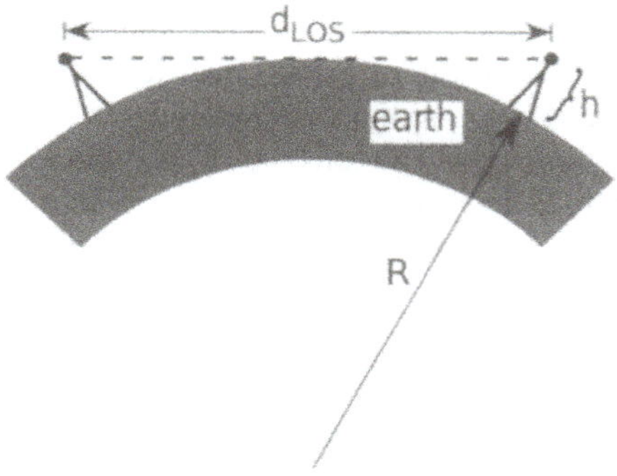

Figure 6.3: Two antennae are shown each having the same height. Line-of-sight transmission means the transmitting and receiving antennae can "see" each other as shown. The maximum distance at which they can see each other, d_{LOS}, occurs when the sighting line just grazes the earth's surface.

At the usual radio frequencies, propagating electromagnetic energy does not follow the earth's surface. **Line-of-sight** communication has the transmitter and receiver antennas in visual contact with each other.

[6]This content is available online at <http://cnx.org/content/m0538/2.13/>.

Assuming both antennas have height h above the earth's surface, maximum line-of-sight distance is

$$d_{LOS} = 2\sqrt{2hR + h^2} \approx 2\sqrt{2Rh} \tag{6.19}$$

where R is the earth's radius ($6.38 \times 10^6 m$).

Exercise 6.5 *(Solution on p. 272.)*

Derive the expression of line-of-sight distance using only the Pythagorean Theorem. Generalize it to the case where the antennas have different heights (as is the case with commercial radio and cellular telephone). What is the range of cellular telephone where the handset antenna has essentially zero height?

Exercise 6.6 *(Solution on p. 273.)*

Can you imagine a situation wherein global wireless communication is possible with only one transmitting antenna? In particular, what happens to wavelength when carrier frequency decreases?

Using a 100 m antenna would provide line-of-sight transmission over a distance of 71.4 km. Using such very tall antennas would provide wireless communication within a town or between closely spaced population centers. Consequently, **networks** of antennas sprinkle the countryside (each located on the highest hill possible) to provide long-distance wireless communications: Each antenna receives energy from one antenna and retransmits to another. This kind of network is known as a **relay network**.

6.6 The Ionosphere and Communications[7]

If we were limited to line-of-sight communications, long distance wireless communication, like ship-to-shore communication, would be impossible. At the turn of the century, Marconi, the inventor of wireless telegraphy, boldly tried such long distance communication without any evidence — either empirical or theoretical — that it was possible. When the experiment worked, but only at night, physicists scrambled to determine why (using Maxwell's equations, of course). It was Oliver Heaviside, a mathematical physicist with strong engineering interests, who hypothesized that an invisible electromagnetic "mirror" surrounded the earth. What he meant was that at optical frequencies (and others as it turned out), the mirror was transparent, but at the frequencies Marconi used, it reflected electromagnetic radiation back to earth. He had predicted the existence of the ionosphere, a plasma that encompasses the earth at altitudes h_i between 80 and 180 km that reacts to solar radiation: It becomes transparent at Marconi's frequencies during the day, but becomes a mirror at night when solar radiation diminishes. The maximum distance along the earth's surface that can be reached by a *single* ionospheric reflection is $2R \arccos\left(\frac{R}{R+h_i}\right)$, which ranges between 2,010 and 3,000 km when we substitute minimum and maximum ionospheric altitudes. This distance does not span the United States or cross the Atlantic; for transatlantic communication, at least two reflections would be required. The communication delay encountered with a single reflection in this channel is $\frac{2\sqrt{2Rh_i+{h_i}^2}}{c}$, which ranges between 6.8 and 10 ms, again a small time interval.

6.7 Communication with Satellites[8]

Global wireless communication relies on satellites. Here, ground stations transmit to orbiting satellites that amplify the signal and retransmit it back to earth. Satellites will move across the sky unless they are in **geosynchronous orbits**, where the time for one revolution about the equator exactly matches the earth's rotation time of one day. TV satellites would require the homeowner to continually adjust his or her antenna if the satellite weren't in geosynchronous orbit. Newton's equations applied to orbiting bodies predict that

[7]This content is available online at <http://cnx.org/content/m0539/2.10/>.
[8]This content is available online at <http://cnx.org/content/m0540/2.10/>.

the time T for one orbit is related to distance from the earth's center R as

$$R = \sqrt[3]{\frac{GMT^2}{4\pi^2}} \tag{6.20}$$

where G is the gravitational constant and M the earth's mass. Calculations yield $R = 42200km$, which corresponds to an altitude of $35700km$. This altitude greatly exceeds that of the ionosphere, requiring satellite transmitters to use frequencies that pass through it. Of great importance in satellite communications is the transmission delay. The time for electromagnetic fields to propagate to a geosynchronous satellite and return is 0.24 s, a significant delay.

Exercise 6.7 *(Solution on p. 273.)*

In addition to delay, the propagation attenuation encountered in satellite communication far exceeds what occurs in ionospheric-mirror based communication. Calculate the attenuation incurred by radiation going to the satellite (one-way loss) with that encountered by Marconi (total going up and down). Note that the attenuation calculation in the ionospheric case, assuming the ionosphere acts like a perfect mirror, is not a straightforward application of the propagation loss formula (p. 215).

6.8 Noise and Interference[9]

We have mentioned that communications are, to varying degrees, subject to interference and noise. It's time to be more precise about what these quantities are and how they differ.

Interference represents man-made signals. Telephone lines are subject to power-line interference (in the United States a distorted 60 Hz sinusoid). Cellular telephone channels are subject to adjacent-cell phone conversations using the same signal frequency. The problem with such interference is that it occupies the same frequency band as the desired communication signal, and has a similar structure.

Exercise 6.8 *(Solution on p. 273.)*

Suppose interference occupied a different frequency band; how would the receiver remove it?

We use the notation $i(t)$ to represent interference. Because interference has man-made structure, we can write an explicit expression for it that may contain some unknown aspects (how large it is, for example).

Noise signals have little structure and arise from both human and natural sources. Satellite channels are subject to deep space noise arising from electromagnetic radiation pervasive in the galaxy. Thermal noise plagues *all* electronic circuits that contain resistors. Thus, in receiving small amplitude signals, receiver amplifiers will most certainly add noise as they boost the signal's amplitude. All channels are subject to noise, and we need a way of describing such signals despite the fact we can't write a formula for the noise signal like we can for interference. The most widely used noise model is **white noise**. It is defined entirely by its frequency-domain characteristics.

- White noise has constant power at all frequencies.
- At each frequency, the phase of the noise spectrum is totally uncertain: It can be any value in between 0 and 2π, and its value at any frequency is unrelated to the phase at any other frequency.
- When noise signals arising from two different sources add, the resultant noise signal has a power equal to the sum of the component powers.

Because of the emphasis here on frequency-domain power, we are lead to define the **power spectrum**. Because of Parseval's Theorem[10], we define the power spectrum $P_s(f)$ of a non-noise signal $s(t)$ to be the magnitude-squared of its Fourier transform.

$$P_s(f) \equiv (|S(f)|)^2 \tag{6.21}$$

[9]This content is available online at <http://cnx.org/content/m0515/2.17/>.
[10]"Parseval's Theorem", (1) <http://cnx.org/content/m0047/latest/#parseval>

Integrating the power spectrum over any range of frequencies equals the power the signal contains in that band. Because signals *must* have negative frequency components that mirror positive frequency ones, we routinely calculate the power in a spectral band as the integral over positive frequencies multiplied by two.

$$\text{Power in } [f_1, f_2] = 2 \int_{f_1}^{f_2} P_s(f)\, df \tag{6.22}$$

Using the notation $n(t)$ to represent a noise signal's waveform, we define noise in terms of its power spectrum. For white noise, the power spectrum equals the constant $\frac{N_0}{2}$. With this definition, the power in a frequency band equals $N_0 (f_2 - f_1)$.

When we pass a signal through a linear, time-invariant system, the output's spectrum equals the product (p. 131) of the system's frequency response and the input's spectrum. Thus, the power spectrum of the system's output is given by

$$P_y(f) = (|H(f)|)^2 P_x(f) \tag{6.23}$$

This result applies to noise signals as well. When we pass white noise through a filter, the output is also a noise signal but with power spectrum $(|H(f)|)^2 \frac{N_0}{2}$.

6.9 Channel Models[11]

Both wireline and wireless channels share characteristics, allowing us to use a common model for how the channel affects transmitted signals.

- The transmitted signal is usually not filtered by the channel.
- The signal can be attenuated.
- The signal propagates through the channel at a speed equal to or less than the speed of light, which means that the channel delays the transmission.
- The channel may introduce additive interference and/or noise.

Letting α represent the attenuation introduced by the channel, the receiver's input signal is related to the transmitted one by

$$r(t) = \alpha x(t - \tau) + i(t) + n(t) \tag{6.24}$$

This expression corresponds to the system model for the channel shown in Figure 6.4. In this book, we shall assume that the noise is white.

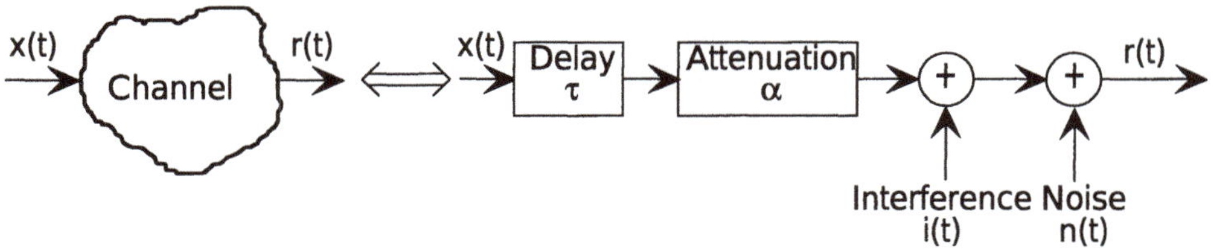

Figure 6.4: The channel component of the fundamental model of communication (Figure 1.4: Fundamental model of communication) has the depicted form. The attenuation is due to propagation loss. Adding the interference and noise is justified by the linearity property of Maxwell's equations.

[11]This content is available online at <http://cnx.org/content/m0516/2.10/>.

Exercise 6.9 *(Solution on p. 273.)*

Is this model for the channel linear?

As expected, the signal that emerges from the channel is corrupted, but does contain the transmitted signal. Communication system design begins with detailing the channel model, then developing the transmitter and receiver that best compensate for the channel's corrupting behavior. We characterize the channel's quality by the signal-to-interference ratio (*SIR*) and the signal-to-noise ratio (*SNR*). The ratios are computed according to the relative power of each *within the transmitted signal's bandwidth.* Assuming the signal $x(t)$'s spectrum spans the frequency interval $[f_l, f_u]$, these ratios can be expressed in terms of power spectra.

$$SIR = \frac{2\alpha^2 \int_0^{\infty} P_x(f)\, df}{2 \int_{f_l}^{f_u} P_i(f)\, df} \qquad (6.25)$$

$$SNR = \frac{2\alpha^2 \int_0^{\infty} P_x(f)\, df}{N_0 (f_u - f_l)} \qquad (6.26)$$

In most cases, the interference and noise powers do not vary for a given receiver. Variations in signal-to-interference and signal-to-noise ratios arise from the attenuation because of transmitter-to-receiver distance variations.

6.10 Baseband Communication[12]

We use analog communication techniques for analog message signals, like music, speech, and television. Transmission and reception of analog signals using analog results in an inherently noisy received signal (assuming the channel adds noise, which it almost certainly does).

The simplest form of analog communication is **baseband communication.**

POINT OF INTEREST: We use analog communication techniques for analog message signals, like music, speech, and television. Transmission and reception of analog signals using analog results in an inherently noisy received signal (assuming the channel adds noise, which it almost certainly does).

Here, the transmitted signal equals the message times a transmitter gain.

$$x(t) = Gm(t) \qquad (6.27)$$

An example, which is somewhat out of date, is the wireline telephone system. You don't use baseband communication in wireless systems simply because low-frequency signals do not radiate well. The receiver in a baseband system can't do much more than filter the received signal to remove out-of-band noise (interference is small in wireline channels). Assuming the signal occupies a bandwidth of W Hz (the signal's spectrum extends from zero to W), the receiver applies a lowpass filter having the same bandwidth, as shown in Figure 6.5.

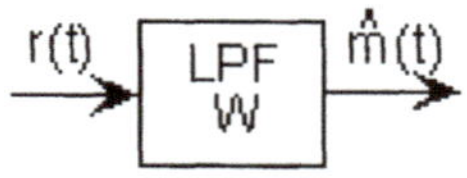

Figure 6.5: The receiver for baseband communication systems is quite simple: a lowpass filter having the same bandwidth as the signal.

[12]This content is available online at <http://cnx.org/content/m0517/2.17/>.

We use the **signal-to-noise ratio** of the receiver's output $\hat{m}(t)$ to evaluate any analog-message communication system. Assume that the channel introduces an attenuation α and white noise of spectral height $\frac{N_0}{2}$. The filter does not affect the signal component—we assume its gain is unity—but does filter the noise, removing frequency components above W Hz. In the filter's output, the received signal power equals $\alpha^2 G^2 power(m)$ and the noise power $N_0 W$, which gives a signal-to-noise ratio of

$$SNR_{baseband} = \frac{\alpha^2 G^2 power(m)}{N_0 W} \tag{6.28}$$

The signal power $power(m)$ will be proportional to the bandwidth W; thus, in baseband communication the signal-to-noise ratio varies only with transmitter gain and channel attenuation and noise level.

6.11 Modulated Communication[13]

Especially for wireless channels, like commercial radio and television, but also for wireline systems like cable television, an analog message signal must be **modulated**: The transmitted signal's spectrum occurs at much higher frequencies than those occupied by the signal.

> Point of Interest: We use analog communication techniques for analog message signals, like music, speech, and television. Transmission and reception of analog signals using analog results in an inherently noisy received signal (assuming the channel adds noise, which it almost certainly does).

The key idea of modulation is to affect the amplitude, frequency or phase of what is known as the **carrier** sinusoid. Frequency modulation (FM) and less frequently used phase modulation (PM) are not discussed here; we focus on amplitude modulation (AM). The amplitude modulated message signal has the form

$$x(t) = A_c (1 + m(t)) \cos(2\pi f_c t) \tag{6.29}$$

where f_c is the **carrier frequency** and A_c the **carrier amplitude**. Also, the signal's amplitude is assumed to be less than one: $|m(t)| < 1$. From our previous exposure to amplitude modulation (see the Fourier Transform example (Example 4.5)), we know that the transmitted signal's spectrum occupies the frequency range $[f_c - W, f_c + W]$, assuming the signal's bandwidth is W Hz (see the figure (Figure 6.6)). The carrier frequency is usually much larger than the signal's highest frequency: $(f_c \gg W)$, which means that the transmitter antenna and carrier frequency are chosen jointly during the design process.

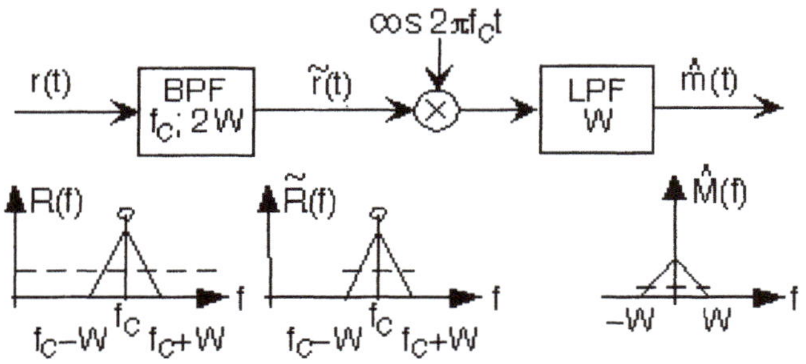

Figure 6.6: The AM coherent receiver along with the spectra of key signals is shown for the case of a triangular-shaped signal spectrum. The dashed line indicates the white noise level. Note that the filters' characteristics — cutoff frequency and center frequency for the bandpass filter — must be match to the modulation and message parameters.

[13]This content is available online at <http://cnx.org/content/m0518/2.24/>.

Ignoring the attenuation and noise introduced by the channel for the moment, reception of an amplitude modulated signal is quite easy (see Problem 4.17). The so-called **coherent** receiver multiplies the input signal by a sinusoid and lowpass-filters the result (Figure 6.6).

$$\begin{aligned}\hat{m}(t) &= LPF(x(t)\cos(2\pi f_c t)) \\ &= LPF\left(A_c(1+m(t))\cos^2(2\pi f_c t)\right)\end{aligned} \quad (6.30)$$

Because of our trigonometric identities, we know that

$$\cos^2(2\pi f_c t) = \frac{1}{2}(1+\cos(2\pi 2 f_c t)) \quad (6.31)$$

At this point, the message signal is multiplied by a constant and a sinusoid at twice the carrier frequency. Multiplication by the constant term returns the message signal to baseband (where we want it to be!) while multiplication by the double-frequency term yields a very high frequency signal. The lowpass filter removes this high-frequency signal, leaving only the baseband signal. Thus, the received signal is

$$\hat{m}(t) = \frac{A_c}{2}(1+m(t)) \quad (6.32)$$

Exercise 6.10 *(Solution on p. 273.)*

This derivation relies solely on the time domain; derive the same result in the frequency domain. You won't need the trigonometric identity with this approach.

Because it is so easy to remove the constant term by electrical means—we insert a capacitor in series with the receiver's output—we typically ignore it and concentrate on the signal portion of the receiver's output when calculating signal-to-noise ratio.

6.12 Signal-to-Noise Ratio of an Amplitude-Modulated Signal[14]

When we consider the much more realistic situation when we have a channel that introduces attenuation and noise, we can make use of the just-described receiver's linear nature to directly derive the receiver's output. The attenuation affects the output in the same way as the transmitted signal: It scales the output signal by the same amount. The white noise, on the other hand, should be filtered from the received signal before demodulation. We must thus insert a bandpass filter having bandwidth $2W$ and center frequency f_c: This filter has no effect on the received signal-related component, but does remove out-of-band noise power. As shown in the triangular-shaped signal spectrum (Figure 6.6), we apply coherent receiver to this filtered signal, with the result that the demodulated output contains noise that cannot be removed: It lies in the same spectral band as the signal.

As we derive the signal-to-noise ratio in the demodulated signal, let's also calculate the signal-to-noise ratio of the bandpass filter's output $\tilde{r}(t)$. The signal component of $\tilde{r}(t)$ equals $\alpha A_c m(t)\cos(2\pi f_c t)$. This signal's Fourier transform equals

$$\frac{\alpha A_c}{2}(M(f+f_c)+M(f-f_c)) \quad (6.33)$$

making the power spectrum,

$$\frac{\alpha^2 {A_c}^2}{4}\left((|M(f+f_c)|)^2+(|M(f-f_c)|)^2\right) \quad (6.34)$$

Exercise 6.11 *(Solution on p. 273.)*

If you calculate the magnitude-squared of the first equation, you don't obtain the second unless you make an assumption. What is it?

[14]This content is available online at <http://cnx.org/content/m0541/2.16/>.

Thus, the total signal-related power in $\tilde{r}(t)$ is $\frac{\alpha^2 {A_c}^2}{2} power(m)$. The noise power equals the integral of the noise power spectrum; because the power spectrum is constant over the transmission band, this integral equals the noise amplitude N_0 times the filter's bandwidth $2W$. The so-called *received signal-to-noise ratio* — the signal-to-noise ratio after the *de rigeur* front-end bandpass filter and before demodulation — equals

$$SNR_r = \frac{\alpha^2 {A_c}^2 power(m)}{4 N_0 W} \tag{6.35}$$

The demodulated signal $\hat{m}(t) = \frac{\alpha A_c m(t)}{2} + n_{\text{out}}(t)$. Clearly, the signal power equals $\frac{\alpha^2 {A_c}^2 power(m)}{4}$. To determine the noise power, we must understand how the coherent demodulator affects the bandpass noise found in $\tilde{r}(t)$. Because we are concerned with noise, we must deal with the power spectrum since we don't have the Fourier transform available to us. Letting $P(f)$ denote the power spectrum of $\tilde{r}(t)$'s noise component, the power spectrum after multiplication by the carrier has the form

$$\frac{P(f+f_c)+P(f-f_c)}{4} \tag{6.36}$$

The delay and advance in frequency indicated here results in two spectral noise bands falling in the low-frequency region of lowpass filter's passband. Thus, the total noise power in this filter's output equals $\left(2 \cdot \frac{N_0}{2} \cdot W \cdot 2 \cdot \frac{1}{4}\right) = \frac{N_0 W}{2}$. The signal-to-noise ratio of the receiver's output thus equals

$$\begin{aligned} SNR_{\hat{m}} &= \frac{\alpha^2 {A_c}^2 power(m)}{2 N_0 W} \\ &= 2 SNR_r \end{aligned} \tag{6.37}$$

Let's break down the components of this signal-to-noise ratio to better appreciate how the channel and the transmitter parameters affect communications performance. Better performance, as measured by the *SNR*, occurs as it increases.

- More transmitter power — increasing A_c — increases the signal-to-noise ratio proportionally.
- The carrier frequency f_c has no effect on SNR, but we have assumed that $(f_c \gg W)$.
- The signal bandwidth W enters the signal-to-noise expression in two places: implicitly through the signal power and explicitly in the expression's denominator. *If the signal spectrum had a constant amplitude* as we increased the bandwidth, signal power would increase proportionally. On the other hand, our transmitter enforced the criterion that signal amplitude was constant (Section 6.7). Signal amplitude essentially equals the integral of the magnitude of the signal's spectrum.

  NOTE: This result isn't exact, but we do know that $m(0) = \int_{-\infty}^{\infty} M(f)\,df$.

  Enforcing the signal amplitude specification means that as the signal's bandwidth increases we must decrease the spectral amplitude, with the result that the signal power remains constant. Thus, increasing signal bandwidth does indeed decrease the signal-to-noise ratio of the receiver's output.
- Increasing channel attenuation — moving the receiver farther from the transmitter — decreases the signal-to-noise ratio as the square. Thus, signal-to-noise ratio decreases as distance-squared between transmitter and receiver.
- Noise added by the channel adversely affects the signal-to-noise ratio.

In summary, amplitude modulation provides an effective means for sending a bandlimited signal from one place to another. For wireline channels, using baseband or amplitude modulation makes little difference in terms of signal-to-noise ratio. For wireless channels, amplitude modulation is the only alternative. The one AM parameter that does not affect signal-to-noise ratio is the carrier frequency f_c: We can choose any value we want so long as the transmitter and receiver use the same value. However, suppose someone else wants to use AM and chooses the same carrier frequency. The two resulting transmissions will add, and *both* receivers will produce the sum of the two signals. What we clearly need to do is talk to the other party, and agree to use separate carrier frequencies. As more and more users wish to use radio, we need a forum for agreeing on carrier frequencies and on signal bandwidth. On earth, this forum is the government. In the United States, the Federal Communications Commission (FCC) strictly controls the use of the electromagnetic spectrum for communications. Separate frequency bands are allocated for commercial AM, FM, cellular telephone (the analog version of which is AM), short wave (also AM), and satellite communications.

Exercise 6.12 *(Solution on p. 273.)*

Suppose all users agree to use the same signal bandwidth. How closely can the carrier frequencies be while avoiding communications crosstalk? What is the signal bandwidth for commercial AM? How does this bandwidth compare to the speech bandwidth?

6.13 Digital Communication[15]

Effective, error-free transmission of a sequence of bits—a **bit stream** $\{b(0), b(1), \dots\}$—is the goal here. We found that analog schemes, as represented by amplitude modulation, always yield a received signal containing noise as well as the message signal when the channel adds noise. Digital communication schemes are very different. Once we decide how to represent bits by analog signals that can be transmitted over wireline (like a computer network) or wireless (like digital cellular telephone) channels, we will then develop a way of tacking on communication bits to the message bits that will reduce channel-induced errors greatly. In theory, digital communication errors can be zero, even though the channel adds noise!

We represent a bit by associating one of two specific analog signals with the bit's value. Thus, if $b(n) = 0$, we transmit the signal $s_0(t)$; if $b(n) = 1$, send $s_1(t)$. These two signals comprise the **signal set** for digital communication and are designed with the channel and bit stream in mind. In virtually every case, these signals have a finite duration T common to both signals; this duration is known as the **bit interval**. Exactly what signals we use ultimately affects how well the bits can be received. Interestingly, baseband and modulated signal sets can yield the same performance. Other considerations determine how signal set choice affects digital communication performance.

Exercise 6.13 *(Solution on p. 273.)*

What is the expression for the signal arising from a digital transmitter sending the bit stream $b(n)$, $n = \{\dots, -1, 0, 1, \dots\}$ using the signal set $s_0(t)$, $s_1(t)$, each signal of which has duration T?

6.14 Binary Phase Shift Keying[16]

A commonly used example of a signal set consists of pulses that are negatives of each other (Figure 6.7).

$$\begin{aligned} s_0(t) &= Ap_T(t) \\ s_1(t) &= -(Ap_T(t)) \end{aligned} \qquad (6.38)$$

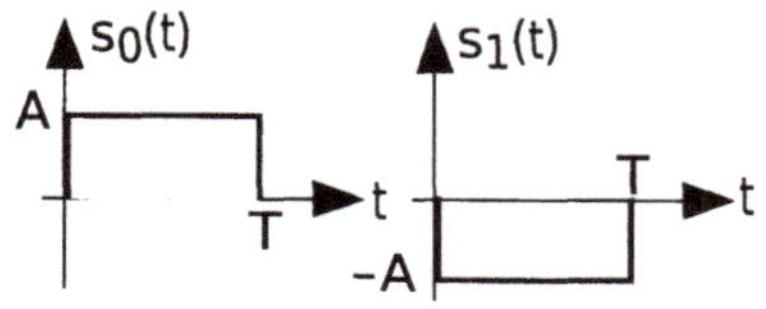

Figure 6.7

[15]This content is available online at <http://cnx.org/content/m0519/2.10/>.
[16]This content is available online at <http://cnx.org/content/m10280/2.13/>.

Here, we have a baseband signal set suitable for wireline transmission. The entire bit stream $b(n)$ is represented by a sequence of these signals. Mathematically, the transmitted signal has the form

$$x(t) = \sum_{n} \left((-1)^{b(n)} A p_T (t - nT) \right) \tag{6.39}$$

and graphically Figure 6.8 shows what a typical transmitted signal might be.

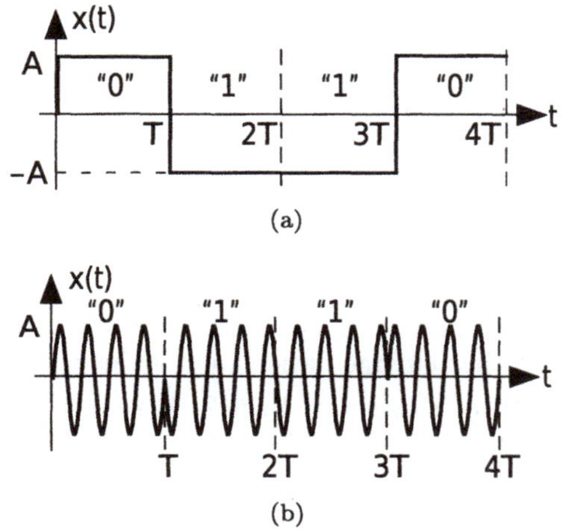

Figure 6.8: The upper plot shows how a baseband signal set for transmitting the bit sequence 0110. The lower one shows an amplitude-modulated variant suitable for wireless channels.

This way of representing a bit stream—changing the bit changes the sign of the transmitted signal—is known as **binary phase shift keying** and abbreviated BPSK. The name comes from concisely expressing this popular way of communicating digital information. The word "binary" is clear enough (one binary-valued quantity is transmitted during a bit interval). Changing the sign of sinusoid amounts to changing—shifting—the phase by π (although we don't have a sinusoid yet). The word "keying" reflects back to the first electrical communication system, which happened to be digital as well: the telegraph.

The **datarate** R of a digital communication system is how frequently an information bit is transmitted. In this example it equals the reciprocal of the bit interval: $R = \frac{1}{T}$. Thus, for a 1 Mbps (megabit per second) transmission, we must have $T = 1\mu s$.

The choice of signals to represent bit values is arbitrary to some degree. Clearly, we do not want to choose signal set members to be the same; we couldn't distinguish bits if we did so. We could also have made the negative-amplitude pulse represent a 0 and the positive one a 1. This choice is indeed arbitrary and will have no effect on performance *assuming* the receiver knows which signal represents which bit. As in all communication systems, we design transmitter and receiver together.

A simple signal set for both wireless and wireline channels amounts to amplitude modulating a baseband signal set (more appropriate for a wireline channel) by a carrier having a frequency harmonic with the bit interval.

$$\begin{aligned} s_0(t) &= A p_T(t) \sin\left(\frac{2\pi k t}{T}\right) \\ s_1(t) &= -\left(A p_T(t) \sin\left(\frac{2\pi k t}{T}\right)\right) \end{aligned} \tag{6.40}$$

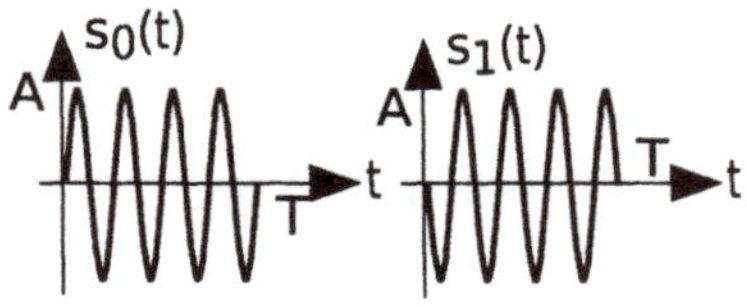

Figure 6.9

Exercise 6.14 *(Solution on p. 273.)*

What is the value of k in this example?

This signal set is also known as a BPSK signal set. We'll show later that indeed both signal sets provide identical performance levels when the signal-to-noise ratios are equal.

Exercise 6.15 *(Solution on p. 273.)*

Write a formula, in the style of the baseband signal set, for the transmitted signal as shown in the plot of the baseband signal set[17] that emerges when we use this modulated signal.

What is the transmission bandwidth of these signal sets? We need only consider the baseband version as the second is an amplitude-modulated version of the first. The bandwidth is determined by the bit sequence. If the bit sequence is constant—always 0 or always 1—the transmitted signal is a constant, which has zero bandwidth. The worst-case—bandwidth consuming—bit sequence is the alternating one shown in Figure 6.10. In this case, the transmitted signal is a square wave having a period of $2T$.

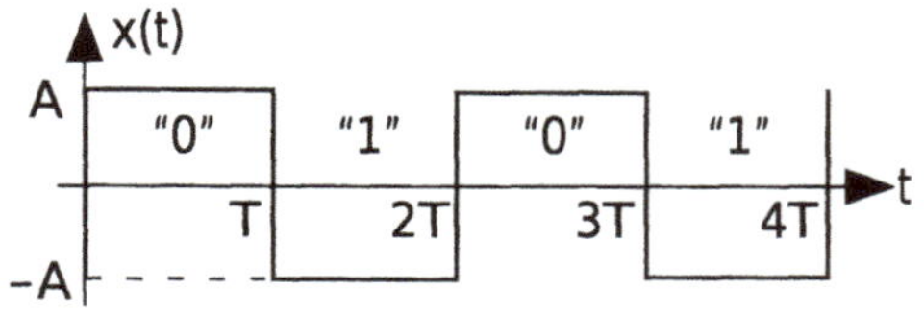

Figure 6.10: Here we show the transmitted waveform corresponding to an alternating bit sequence.

From our work in Fourier series, we know that this signal's spectrum contains odd-harmonics of the fundamental, which here equals $\frac{1}{2T}$. Thus, strictly speaking, the signal's bandwidth is infinite. In practical terms, we use the 90%-power bandwidth to assess the effective range of frequencies consumed by the signal. The first and third harmonics contain that fraction of the total power, meaning that the effective bandwidth of our baseband signal is $\frac{3}{2T}$ or, expressing this quantity in terms of the datarate, $\frac{3R}{2}$. Thus, a digital communications signal requires more bandwidth than the datarate: a 1 Mbps baseband system requires a bandwidth of at least 1.5 MHz. Listen carefully when someone describes the transmission bandwidth of digital communication systems: Did they say "megabits" or "megahertz"?

Exercise 6.16 *(Solution on p. 274.)*

Show that indeed the first and third harmonics contain 90% of the transmitted power. If the receiver uses a front-end filter of bandwidth $\frac{3}{2T}$, what is the total harmonic distortion of the received signal?

Exercise 6.17 *(Solution on p. 274.)*

What is the 90% transmission bandwidth of the modulated signal set?

[17] "Signal Sets", Figure 2 <http://cnx.org/content/m0542/latest/#fig1001>

6.15 Frequency Shift Keying[18]

In **frequency-shift keying** (FSK), the bit affects the frequency of a carrier sinusoid.

$$\begin{aligned} s_0(t) &= Ap_T(t)\sin(2\pi f_0 t) \\ s_1(t) &= Ap_T(t)\sin(2\pi f_1 t) \end{aligned} \quad (6.41)$$

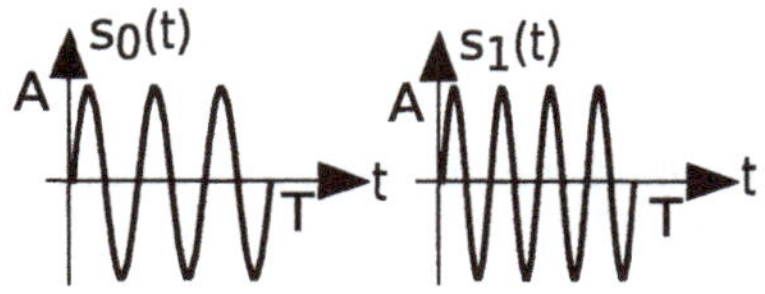

Figure 6.11

The frequencies f_0, f_1 are usually harmonically related to the bit interval. In the depicted example, $f_0 = \frac{3}{T}$ and $f_1 = \frac{4}{T}$. As can be seen from the transmitted signal for our example bit stream (Figure 6.12), the transitions at bit interval boundaries are smoother than those of BPSK.

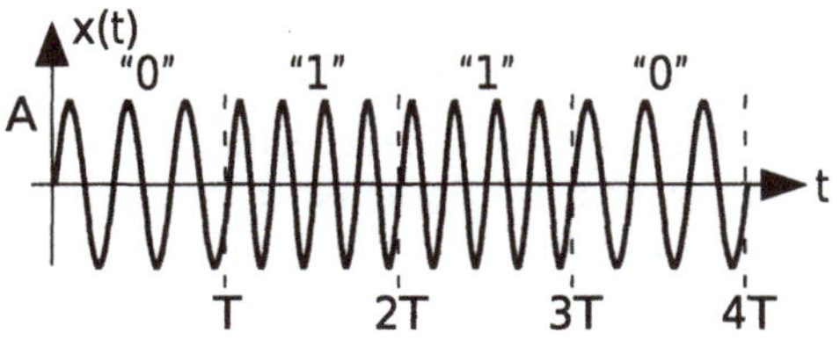

Figure 6.12: This plot shows the FSK waveform for same bitstream used in the BPSK example (Figure 6.8).

To determine the bandwidth required by this signal set, we again consider the alternating bit stream. Think of it as two signals added together: The first comprised of the signal $s_0(t)$, the zero signal, $s_0(t)$, zero, *etc.*, and the second having the same structure but interleaved with the first and containing $s_1(t)$ (Figure 6.13).

Each component can be thought of as a fixed-frequency sinusoid multiplied by a square wave of period $2T$ that alternates between one and zero. This baseband square wave has the same Fourier spectrum as our BPSK example, but with the addition of the constant term c_0. This quantity's presence changes the number of Fourier series terms required for the 90% bandwidth: Now we need only include the zero and first harmonics to achieve it. The bandwidth thus equals, with $f_0 < f_1$, $f_1 + \frac{1}{2T} - \left(f_0 - \frac{1}{2T}\right) = f_1 - f_0 + \frac{1}{T}$. If the two frequencies are harmonics of the bit-interval duration, $f_0 = \frac{k_0}{T}$ and $f_1 = \frac{k_1}{T}$ with $k_1 > k_0$, the bandwidth equals $\frac{k_1 + (-k_0) + 1}{T}$. If the difference between harmonic numbers is 1, then the FSK bandwidth is

[18] This content is available online at <http://cnx.org/content/m0545/2.11/>.

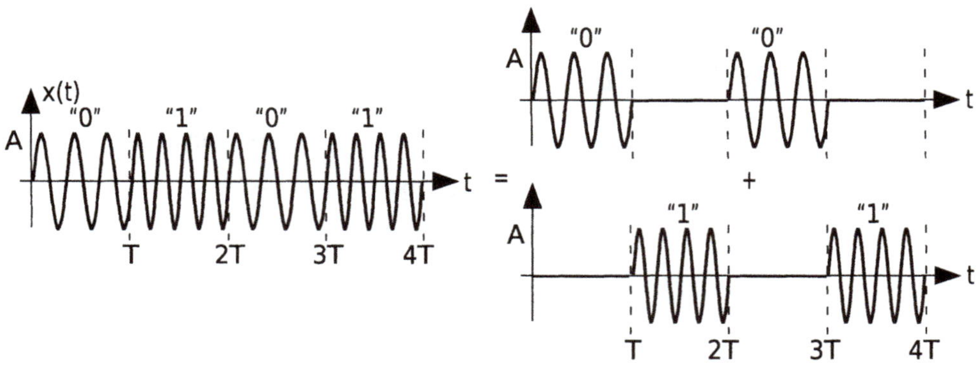

Figure 6.13: The depicted decomposition of the FSK-modulated alternating bit stream into its frequency components simplifies the calculation of its bandwidth.

smaller than the BPSK bandwidth. If the difference is 2, the bandwidths are equal and larger differences produce a transmission bandwidth larger than that resulting from using a BPSK signal set.

6.16 Digital Communication Receivers[19]

The receiver interested in the transmitted bit stream must perform two tasks when received waveform $r(t)$ begins.

- It must determine when bit boundaries occur: The receiver needs to **synchronize** with the transmitted signal. Because transmitter and receiver are designed in concert, both use the same value for the bit interval T. Synchronization can occur because the transmitter begins sending with a reference bit sequence, known as the **preamble**. This reference bit sequence is usually the alternating sequence as shown in the square wave example [20] and in the FSK example (Figure 6.13). The receiver knows what the preamble bit sequence is and uses it to determine when bit boundaries occur. This procedure amounts to what in digital hardware as **self-clocking signaling**: The receiver of a bit stream must derive the clock — when bit boundaries occur — from its input signal. Because the receiver usually does not determine which bit was sent until synchronization occurs, it does not know when during the preamble it obtained synchronization. The transmitter signals the end of the preamble by switching to a second bit sequence. The second preamble phase informs the receiver that data bits are about to come and that the preamble is almost over.
- Once synchronized and data bits are transmitted, the receiver must then determine every T seconds what bit was transmitted during the previous bit interval. We focus on this aspect of the digital receiver because this strategy is also used in synchronization.

The receiver for digital communication is known as a **matched filter**.

[19]This content is available online at <http://cnx.org/content/m0520/2.17/>.
[20]"Transmission Bandwidth", Figure 1 <http://cnx.org/content/m0544/latest/#fig1003>

Optimal receiver structure

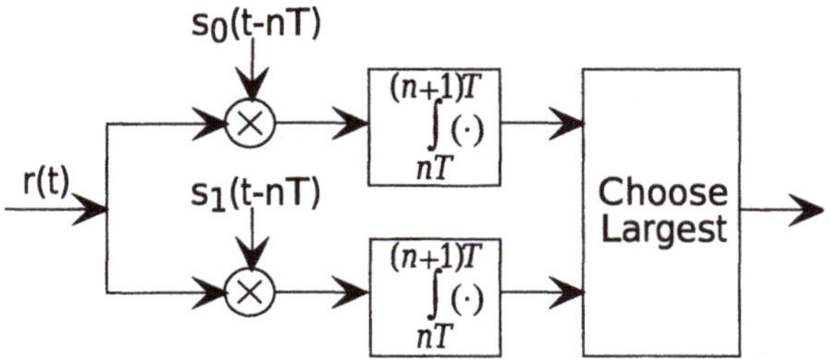

Figure 6.14: The optimal receiver structure for digital communication faced with additive white noise channels is the depicted matched filter.

This receiver, shown in Figure 6.14 (Optimal receiver structure), multiplies the received signal by each of the possible members of the transmitter signal set, integrates the product over the bit interval, and compares the results. Whichever path through the receiver yields the largest value corresponds to the receiver's decision as to what bit was sent during the previous bit interval. For the next bit interval, the multiplication and integration begins again, with the next bit decision made at the end of the bit interval. Mathematically, the received value of $b(n)$, which we label $\hat{b}(n)$, is given by

$$\hat{b}(n) = \arg\max_{i} \int_{nT}^{(n+1)T} r(t)\, s_i(t)\, dt \tag{6.42}$$

You may not have seen the arg max notation before. $\max_i \{\cdot\}$ yields the maximum value of its argument with respect to the index i. $\arg\max_i$ equals the value of the index that yields the maximum. Note that the precise numerical value of the integrator's output does not matter; what does matter is its value relative to the other integrator's output.

Let's assume a perfect channel for the moment: The received signal equals the transmitted one. If bit 0 were sent using the baseband BPSK signal set, the integrator outputs would be

$$\int_{nT}^{(n+1)T} r(t)\, s_0(t)\, dt = A^2T \tag{6.43}$$

$$\int_{nT}^{(n+1)T} r(t)\, s_1(t)\, dt = -\left(A^2T\right)$$

If bit 1 were sent,

$$\int_{nT}^{(n+1)T} r(t)\, s_0(t)\, dt = -\left(A^2T\right) \tag{6.44}$$

$$\int_{nT}^{(n+1)T} r(t)\, s_1(t)\, dt = A^2T$$

Exercise 6.18 *(Solution on p. 274.)*

Can you develop a receiver for BPSK signal sets that requires only one multiplier-integrator combination?

Exercise 6.19 *(Solution on p. 274.)*

What is the corresponding result when the amplitude-modulated BPSK signal set is used?

Clearly, this receiver would always choose the bit correctly. Channel attenuation would not affect this correctness; it would only make the values smaller, but all that matters is which is largest.

6.17 Digital Communication in the Presence of Noise[21]

When we incorporate additive noise into our channel model, so that $r(t) = \alpha s_i(t) + n(t)$, errors can creep in. If the transmitter sent bit 0 using a BPSK signal set (Section 6.14), the integrators' outputs in the matched filter receiver (Figure 6.14: Optimal receiver structure) would be:

$$\int_{nT}^{(n+1)T} r(t)\, s_0(t)\, dt = \alpha A^2 T + \int_{nT}^{(n+1)T} n(t)\, s_0(t)\, dt \tag{6.45}$$

$$\int_{nT}^{(n+1)T} r(t)\, s_1(t)\, dt = (-\alpha) A^2 T + \int_{nT}^{(n+1)T} n(t)\, s_1(t)\, dt$$

It is the quantities containing the noise terms that cause errors in the receiver's decision-making process. Because they involve noise, the values of these integrals are random quantities drawn from some probability distribution that vary erratically from bit interval to bit interval. Because the noise has zero average value and has an equal amount of power in all frequency bands, the values of the integrals will hover about zero. What is important is how much they vary. If the noise is such that its integral term is more negative than $\alpha A^2 T$, then the receiver will make an error, deciding that the transmitted zero-valued bit was indeed a one. The probability that this situation occurs depends on three factors:

- *Signal Set Choice* — The difference between the signal-dependent terms in the integrators' outputs (equations (6.45)) defines how large the noise term must be for an incorrect receiver decision to result. What affects the probability of such errors occurring is the square of this difference in comparison to the noise term's variability. For our BPSK baseband signal set, the signal-related value is $4\alpha^2 A^4 T^2$.
- *Variability of the Noise Term* — We quantify variability by the average value of its square, which is essentially the noise term's power. This calculation is best performed in the frequency domain and equals
$$power\left(\int_{nT}^{(n+1)T} n(t)\, s_0(t)\, dt\right) = \int_{-\infty}^{\infty} \frac{N_0}{2} (|S_0(f)|)^2 df$$
Because of Parseval's Theorem, we know that $\int_{-\infty}^{\infty} (|S_0(f)|)^2 df = \int_{-\infty}^{\infty} {s_0}^2(t)\, dt$, which for the baseband signal set equals $A^2 T$. Thus, the noise term's power is $\frac{N_0 A^2 T}{2}$.
- *Probability Distribution of the Noise Term* — The value of the noise terms relative to the signal terms and the probability of their occurrence directly affect the likelihood that a receiver error will occur. For the white noise we have been considering, the underlying distributions are Gaussian. The probability the receiver makes an error on any bit transmission equals:
$$p_e = Q\left(\frac{1}{2}\sqrt{\frac{\text{square of signal difference}}{\text{noise term power}}}\right) \tag{6.46}$$
$$p_e = Q\left(\sqrt{\frac{2\alpha^2 A^2 T}{N_0}}\right)$$
Here $Q(\cdot)$ is the integral $Q(x) = \frac{1}{\sqrt{2\pi}} \int_x^{\infty} e^{-\left(\frac{\alpha^2}{2}\right)} d\alpha$. This integral has no closed form expression, but it can be accurately computed. As Figure 6.15 illustrates, $Q(\cdot)$ is a decreasing, very nonlinear function.

[21] This content is available online at <http://cnx.org/content/m0546/2.12/>.

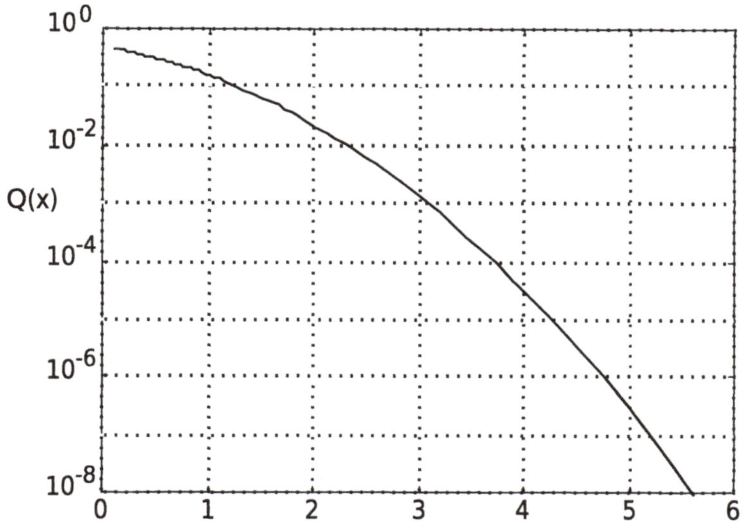

Figure 6.15: The function $Q(x)$ is plotted in semilogarithmic coordinates. Note that it decreases very rapidly for small increases in its arguments. For example, when x increases from 4 to 5, $Q(x)$ decreases by a factor of 100.

The term A^2T equals the energy expended by the transmitter in sending the bit; we label this term E_b. We arrive at a concise expression for the probability the matched filter receiver makes a bit-reception error.

$$p_e = Q\left(\sqrt{\frac{2\alpha^2 E_b}{N_0}}\right) \tag{6.47}$$

Figure 6.16 shows how the receiver's error rate varies with the signal-to-noise ratio $\frac{\alpha^2 E_b}{N_0}$.

Exercise 6.20 ***(Solution on p. 274.)***

Derive the probability of error expression for the modulated BPSK signal set, and show that its performance identically equals that of the baseband BPSK signal set.

6.18 Digital Communication System Properties[22]

Results from the Receiver Error module (Section 6.17) reveals several properties about digital communication systems.

- As the received signal becomes increasingly noisy, whether due to increased distance from the transmitter (smaller α) or to increased noise in the channel (larger N_0), the probability the receiver makes an error approaches 1/2. In such situations, the receiver performs only slightly better than the "receiver" that ignores what was transmitted and merely guesses what bit was transmitted. Consequently, it becomes almost impossible to communicate information when digital channels become noisy.

[22]This content is available online at <http://cnx.org/content/m10282/2.9/>.

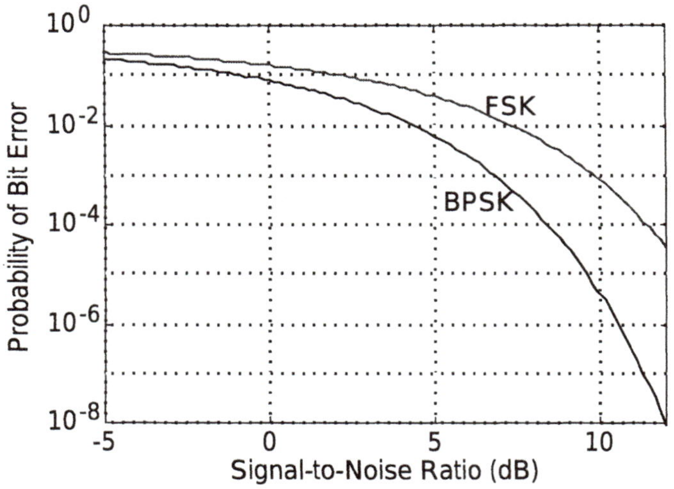

Figure 6.16: The probability that the matched-filter receiver makes an error on any bit transmission is plotted against the signal-to-noise ratio of the received signal. The upper curve shows the performance of the FSK signal set, the lower (and therefore better) one the BPSK signal set.

- As the signal-to-noise ratio increases, performance gains–smaller probability of error p_e – can be easily obtained. At a signal-to-noise ratio of 12 dB, the probability the receiver makes an error equals 10^{-8}. In words, one out of one hundred million bits will, on the average, be in error.
- Once the signal-to-noise ratio exceeds about 5 dB, the error probability decreases dramatically. Adding 1 dB improvement in signal-to-noise ratio can result in a factor of 10 smaller p_e.
- Signal set choice can make a significant difference in performance. All BPSK signal sets, baseband or modulated, yield the same performance for the same bit energy. The BPSK signal set does perform much better than the FSK signal set once the signal-to-noise ratio exceeds about 5 dB.

Exercise 6.21 ***(Solution on p. 274.)***

Derive the expression for the probability of error that would result if the FSK signal set were used.

The matched-filter receiver provides impressive performance once adequate signal-to-noise ratios occur. You might wonder whether another receiver might be better. The answer is that the matched-filter receiver is optimal: *No other receiver can provide a smaller probability of error than the matched filter regardless of the SNR.* Furthermore, no signal set can provide better performance than the BPSK signal set, where the signal representing a bit is the negative of the signal representing the other bit. The reason for this result rests in the dependence of probability of error p_e on the difference between the noise-free integrator outputs: For a given E_b, no other signal set provides a greater difference.

How small should the error probability be? Out of N transmitted bits, on the average Np_e bits will be received in error. Do note the phrase "on the average" here: Errors occur randomly because of the noise introduced by the channel, and we can only predict the probability of occurrence. Since bits are transmitted at a rate R, errors occur at an average frequency of Rp_e. Suppose the error probability is an impressively small number like 10^{-6}. Data on a computer network like Ethernet is transmitted at a rate $R = 100Mbps$, which means that errors would occur roughly 100 per second. This error rate is very high, requiring a much smaller p_e to achieve a more acceptable average occurrence rate for errors occurring. Because Ethernet is a wireline channel, which means the channel noise is small and the attenuation low, obtaining very small error probabilities is not difficult. We do have some tricks up our sleeves, however, that can essentially reduce the error rate to zero *without* resorting to expending a large amount of energy at the transmitter. We need to understand digital channels (Section 6.19) and Shannon's Noisy Channel Coding Theorem (Section 6.30).

6.19 Digital Channels[23]

Let's review how digital communication systems work within the Fundamental Model of Communication (Figure 1.4: Fundamental model of communication). As shown in Figure 6.17 (DigMC), the message is a single bit. The entire analog transmission/reception system, which is discussed in Digital Communication (Section 6.13), Signal Sets[24], BPSK Signal Set[25], Transmission Bandwidth[26], Frequency Shift Keying (Section 6.15), Digital Communication Receivers (Section 6.16), Factors in Receiver Error (Section 6.17), Digital Communication System Properties[27], and Error Probability[28], can be lumped into a single system known as the digital channel.

DigMC

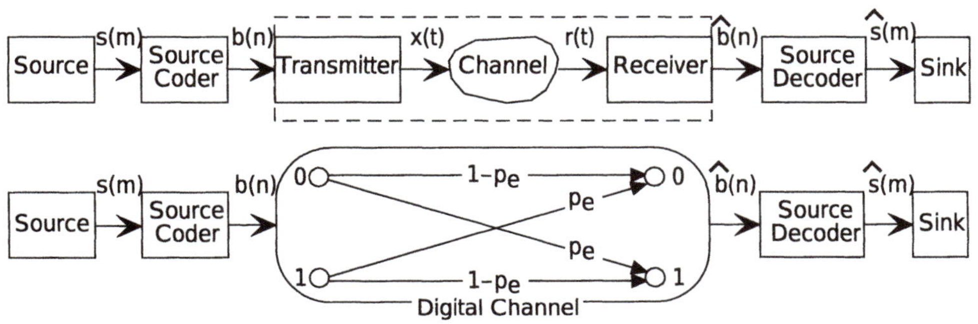

Figure 6.17: The steps in transmitting digital information are shown in the upper system, the Fundamental Model of Communication. The symbolic-valued signal $s(m)$ forms the message, and it is encoded into a bit sequence $b(n)$. The indices differ because more than one bit/symbol is usually required to represent the message by a bitstream. Each bit is represented by an analog signal, transmitted through the (unfriendly) channel, and received by a matched-filter receiver. From the received bitstream $\hat{b}(n)$ the received symbolic-valued signal $\hat{s}(m)$ is derived. The lower block diagram shows an equivalent system wherein the analog portions are combined and modeled by a transition diagram, which shows how each transmitted bit could be received. For example, transmitting a 0 results in the reception of a 1 with probability p_e (an error) or a 0 with probability $1 - p_e$ (no error).

Digital channels are described by **transition diagrams**, which indicate the output alphabet symbols that result for each possible transmitted symbol and the probabilities of the various reception possibilities. The probabilities on transitions coming from the same symbol must sum to one. For the matched-filter receiver and the signal sets we have seen, the depicted transition diagram, known as a **binary symmetric channel**, captures how transmitted bits are received. The probability of error p_e is the sole parameter of the digital channel, and it encapsulates signal set choice, channel properties, and the matched-filter receiver. With this simple but entirely accurate model, we can concentrate on how bits are received.

[23] This content is available online at <http://cnx.org/content/m0102/2.13/>.
[24] "Signal Sets" <http://cnx.org/content/m0542/latest/>
[25] "BPSK signal set" <http://cnx.org/content/m0543/latest/>
[26] "Transmission Bandwidth" <http://cnx.org/content/m0544/latest/>
[27] "Digital Communcation System Properties" <http://cnx.org/content/m0547/latest/>
[28] "Error Probability" <http://cnx.org/content/m0548/latest/>

6.20 Entropy[29]

Communication theory has been formulated best for symbolic-valued signals. Claude Shannon[30] published in 1948 *The Mathematical Theory of Communication*, which became the cornerstone of digital communication. He showed the power of **probabilistic models** for symbolic-valued signals, which allowed him to quantify the information present in a signal. In the simplest signal model, each symbol can occur at index n with a probability $Pr[a_k]$, $k = \{0, \dots, K-1\}$. What this model says is that for each signal value a K-sided coin is flipped (note that the coin need not be fair). For this model to make sense, the probabilities must be numbers between zero and one and must sum to one.

$$0 \leq Pr[a_k] \leq 1 \tag{6.48}$$

$$\sum_{k=1}^{K} (Pr[a_k]) = 1 \tag{6.49}$$

This coin-flipping model assumes that symbols occur without regard to what preceding or succeeding symbols were, a false assumption for typed text. Despite this probabilistic model's over-simplicity, the ideas we develop here also work when more accurate, but still probabilistic, models are used. The key quantity that characterizes a symbolic-valued signal is the **entropy** of its alphabet.

$$H(A) = -\left(\sum_k (Pr[a_k] \log_2 (Pr[a_k]))\right) \tag{6.50}$$

Because we use the base-2 logarithm, entropy has units of bits. For this definition to make sense, we must take special note of symbols having probability zero of occurring. A zero-probability symbol never occurs; thus, we define $0 \log_2 0 = 0$ so that such symbols do not affect the entropy. The maximum value attainable by an alphabet's entropy occurs when the symbols are equally likely ($Pr[a_k] = Pr[a_l]$). In this case, the entropy equals $\log_2 K$. The minimum value occurs when only one symbol occurs; it has probability one of occurring and the rest have probability zero.

Exercise 6.22 *(Solution on p. 274.)*

Derive the maximum-entropy results, both the numeric aspect (entropy equals $\log_2 K$) and the theoretical one (equally likely symbols maximize entropy). Derive the value of the minimum entropy alphabet.

Example 6.1

A four-symbol alphabet has the following probabilities.

$$\begin{array}{cc} \Pr[a_0] = \frac{1}{2} & \Pr[a_1] = \frac{1}{4} \\ \Pr[a_2] = \frac{1}{8} & \Pr[a_3] = \frac{1}{8} \end{array}$$

Note that these probabilities sum to one as they should. As $\frac{1}{2} = 2^{-1}$, $\log_2\left(\frac{1}{2}\right) = -1$. The entropy of this alphabet equals

$$\begin{aligned} H(A) &= -\left(\tfrac{1}{2}\log_2\left(\tfrac{1}{2}\right) + \tfrac{1}{4}\log_2\left(\tfrac{1}{4}\right) + \tfrac{1}{8}\log_2\left(\tfrac{1}{8}\right) + \tfrac{1}{8}\log_2\left(\tfrac{1}{8}\right)\right) \\ &= -\left(\tfrac{1}{2}(-1) + \tfrac{1}{4}(-2) + \tfrac{1}{8}(-3) + \tfrac{1}{8}(-3)\right) \\ &= 1.75 \text{ bits} \end{aligned} \tag{6.51}$$

[29]This content is available online at <http://cnx.org/content/m0070/2.13/>.
[30]http://www.lucent.com/minds/infotheory/

6.21 Source Coding Theorem[31]

The significance of an alphabet's entropy rests in how we can represent it with a sequence of **bits**. Bit sequences form the "coin of the realm" in digital communications: they are the universal way of representing symbolic-valued signals. We convert back and forth between symbols to bit-sequences with what is known as a **codebook**: a table that associates symbols to bit sequences. In creating this table, we must be able to assign a *unique* bit sequence to each symbol so that we can go between symbol and bit sequences without error.

> POINT OF INTEREST: You may be conjuring the notion of hiding information from others when we use the name codebook for the symbol-to-bit-sequence table. There is no relation to cryptology, which comprises mathematically provable methods of securing information. The codebook terminology was developed during the beginnings of information theory just after World War II.

As we shall explore in some detail elsewhere, digital communication (Section 6.13) is the transmission of symbolic-valued signals from one place to another. When faced with the problem, for example, of sending a file across the Internet, we must first represent each character by a bit sequence. Because we want to send the file quickly, we want to use as few bits as possible. However, we don't want to use so few bits that the receiver cannot determine what each character was from the bit sequence. For example, we could use one bit for every character: File transmission would be fast but useless because the codebook creates errors. Shannon[32] proved in his monumental work what we call today the **Source Coding Theorem**. Let $B(a_k)$ denote the number of bits used to represent the symbol a_k. The average number of bits $\overline{B(A)}$ required to represent the entire alphabet equals $\sum_{k=1}^{K} (B(a_k) Pr[a_k])$. *The Source Coding Theorem states that the average number of bits needed to* accurately *represent the alphabet need only to satisfy*

$$H(A) \leq \overline{B(A)} < H(A) + 1 \qquad (6.52)$$

Thus, the alphabet's entropy specifies to within one bit how many bits on the average need to be used to send the alphabet. The smaller an alphabet's entropy, the fewer bits required for digital transmission of files expressed in that alphabet.

Example 6.2

A four-symbol alphabet has the following probabilities.

$$\Pr[a_0] = \frac{1}{2} \quad \Pr[a_1] = \frac{1}{4}$$
$$\Pr[a_2] = \frac{1}{8} \quad \Pr[a_3] = \frac{1}{8}$$

and an entropy of 1.75 bits (Example 6.1). Let's see if we can find a codebook for this four-letter alphabet that satisfies the Source Coding Theorem. The simplest code to try is known as the **simple binary code**: convert the symbol's index into a binary number and use the same number of bits for each symbol by including leading zeros where necessary.

$$(a_0 \leftrightarrow 00)\ (a_1 \leftrightarrow 01)\ (a_2 \leftrightarrow 10)\ (a_3 \leftrightarrow 11) \qquad (6.53)$$

Whenever the number of symbols in the alphabet is a power of two (as in this case), the average number of bits $\overline{B(A)}$ equals $\log_2 K$, which equals 2 in this case. Because the entropy equals 1.75bits, the simple binary code indeed satisfies the Source Coding Theorem—we are within one bit of the entropy limit—but you might wonder if you can do better. If we chose a codebook with differing number of bits for the symbols, a smaller average number of bits can indeed be obtained. The idea is to use shorter bit sequences for the symbols that occur more often. One codebook like this is

$$(a_0 \leftrightarrow 0)\ (a_1 \leftrightarrow 10)\ (a_2 \leftrightarrow 110)\ (a_3 \leftrightarrow 111) \qquad (6.54)$$

[31]This content is available online at <http://cnx.org/content/m0091/2.13/>.
[32]http://www.lucent.com/minds/infotheory/

Now $\overline{B(A)} = 1\frac{1}{2} + 2\frac{1}{4} + 3\frac{1}{8} + 3\frac{1}{8} = 1.75$. We can reach the entropy limit! The simple binary code is, in this case, less efficient than the unequal-length code. Using the efficient code, we can transmit the symbolic-valued signal having this alphabet 12.5% faster. Furthermore, we know that no more efficient codebook can be found because of Shannon's Theorem.

6.22 Compression and the Huffman Code[33]

Shannon's Source Coding Theorem (6.52) has additional applications in **data compression**. Here, we have a symbolic-valued signal source, like a computer file or an image, that we want to represent with as few bits as possible. Compression schemes that assign symbols to bit sequences are known as **lossless** if they obey the Source Coding Theorem; they are **lossy** if they use fewer bits than the alphabet's entropy. Using a lossy compression scheme means that you cannot recover a symbolic-valued signal from its compressed version without incurring some error. You might be wondering why anyone would want to intentionally create errors, but lossy compression schemes are frequently used where the efficiency gained in representing the signal outweighs the significance of the errors.

Shannon's Source Coding Theorem states that symbolic-valued signals require *on the average* at least $H(A)$ number of bits to represent each of its values, which are symbols drawn from the alphabet A. In the module on the Source Coding Theorem (Section 6.21) we find that using a so-called **fixed rate** source coder, one that produces a fixed number of bits/symbol, may not be the most efficient way of encoding symbols into bits. What is not discussed there is a procedure for designing an efficient source coder: one *guaranteed* to produce the fewest bits/symbol on the average. That source coder is not unique, and one approach that does achieve that limit is the **Huffman source coding algorithm**.

POINT OF INTEREST: In the early years of information theory, the race was on to be the first to find a *provably* maximally efficient source coding algorithm. The race was won by then MIT graduate student David Huffman in 1954, who worked on the problem as a project in his information theory course. We're pretty sure he received an "A."

- Create a vertical table for the symbols, the best ordering being in decreasing order of probability.
- Form a binary tree to the right of the table. A binary tree always has two branches at each node. Build the tree by merging the two lowest probability symbols at each level, making the probability of the node equal to the sum of the merged nodes' probabilities. If more than two nodes/symbols share the lowest probability at a given level, pick any two; your choice won't affect $\overline{B(A)}$.
- At each node, label each of the emanating branches with a binary number. The bit sequence obtained from passing from the tree's root to the symbol is its Huffman code.

Example 6.3

The simple four-symbol alphabet used in the Entropy (Example 6.1) and Source Coding (Example 6.2) modules has a four-symbol alphabet with the following probabilities,

$$\Pr[a_0] = \frac{1}{2} \quad \Pr[a_1] = \frac{1}{4}$$
$$\Pr[a_2] = \frac{1}{8} \quad \Pr[a_3] = \frac{1}{8}$$

and an entropy of 1.75 bits (Example 6.1). This alphabet has the Huffman coding tree shown in Figure 6.18 (Huffman Coding Tree).

[33]This content is available online at <http://cnx.org/content/m0092/2.17/>.

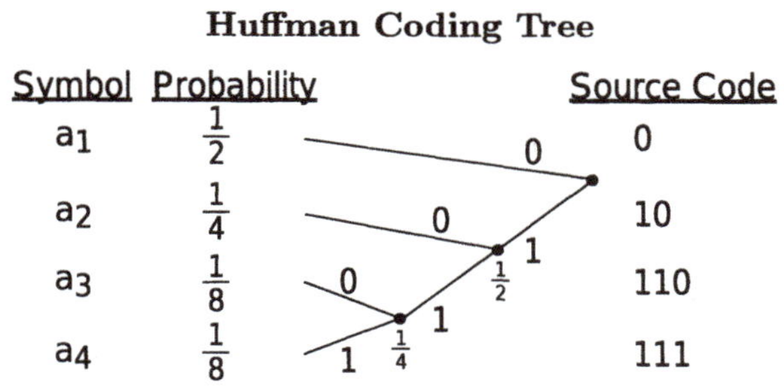

Figure 6.18: We form a Huffman code for a four-letter alphabet having the indicated probabilities of occurrence. The binary tree created by the algorithm extends to the right, with the root node (the one at which the tree begins) defining the codewords. The bit sequence obtained by traversing the tree from the root to the symbol defines that symbol's binary code.

The code thus obtained is not unique as we could have labeled the branches coming out of each node differently. The average number of bits required to represent this alphabet equals 1.75 bits, which is the Shannon entropy limit for this source alphabet. If we had the symbolic-valued signal $s(m) = \{a_2, a_3, a_1, a_4, a_1, a_2, \dots\}$, our Huffman code would produce the bitstream $b(n) = 101100111010\dots$.

If the alphabet probabilities were different, clearly a different tree, and therefore different code, could well result. Furthermore, we may not be able to achieve the entropy limit. If our symbols had the probabilities $Pr[a_1] = \frac{1}{2}$, $Pr[a_2] = \frac{1}{4}$, $Pr[a_3] = \frac{1}{5}$, and $Pr[a_4] = \frac{1}{20}$, the average number of bits/symbol resulting from the Huffman coding algorithm would equal 1.75 bits. However, the entropy limit is 1.68 bits. The Huffman code does satisfy the Source Coding Theorem—its average length is within one bit of the alphabet's entropy—but you might wonder if a better code existed. David Huffman showed mathematically that no other code could achieve a shorter average code than his. We can't do better.

Exercise 6.23 *(Solution on p. 274.)*

Derive the Huffman code for this second set of probabilities, and verify the claimed average code length and alphabet entropy.

6.23 Subtlies of Coding[34]

In the Huffman code, the bit sequences that represent individual symbols can have differing lengths so the bitstream index m does not increase in lock step with the symbol-valued signal's index n. To capture how often bits must be transmitted to keep up with the source's production of symbols, we can only compute averages. If our source code averages $\overline{B(A)}$ bits/symbol and symbols are produced at a rate R, the average bit rate equals $\overline{B(A)}R$, and this quantity determines the bit interval duration T.

Exercise 6.24 *(Solution on p. 274.)*

Calculate what the relation between T and the average bit rate $\overline{B(A)}R$ is.

A subtlety of source coding is whether we need "commas" in the bitstream. When we use an unequal number of bits to represent symbols, how does the receiver determine when symbols begin and end? If you created a source code that *required* a separation marker in the bitstream between symbols, it would be very inefficient since you are essentially requiring an extra symbol in the transmission stream.

[34]This content is available online at <http://cnx.org/content/m0093/2.16/>.

POINT OF INTEREST: A good example of this need is the Morse Code: Between each letter, the telegrapher needs to insert a pause to inform the receiver when letter boundaries occur.

As shown in this example (Example 6.3), no commas are placed in the bitstream, but you can unambiguously decode the sequence of symbols from the bitstream. Huffman showed that his (maximally efficient) code had the **prefix** property: No code for a symbol began another symbol's code. Once you have the prefix property, the bitstream is *partially* self-synchronizing: Once the receiver knows where the bitstream starts, we can assign a unique and correct symbol sequence to the bitstream.

Exercise 6.25 *(Solution on p. 274.)*

Sketch an argument that prefix coding, whether derived from a Huffman code or not, will provide unique decoding when an unequal number of bits/symbol are used in the code.

However, having a prefix code does not guarantee total synchronization: After hopping into the middle of a bitstream, can we always find the correct symbol boundaries? The self-synchronization issue does mitigate the use of efficient source coding algorithms.

Exercise 6.26 *(Solution on p. 274.)*

Show by example that a bitstream produced by a Huffman code is not necessarily self-synchronizing. Are fixed-length codes self synchronizing?

Another issue is bit errors induced by the digital channel; if they occur (and they will), synchronization can easily be lost even if the receiver started "in synch" with the source. Despite the small probabilities of error offered by good signal set design and the matched filter, an infrequent error can devastate the ability to translate a bitstream into a symbolic signal. We need ways of reducing reception errors *without* demanding that p_e be smaller.

Example 6.4

The first electrical communications system—the telegraph—was digital. When first deployed in 1844, it communicated text over wireline connections using a binary code—the Morse code—to represent individual letters. To send a message from one place to another, telegraph operators would tap the message using a telegraph key to another operator, who would relay the message on to the next operator, presumably getting the message closer to its destination. In short, the telegraph relied on a **network** not unlike the basics of modern computer networks. To say it presaged modern communications would be an understatement. It was also far ahead of some needed technologies, namely the Source Coding Theorem. The Morse code, shown in Figure 6.19, was not a prefix code. To separate codes for each letter, Morse code required that a space—a pause—be inserted between each letter. In information theory, that space counts as another code letter, which means that the Morse code encoded text with a three-letter source code: dots, dashes and space. The resulting source code is not within a bit of entropy, and is grossly inefficient (about 25%). Figure 6.19 shows a Huffman code for English text, which as we know *is* efficient.

6.24 Channel Coding[35]

We can, to some extent, *correct* errors made by the receiver with only the error-filled bit stream emerging from the digital channel available to us. The idea is for the transmitter to send not only the symbol-derived bits emerging from the source coder but also additional bits derived from the coder's bit stream. These additional bits, *the error correcting bits*, help the receiver determine if an error has occurred in the data bits (the important bits) or in the error-correction bits. Instead of the communication model (Figure 6.17: DigMC) shown previously, the transmitter inserts a **channel coder** before analog modulation, and the receiver the corresponding channel decoder (Figure 6.20). This block diagram shown there forms the **Fundamental Model of Digital Communication**.

[35]This content is available online at <http://cnx.org/content/m10782/2.4/>.

Morse and Huffman Code Table

	%	Morse Code	Huffman Code
A	6.22	.-	1011
B	1.32	-...	010100
C	3.11	-.-.	10101
D	2.97	-..	01011
E	10.53	.	001
F	1.68	..-.	110001
G	1.65	–.	110000
H	3.63	....	11001
I	6.14	..	1001
J	0.06	.—	01010111011
K	0.31	-.-	01010110
L	3.07	.-..	10100
M	2.48	–	00011
N	5.73	-.	0100
O	6.06	—	1000
P	1.87	.–.	00000
Q	0.10	–.-	0101011100
R	5.87	.-.	0111
S	5.81	...	0110
T	7.68	-	1101
U	2.27	..-	00010
V	0.70	...-	0101010
W	1.13	.–	000011
X	0.25	-..-	010101111
Y	1.07	-.–	000010
Z	0.06	–..	0101011101011

Figure 6.19: Morse and Huffman Codes for American-Roman Alphabet. The % column indicates the average probability (expressed in percent) of the letter occurring in English. The entropy $H(A)$ of the this source is 4.14 bits. The average Morse codeword length is 2.5 symbols. Adding one more symbol for the letter separator and converting to bits yields an average codeword length of 5.56 bits. The average Huffman codeword length is 4.35 bits.

Figure 6.20: To correct errors that occur in the digital channel, a channel coder and decoder are added to the communication system. Properly designed channel coding can greatly reduce the probability (from the uncoded value of p_e) that a data bit $b(n)$ is received incorrectly even when the probability of $c(l)$ be received in error remains p_e or becomes larger. This system forms the Fundamental Model of Digital Communication.

Shannon's Noisy Channel Coding Theorem (Section 6.30) says that if the data aren't transmitted too quickly, that error correction codes exist that can correct *all* the bit errors introduced by the channel. Unfortunately, Shannon did not demonstrate an error correcting code that would achieve this remarkable feat; in fact, no one has found such a code. Shannon's result proves it exists; seems like there is always more work to do. In any case, that should not prevent us from studying commonly used error correcting codes that not only find their way into all digital communication systems, but also into CDs and bar codes used on merchandise.

6.25 Repetition Codes[36]

Perhaps the simplest error correcting code is the **repetition code**.

Repetition Code

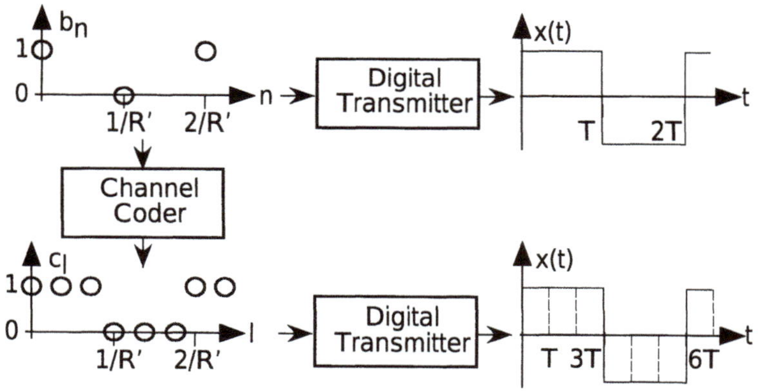

Figure 6.21: The upper portion depicts the result of directly modulating the bit stream $b(n)$ into a transmitted signal $x(t)$ using a baseband BPSK signal set. R' is the datarate produced by the source coder. If that bit stream passes through a (3,1) channel coder to yield the bit stream $c(l)$, the resulting transmitted signal requires a bit interval T three times smaller than the uncoded version. This reduction in the bit interval means that the transmitted energy/bit decreases by a factor of three, which results in an increased error probability in the receiver.

Here, the transmitter sends the data bit several times, an odd number of times in fact. Because the

[36]This content is available online at <http://cnx.org/content/m0071/2.21/>.

error probability p_e is always less than $\frac{1}{2}$, we know that more of the bits should be correct rather than in error. Simple majority voting of the received bits (hence the reason for the odd number) determines the transmitted bit more accurately than sending it alone. For example, let's consider the three-fold repetition code: for every bit $b(n)$ emerging from the source coder, the channel coder produces three. Thus, the bit stream emerging from the channel coder $c(l)$ has a data rate three times higher than that of the original bit stream $b(n)$. The coding table illustrates when errors can be corrected and when they can't by the majority-vote decoder.

Coding Table

Code	Probability	Bit
000	$(1-p_e)^3$	0
001	$p_e(1-p_e)^2$	0
010	$p_e(1-p_e)^2$	0
011	${p_e}^2(1-p_e)$	1
100	$p_e(1-p_e)^2$	0
101	${p_e}^2(1-p_e)$	1
110	${p_e}^2(1-p_e)$	1
111	${p_e}^3$	1

Figure 6.22: In this example, the transmitter encodes 0 as 000. The channel creates an error (changing a 0 into a 1) that with probability p_e. The first column lists all possible received datawords and the second the probability of each dataword being received. The last column shows the results of the majority-vote decoder. When the decoder produces 0, it successfully corrected the errors introduced by the channel (if there were any; the top row corresponds to the case in which no errors occurred). The error probability of the decoders is the sum of the probabilities when the decoder produces 1.

Thus, if one bit of the three bits is received in error, the receiver can correct the error; if more than one error occurs, the channel decoder announces the bit is *1* instead of transmitted value of *0*. Using this repetition code, the probability of $\hat{b}(n) \neq 0$ equals $3{p_e}^2(1-p_e)+{p_e}^3$. This probability of a decoding error is always less than p_e, the uncoded value, so long as $p_e < \frac{1}{2}$.

Exercise 6.27 ***(Solution on p. 275.)***

Demonstrate mathematically that this claim is indeed true. Is $3{p_e}^2(1-p_e)+{p_e}^3 \leq p_e$?

6.26 Block Channel Coding[37]

Because of the higher datarate imposed by the channel coder, the probability of bit error occurring in the digital channel *increases* relative to the value obtained when no channel coding is used. The bit interval duration must be reduced by $\frac{K}{N}$ in comparison to the no-channel-coding situation, which means the energy

[37]This content is available online at <http://cnx.org/content/m0094/2.14/>.

per bit E_b goes *down* by the same amount. The bit interval must decrease by a factor of three if the transmitter is to keep up with the data stream, as illustrated here (Figure 6.21: Repetition Code).

POINT OF INTEREST: It is unlikely that the transmitter's power could be increased to compensate. Such is the sometimes-unfriendly nature of the real world.

Because of this reduction, the error probability p_e of the digital channel goes up. The question thus becomes does channel coding *really* help: Is the effective error probability lower with channel coding even though the error probability for each transmitted bit is larger? The answer is *no*: Using a repetition code for channel coding cannot ultimately reduce the probability that a data bit is received in error. The ultimate reason is the repetition code's inefficiency: transmitting one data bit for every three transmitted is too inefficient for the amount of error correction provided.

Exercise 6.28 *(Solution on p. 275.)*

Using MATLAB, calculate the probability a bit is received incorrectly with a three-fold repetition code. Show that when the energy per bit E_b is reduced by 1/3 that this probability is larger than the no-coding probability of error.

The repetition code (p. 240) represents a special case of what is known as **block channel coding**. For every K bits that enter the block channel coder, it inserts an additional $N-K$ error-correction bits to produce a block of N bits for transmission. We use the notation (N,K) to represent a given block code's parameters. In the three-fold repetition code (p. 240), $K=1$ and $N=3$. A block code's **coding efficiency** E equals the ratio $\frac{K}{N}$, and quantifies the overhead introduced by channel coding. The rate at which bits must be transmitted again changes: So-called data bits $b(n)$ emerge from the source coder at an average rate $\overline{B(A)}$ and exit the channel at a rate $\frac{1}{E}$ higher. We represent the fact that the bits sent through the digital channel operate at a different rate by using the index l for the channel-coded bit stream $c(l)$. Note that the blocking (framing) imposed by the channel coder does not correspond to symbol boundaries in the bit stream $b(n)$, especially when we employ variable-length source codes.

Does any error-correcting code reduce communication errors when real-world constraints are taken into account? The answer now is yes. To understand channel coding, we need to develop first a general framework for channel coding, and discover what it takes for a code to be maximally efficient: Correct as many errors as possible using the fewest error correction bits as possible (making the efficiency $\frac{K}{N}$ as large as possible).

6.27 Error-Correcting Codes: Hamming Distance[38]

So-called **linear codes** create error-correction bits by combining the data bits linearly. The phrase "linear combination" means here single-bit binary arithmetic.

$(0\oplus 0)=0$	$(1\oplus 1)=0$	$(0\oplus 1)=1$	$(1\oplus 0)=1$
$(0\cdot 0)=0$	$(1\cdot 1)=1$	$(0\cdot 1)=0$	$(1\cdot 0)=0$

Figure 6.23

For example, let's consider the specific (3, 1) error correction code described by the following coding table and, more concisely, by the succeeding matrix expression.

$$c(1)=b(1)$$
$$c(2)=b(1)$$
$$c(3)=b(1)$$

[38]This content is available online at <http://cnx.org/content/m10283/2.28/>.

or

$$\mathbf{c} = G\mathbf{b}$$

where

$$G = \begin{pmatrix} 1 \\ 1 \\ 1 \end{pmatrix}$$

$$\mathbf{c} = \begin{pmatrix} c(1) \\ c(2) \\ c(3) \end{pmatrix}$$

$$\mathbf{b} = \begin{pmatrix} b(1) \end{pmatrix}$$

The length-K (in this simple example $K = 1$) block of data bits is represented by the vector **b**, and the length-N output block of the channel coder, known as a **codeword**, by c. The **generator matrix** G defines all block-oriented linear channel coders.

As we consider other block codes, the simple idea of the decoder taking a majority vote of the received bits won't generalize easily. We need a broader view that takes into account the *distance* between codewords. A length-N codeword means that the receiver must decide among the 2^N possible datawords to select which of the 2^K codewords was actually transmitted. As shown in Figure 6.24, we can think of the datawords geometrically. We define the **Hamming distance** between binary datawords c_1 and c_2, denoted by $d(c_1, c_2)$ to be the minimum number of bits that must be "flipped" to go from one word to the other. For example, the distance between codewords is 3 bits. In our table of binary arithmetic, we see that adding a 1 corresponds to flipping a bit. Furthermore, subtraction and addition are equivalent. We can express the Hamming distance as

$$d(c_1, c_2) = \text{sum}((c_1 \oplus c_2)) \tag{6.55}$$

Exercise 6.29 *(Solution on p. 275.)*

Show that adding the error vector col[1,0,...,0] to a codeword flips the codeword's leading bit and leaves the rest unaffected.

The probability of one bit being flipped anywhere in a codeword is $Np_e(1 - p_e)^{N-1}$. The number of errors the channel introduces equals the number of ones in e; the probability of any particular error vector decreases with the number of errors.

To perform decoding when errors occur, we want to find the codeword (one of the filled circles in Figure 6.24) that has the highest probability of occurring: the one closest to the one received. Note that if a dataword lies a distance of 1 from two codewords, it is *impossible* to determine which codeword was actually sent. This criterion means that if any two codewords are two bits apart, then the code *cannot* correct the channel-induced error. *Thus, to have a code that can correct all single-bit errors, codewords must have a minimum separation of three.* Our repetition code has this property.

Introducing code bits increases the probability that any bit arrives in error (because bit interval durations decrease). However, using a well-designed error-correcting code corrects bit reception errors. Do we win or lose by using an error-correcting code? The answer is that we can win *if* the code is well-designed. The (3,1) repetition code demonstrates that we can lose (Exercise 6.28). To develop good channel coding, we need to develop first a general framework for channel codes and discover what it takes for a code to be maximally efficient: Correct as many errors as possible using the fewest error correction bits as possible (making the efficiency $\frac{K}{N}$ as large as possible.) We also need a systematic way of finding the codeword closest to any

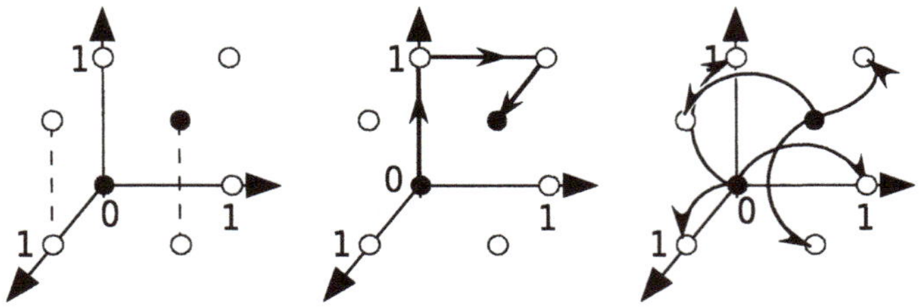

Figure 6.24: In a (3,1) repetition code, only 2 of the possible 8 three-bit data blocks are codewords. We can represent these bit patterns geometrically with the axes being bit positions in the data block. In the left plot, the filled circles represent the codewords [0 0 0] and [1 1 1], the only possible codewords. The unfilled ones correspond to the transmission. The center plot shows that the distance between codewords is 3. Because distance corresponds to flipping a bit, calculating the Hamming distance geometrically means following the axes rather than going "as the crow flies". The right plot shows the datawords that result when one error occurs as the codeword goes through the channel. The three datawords are unit distance from the original codeword. Note that the received dataword groups do not overlap, which means the code can correct all single-bit errors.

received dataword. A much better code than our (3,1) repetition code is the following (7,4) code.

$$\begin{aligned} c(1) &= b(1) \\ c(2) &= b(2) \\ c(3) &= b(3) \\ c(4) &= b(4) \\ c(5) &= (b(1) \oplus b(2) \oplus b(3)) \\ c(6) &= (b(2) \oplus b(3) \oplus b(4)) \\ c(7) &= (b(1) \oplus b(2) \oplus b(4)) \end{aligned}$$

where the generator matrix is

$$G = \begin{pmatrix} 1 & 0 & 0 & 0 \\ 0 & 1 & 0 & 0 \\ 0 & 0 & 1 & 0 \\ 0 & 0 & 0 & 1 \\ 1 & 1 & 1 & 0 \\ 0 & 1 & 1 & 1 \\ 1 & 1 & 0 & 1 \end{pmatrix}$$

In this (7,4) code, $2^4 = 16$ of the $2^7 = 128$ possible blocks at the channel decoder correspond to error-free transmission and reception.

Error correction amounts to searching for the codeword $\mathbf{c}$ closest to the received block $\hat{c}$ in terms of the Hamming distance between the two. The error correction capability of a channel code is limited by how close together any two error-free blocks are. Bad codes would produce blocks close together, which would result in ambiguity when assigning a block of data bits to a received block. The quantity to examine, therefore, in designing code error correction codes is the minimum distance between codewords.

$$d_{min} = min\left(d\left(c_i, c_j\right)\right) \quad , \quad c_i \neq c_j \tag{6.56}$$

To have a channel code that can correct all single-bit errors, $d_{min} \geq 3$.

Exercise 6.30 ***(Solution on p. 276.)***

Suppose we want a channel code to have an error-correction capability of n bits. What must the minimum Hamming distance between codewords d_{min} be?

How do we calculate the minimum distance between codewords? Because we have 2^K codewords, the number of possible unique pairs equals $2^{K-1}\left(2^K - 1\right)$, which can be a large number. Recall that our channel coding procedure is linear, with $\mathbf{c} = G\mathbf{b}$. Therefore $(c_i \oplus c_j) = G\left((b_i \oplus b_j)\right)$. Because $b_i \oplus b_j$ always yields another block of *data* bits, we find that the difference between any two codewords is another codeword! Thus, to find d_{min} we need only compute the number of ones that comprise all non-zero codewords. Finding these codewords is easy once we examine the coder's generator matrix. Note that the columns of G are codewords (why is this?), and that all codewords can be found by all possible pairwise sums of the columns. To find d_{min} , we need only count the number of bits in each column and sums of columns. For our example (7, 4), G's first column has three ones, the next one four, and the last two three. Considering sums of column pairs next, note that because the upper portion of G is an identity matrix, the corresponding upper portion of all column sums must have exactly two bits. Because the bottom portion of each column differs from the other columns in at least one place, the bottom portion of a sum of columns must have at least one bit. Triple sums will have at least three bits because the upper portion of G is an identity matrix. Thus, no sum of columns has fewer than three bits, which means that $d_{min} = 3$, and we have a channel coder that can correct all occurrences of one error within a received 7-bit block.

6.28 Error-Correcting Codes: Channel Decoding[39]

Because the idea of channel coding has merit (so long as the code is efficient), let's develop a systematic procedure for performing channel decoding. One way of checking for errors is to try recreating the error correction bits from the data portion of the received block $\hat{c}$. Using matrix notation, we make this calculation by multiplying the received block $\hat{c}$ by the matrix H known as the **parity check matrix**. It is formed from the generator matrix G by taking the bottom, error-correction portion of G and attaching to it an identity matrix. For our (7,4) code,

$$H = \left[\begin{array}{c} \underbrace{\begin{array}{cccc} 1 & 1 & 1 & 0 \\ 0 & 1 & 1 & 1 \\ 1 & 1 & 0 & 1 \end{array}}_{\text{Lower portion of } G} \quad \underbrace{\begin{array}{ccc} 1 & 0 & 0 \\ 0 & 1 & 0 \\ 0 & 0 & 1 \end{array}}_{\text{Identity}} \end{array}\right] \tag{6.57}$$

The parity check matrix thus has size $(N-K) \times N$, and the result of multiplying this matrix with a received word is a length- $(N-K)$ binary vector. If no digital channel errors occur—we receive a codeword so that $\hat{c} = \mathbf{c}$ — then $H\hat{c} = 0$. For example, the first column of G, $(1,0,0,0,1,0,1)^T$, is a codeword. Simple calculations show that multiplying this vector by H results in a length-$(N-K)$ zero-valued vector.

Exercise 6.31 ***(Solution on p. 276.)***

Show that $H\mathbf{c} = 0$ for all the columns of G. In other words, show that $HG = 0$ an $(N-K) \times K$ matrix of zeroes. Does this property guarantee that all codewords also satisfy $H\mathbf{c} = 0$?

When the received bits $\hat{c}$ do *not* form a codeword, $H\hat{c}$ does not equal zero, indicating the presence of one or more errors induced by the digital channel. Because the presence of an error can be mathematically written as $\hat{c} = (\mathbf{c} \oplus \mathbf{e})$, with $\mathbf{e}$ a vector of binary values having a 1 in those positions where a bit error occurred.

[39]This content is available online at <http://cnx.org/content/m0072/2.20/>.

Exercise 6.32 *(Solution on p. 276.)*

Show that adding the error vector $(1, 0, \dots, 0)^T$ to a codeword flips the codeword's leading bit and leaves the rest unaffected.

Consequently, $H\hat{c} = H(\mathbf{c} \oplus \mathbf{e}) = H\mathbf{e}$. Because the result of the product is a length- $(N-K)$ vector of binary values, we can have $2^{N-K} - 1$ non-zero values that correspond to non-zero error patterns $\mathbf{e}$. To perform our channel decoding,

1. compute (conceptually at least) $H\hat{c}$;
2. if this result is zero, no detectable or correctable error occurred;
3. if non-zero, consult a table of length-$(N-K)$ binary vectors to associate them with the *minimal* error pattern that could have resulted in the non-zero result; then
4. add the error vector thus obtained to the received vector $\hat{c}$ to correct the error (because $(\mathbf{c} \oplus \mathbf{e} \oplus \mathbf{e}) = \mathbf{c}$).
5. Select the data bits from the corrected word to produce the received bit sequence $\hat{b}(n)$.

The phrase *minimal* in the third item raises the point that a double (or triple or quadruple...) error occurring during the transmission/reception of one codeword can create the same received word as a single-bit error or no error in *another* codeword. For example, $(1,0,0,0,1,0,1)^T$ and $(0,1,0,0,1,1,1)^T$ are both codewords in the example (7,4) code. The second results when the first one experiences three bit errors (first, second, and sixth bits). Such an error pattern cannot be detected by our coding strategy, but such multiple error patterns are very unlikely to occur. Our receiver uses the principle of maximum probability: An error-free transmission is much more likely than one with three errors if the bit-error probability p_e is small enough.

Exercise 6.33 *(Solution on p. 276.)*

How small must p_e be so that a single-bit error is more likely to occur than a triple-bit error?

6.29 Error-Correcting Codes: Hamming Codes[40]

For the (7,4) example, we have $2^{N-K} - 1 = 7$ error patterns that can be corrected. We start with single-bit error patterns, and multiply them by the parity check matrix. If we obtain unique answers, we are done; if two or more error patterns yield the same result, we can try double-bit error patterns. In our case, single-bit error patterns give a unique result.

Parity Check Matrix

e	He
1000000	101
0100000	111
0010000	110
0001000	011
0000100	100
0000010	010
0000001	001

This corresponds to our decoding table: We associate the parity check matrix multiplication result with the error pattern and add this to the received word. If more than one error occurs (unlikely though it may be), this "error correction" strategy usually makes the error worse in the sense that more bits are changed from what was transmitted.

[40]This content is available online at <http://cnx.org/content/m0097/2.24/>.

As with the repetition code, we must question whether our (7,4) code's error correction capability compensates for the increased error probability due to the necessitated reduction in bit energy. Figure 6.25 (Probability of error occurring) shows that if the signal-to-noise ratio is large enough channel coding yields a smaller error probability. Because the bit stream emerging from the source decoder is segmented into four-bit blocks, the fair way of comparing coded and uncoded transmission is to compute the probability of **block** error: the probability that any bit in a block remains in error despite error correction and regardless of whether the error occurs in the data or in coding buts. Clearly, our (7,4) channel code does yield smaller error rates, and is worth the additional systems required to make it work.

Probability of error occurring

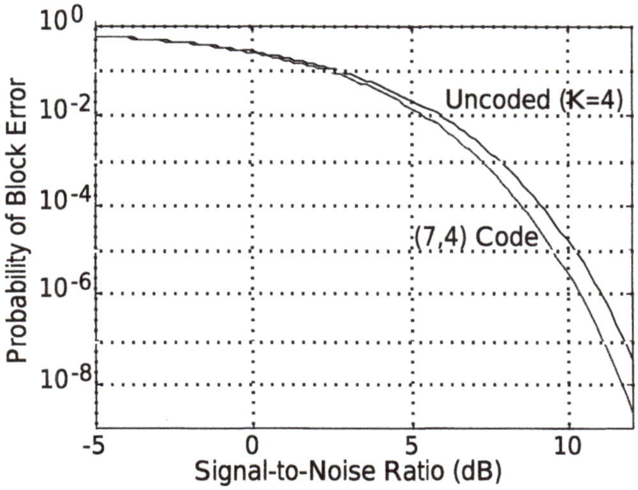

Figure 6.25: The probability of an error occurring in transmitted $K = 4$ data bits equals $1-(1-p_e)^4$ as $(1-p_e)^4$ equals the probability that the four bits are received without error. The upper curve displays how this probability of an error anywhere in the four-bit block varies with the signal-to-noise ratio. When a (7,4) single-bit error correcting code is used, the transmitter reduced the energy it expends during a single-bit transmission by 4/7, appending three extra bits for error correction. Now the probability of any bit in the *seven*-bit block being in error after error correction equals $1-(1-p_e)^7-(7p_e')(1-p_e')^6$, where p_e' is the probability of a bit error occurring in the channel when channel coding occurs. Here $(7p_e')(1-p_e')^6$ equals the probability of exactly on in seven bits emerging from the channel in error; The channel decoder corrects this type of error, and all data bits in the block are received correctly.

Note that our (7,4) code has the length and number of data bits that perfectly fits correcting single bit errors. This pleasant property arises because the number of error patterns that can be corrected, $2^{N-K}-1$, equals the codeword length N. Codes that have $2^{N-K}-1=N$ are known as **Hamming codes**, and the following table (Hamming Codes, p. 247) provides the parameters of these codes. Hamming codes are the simplest single-bit error correction codes, and the generator/parity check matrix formalism for channel coding and decoding works for them.

Hamming Codes

N	K	E (efficiency)
3	1	0.33
7	4	0.57
15	11	0.73
31	26	0.84
63	57	0.90
127	120	0.94

Unfortunately, for such large blocks, the probability of multiple-bit errors can exceed the number of single-bit errors unless the channel single-bit error probability p_e is very small. Consequently, we need to enhance the code's error correcting capability by adding double as well as single-bit error correction.

Exercise 6.34 *(Solution on p. 276.)*

What must the relation between N and K be for a code to correct all single- and double-bit errors with a "perfect fit"?

6.30 Noisy Channel Coding Theorem[41]

As the block length becomes larger, more error correction will be needed. Do codes exist that can correct *all* errors? Perhaps the crowning achievement of Claude Shannon's[42] creation of information theory answers this question. His result comes in two complementary forms: the Noisy Channel Coding Theorem and its converse.

6.30.1 Noisy Channel Coding Theorem

Let E denote the efficiency of an error-correcting code: the ratio of the number of data bits to the total number of bits used to represent them. If the efficiency is less than the **capacity** of the digital channel, an error-correcting code exists that has the property that as the length of the code increases, the probability of an error occurring in the decoded block approaches zero.

$$\lim_{N \to \infty} \Pr[\text{block error}] = 0 \quad , \quad E < C \tag{6.58}$$

6.30.2 Converse to the Noisy Channel Coding Theorem

If $E > C$, the probability of an error in a decoded block must approach one regardless of the code that might be chosen.

$$\lim_{N \to \infty} \Pr[\text{block error}] = 1 \tag{6.59}$$

These results mean that it is possible to transmit digital information over a noisy channel (one that introduces errors) and receive the information without error *if* the code is sufficiently *inefficient* compared to the channel's characteristics. Generally, a channel's capacity changes with the signal-to-noise ratio: As one increases or decreases, so does the other. The capacity measures the overall error characteristics of a

[41] This content is available online at <http://cnx.org/content/m0073/2.11/>.

[42] http://www.lucent.com/minds/infotheory/

channel—the smaller the capacity the more frequently errors occur—and an overly efficient error-correcting code will not build in enough error correction capability to counteract channel errors.

This result astounded communication engineers when Shannon published it in 1948. Analog communication always yields a noisy version of the transmitted signal; in digital communication, error correction can be powerful enough to correct all errors as the block length increases. The key for this capability to exist is that the code's efficiency be less than the channel's capacity. For a binary symmetric channel, the capacity is given by

$$C = 1 + p_e \log_2 p_e + (1 - p_e) \log_2 (1 - p_e)\ bits/transmission \tag{6.60}$$

Figure 6.26 (capacity of a channel) shows how capacity varies with error probability. For example, our (7,4) Hamming code has an efficiency of 0.57, and codes having the same efficiency but longer block sizes can be used on additive noise channels where the signal-to-noise ratio exceeds $0dB$.

capacity of a channel

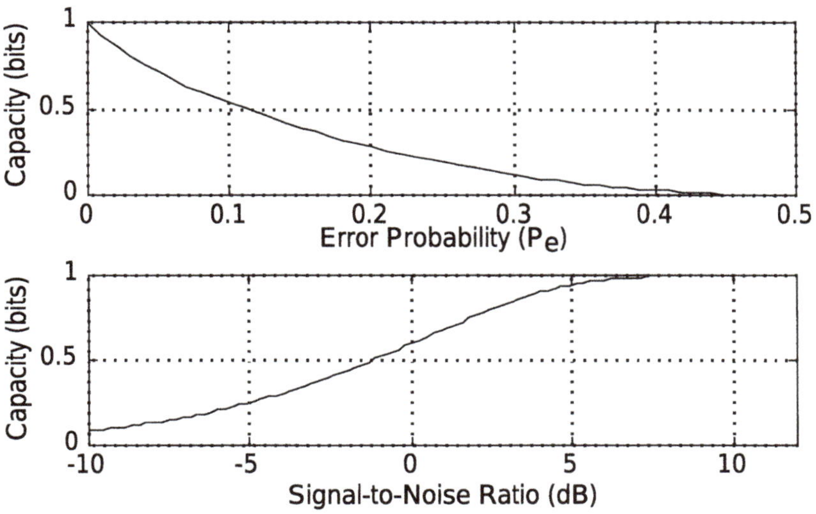

Figure 6.26: The capacity per transmission through a binary symmetric channel is plotted as a function of the digital channel's error probability (upper) and as a function of the signal-to-noise ratio for a BPSK signal set (lower).

6.31 Capacity of a Channel[43]

In addition to the Noisy Channel Coding Theorem and its converse (Section 6.30), Shannon also derived the capacity for a bandlimited (to W Hz) additive white noise channel. For this case, the signal set is unrestricted, even to the point that more than one bit can be transmitted each "bit interval." Instead of constraining channel code efficiency, the revised Noisy Channel Coding Theorem states that some error-correcting code exists such that as the block length increases, error-free transmission is possible if the source coder's datarate, $\overline{B(A)}R$, is less than capacity.

$$C = W \log_2 (1 + SNR)\ \text{bits/s} \tag{6.61}$$

[43]This content is available online at <http://cnx.org/content/m0098/2.13/>.

This result sets the maximum datarate of the source coder's output that can be transmitted through the bandlimited channel with no error. [44] Shannon's proof of his theorem was very clever, and did not indicate what this code might be; it has never been found. Codes such as the Hamming code work quite well in practice to keep error rates low, but they remain greater than zero. Until the "magic" code is found, more important in communication system design is the converse. It states that if your data rate exceeds capacity, errors will overwhelm you no matter what channel coding you use. For this reason, capacity calculations are made to understand the fundamental limits on transmission rates.

Exercise 6.35 ***(Solution on p. 276.)***

The first definition of capacity applies only for binary symmetric channels, and represents the number of bits/transmission. The second result states capacity more generally, having units of bits/second. How would you convert the first definition's result into units of bits/second?

Example 6.5

The telephone channel has a bandwidth of 3 kHz and a signal-to-noise ratio exceeding 30 dB (at least they promise this much). The maximum data rate a modem can produce for this wireline channel and hope that errors will not become rampant is the capacity.

$$\begin{aligned} C &= 3 \times 10^3 \log_2 \left(1 + 10^3\right) \\ &= 29.901 \text{ kbps} \end{aligned} \tag{6.62}$$

Thus, the so-called 33 kbps modems operate right at the capacity limit.

Note that the data rate allowed by the capacity can exceed the bandwidth when the signal-to-noise ratio exceeds 0 dB. Our results for BPSK and FSK indicated the bandwidth they require exceeds $\frac{1}{T}$. What kind of signal sets might be used to achieve capacity? Modem signal sets send more than one bit/transmission using a number, one of the most popular of which is **multi-level signaling**. Here, we can transmit several bits during one transmission interval by representing bit by some signal's amplitude. For example, two bits can be sent with a signal set comprised of a sinusoid with amplitudes of $\pm A$ and $\pm\left(\frac{A}{2}\right)$.

6.32 Comparison of Analog and Digital Communication[45]

Analog communication systems, amplitude modulation (AM) radio being a typifying example, can inexpensively communicate a bandlimited analog signal from one location to another (point-to-point communication) or from one point to many (broadcast). Although it is not shown here, the coherent receiver (Figure 6.6) provides the largest possible signal-to-noise ratio for the demodulated message. An analysis (Section 6.12) of this receiver thus indicates that some residual error will *always* be present in an analog system's output.

Although analog systems are less expensive in many cases than digital ones for the same application, digital systems offer much more efficiency, better performance, and much greater flexibility.

- **Efficiency**: The Source Coding Theorem allows quantification of just how complex a given message source is and allows us to exploit that complexity by source coding (compression). In analog communication, the only parameters of interest are message bandwidth and amplitude. We cannot exploit signal structure to achieve a more efficient communication system.
- **Performance**: Because of the Noisy Channel Coding Theorem, we have a specific criterion by which to formulate error-correcting codes that can bring us as close to error-free transmission as we might want. Even though we may send information by way of a noisy channel, digital schemes are capable of error-free transmission while analog ones cannot overcome channel disturbances; see this problem (Problem 6.15) for a comparison.

[44] The bandwidth restriction arises not so much from channel properties, but from spectral regulation, especially for wireless channels.

[45] This content is available online at <http://cnx.org/content/m0074/2.11/>.

- **Flexibility**: Digital communication systems can transmit real-valued discrete-time signals, which could be analog ones obtained by analog-to-digital conversion, *and* symbolic-valued ones (computer data, for example). Any signal that can be transmitted by analog means can be sent by digital means, with the only issue being the number of bits used in A/D conversion (how accurately do we need to represent signal amplitude). Images can be sent by analog means (commercial television), but better communication performance occurs when we use digital systems (HDTV). In addition to digital communication's ability to transmit a wider variety of signals than analog systems, point-to-point digital systems can be organized into global (and beyond as well) systems that provide efficient and flexible information transmission. **Computer networks**, explored in the next section, are what we call such systems today. Even analog-based networks, such as the telephone system, employ modern computer networking ideas rather than the purely analog systems of the past.

Consequently, with the increased speed of digital computers, the development of increasingly efficient algorithms, and the ability to interconnect computers to form a communications infrastructure, digital communication is now the best choice for many situations.

6.33 Communication Networks[46]

Communication networks elaborate the Fundamental Model of Communications (Figure 1.4: Fundamental model of communication). The model shown in Figure 6.27 describes **point-to-point** communications well, wherein the link between transmitter and receiver is straightforward, and they have the channel to themselves. One modern example of this communications mode is the modem that connects a personal computer with an information server via a telephone line. The key aspect, some would say flaw, of this model is that the channel is **dedicated**: Only one communications link through the channel is allowed for all time. Regardless whether we have a wireline or wireless channel, communication bandwidth is precious, and if it could be shared without significant degradation in communications performance (measured by signal-to-noise ratio for analog signal transmission and by bit-error probability for digital transmission) so much the better.

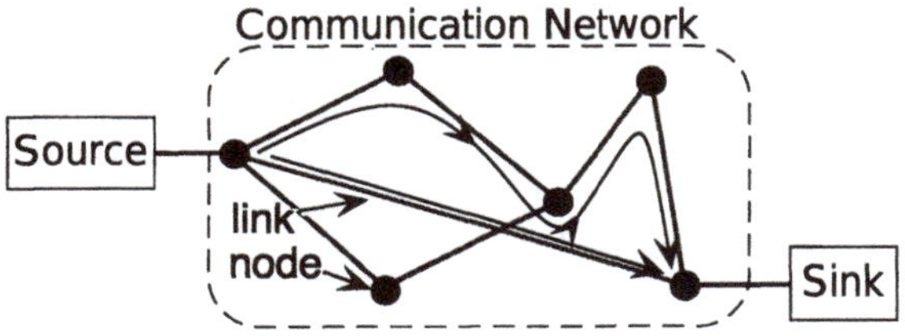

Figure 6.27: The prototypical communications network—whether it be the postal service, cellular telephone, or the Internet—consists of nodes interconnected by links. Messages formed by the source are transmitted within the network by dynamic routing. Two routes are shown. The longer one would be used if the direct link were disabled or congested.

The idea of a network first emerged with perhaps the oldest form of organized communication: the postal service. Most communication networks, even modern ones, share many of its aspects.

- A user writes a letter, serving in the communications context as the message source.
- This message is sent to the network by delivery to one of the network's public entry points. Entry points in the postal case are mailboxes, post offices, or your friendly mailman or mailwoman picking up the letter.

[46]This content is available online at <http://cnx.org/content/m0075/2.10/>.

- The communications network delivers the message in the most efficient (timely) way possible, trying not to corrupt the message while doing so.
- The message arrives at one of the network's exit points, and is delivered to the recipient (what we have termed the message sink).

Exercise 6.36 *(Solution on p. 276.)*

Develop the network model for the telephone system, making it as analogous as possible with the postal service-communications network metaphor.

What is most interesting about the network system is the ambivalence of the message source and sink about how the communications link is made. What they do care about is message integrity and communications efficiency. Furthermore, today's networks use heterogeneous links. Communication paths that form the Internet use wireline, optical fiber, and satellite communication links.

The first **electrical** communications network was the telegraph. Here the network consisted of telegraph operators who transmitted the message efficiently using Morse code and *routed* the message so that it took the shortest possible path to its destination while taking into account internal network failures (downed lines, drunken operators). From today's perspective, the fact that this nineteenth century system handled digital communications is astounding. Morse code, which assigned a sequence of dots and dashes to each letter of the alphabet, served as the source coding algorithm. The signal set consisted of a short and a long pulse. Rather than a matched filter, the receiver was the operator's ear, and he wrote the message (translating from received bits to symbols).

> NOTE: Because of the need for a comma between dot-dash sequences to define letter (symbol) boundaries, the average number of bits/symbol, as described in Subtleties of Coding (Example 6.4), exceeded the Source Coding Theorem's *upper* bound.

Internally, communication networks do have point-to-point communication links between network **nodes** well described by the Fundamental Model of Communications. However, many messages share the communications channel between nodes using what we call **time-domain multiplexing**: Rather than the continuous communications mode implied in the Model as presented, message sequences are sent, sharing in time the channel's capacity. At a grander viewpoint, the network must **route** messages—decide what nodes and links to use—based on destination information—the **address**—that is usually separate from the message information. Routing in networks is necessarily dynamic: The complete route taken by messages is formed as the network handles the message, with nodes relaying the message having some notion of the best possible path at the time of transmission. Note that no omnipotent router views the network as a whole and pre-determines every message's route. Certainly in the case of the postal system dynamic routing occurs, and can consider issues like inoperative and overly busy links. In the telephone system, routing takes place when you place the call; the route is fixed once the phone starts ringing. Modern communication networks strive to achieve the most efficient (timely) and most reliable information delivery system possible.

6.34 Message Routing[47]

Focusing on electrical networks, most analog ones make inefficient use of communication links because truly dynamic routing is difficult, if not impossible, to obtain. In radio networks, such as commercial television, each station has a dedicated portion of the electromagnetic spectrum, and this spectrum cannot be shared with other stations or used in any other than the regulated way. The telephone network is more dynamic, but once it establishes a call the path through the network is fixed. The users of that path control its use, and may not make efficient use of it (long pauses while one person thinks, for example). Telephone network customers would be quite upset if the telephone company momentarily disconnected the path so that someone else could use it. This kind of connection through a network—fixed for the duration of the communication session—is known as a **circuit-switched** connection.

[47]This content is available online at <http://cnx.org/content/m0076/2.8/>.

During the 1960s, it was becoming clear that not only was digital communication technically superior, but also that the wide variety of communication modes—computer login, file transfer, and electronic mail—needed a different approach than point-to-point. The notion of computer networks was born then, and what was then called the ARPANET, now called the Internet, was born. Computer networks elaborate the basic network model by subdividing messages into smaller chunks called **packets** (Figure 6.28). The rationale for the network enforcing smaller transmissions was that large file transfers would consume network resources all along the route, and, because of the long transmission time, a communication failure might require retransmission of the entire file. By creating packets, each of which has its own address and is routed independently of others, the network can better manage congestion. The analogy is that the postal service, rather than sending a long letter in the envelope you provide, opens the envelope, places each page in a separate envelope, and using the address on your envelope, addresses each page's envelope accordingly, and mails them separately. The network does need to make sure packet sequence (page numbering) is maintained, and the network exit point must reassemble the original message accordingly.

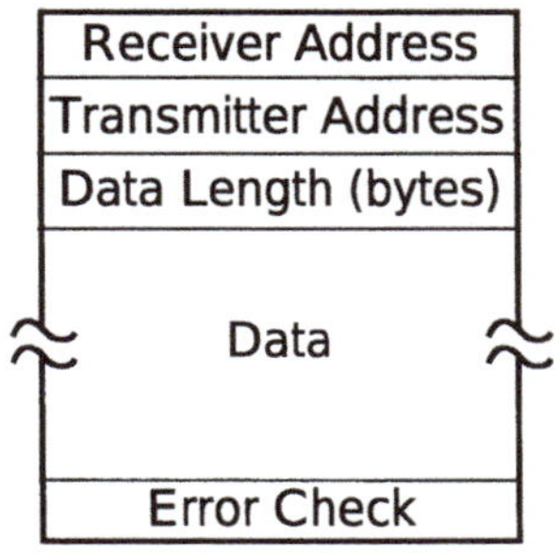

Figure 6.28: Long messages, such as files, are broken into separate packets, then transmitted over computer networks. A packet, like a letter, contains the destination address, the return address (transmitter address), and the data. The data includes the message part and a sequence number identifying its order in the transmitted message.

Communications networks are now categorized according to whether they use packets or not. A system like the telephone network is said to be **circuit switched**: The network establishes a **fixed** route that lasts the entire duration of the message. Circuit switching has the advantage that once the route is determined, the users can use the capacity provided them however they like. Its main disadvantage is that the users may not use their capacity efficiently, clogging network links and nodes along the way. **Packet-switched** networks continuously monitor network utilization, and route messages accordingly. Thus, messages can, on the average, be delivered efficiently, but the network cannot guarantee a specific amount of capacity to the users.

6.35 Network architectures and interconnection[48]

The network structure—its architecture (Figure 6.27)—typifies what are known as **wide area networks** (WANs). The nodes, and users for that matter, are spread geographically over long distances. "Long" has no precise definition, and is intended to suggest that the communication links vary widely. The Internet is certainly the largest WAN, spanning the entire earth and beyond. **Local area networks**, LANs, employ a single communication link and special routing. Perhaps the best known LAN is Ethernet[49]. LANs con-

[48]This content is available online at <http://cnx.org/content/m0077/2.9/>.

[49]"Ethernet" <http://cnx.org/content/m0078/latest/>

nect to other LANs and to wide area networks through special nodes known as **gateways** (Figure 6.29). In the Internet, a computer's address consists of a four byte sequence, which is known as its **IP address** (Internet Protocol address). An example address is *128.42.4.32*: each byte is separated by a period. The first two bytes specify the computer's **domain** (here Rice University). Computers are also addressed by a more human-readable form: a sequence of alphabetic abbreviations representing institution, type of institution, and computer name. A given computer has both names (*128.42.4.32* is the same as *soma.rice.edu*). Data transmission on the Internet requires the numerical form. So-called **name servers** translate between alphabetic and numerical forms, and the transmitting computer requests this translation before the message is sent to the network.

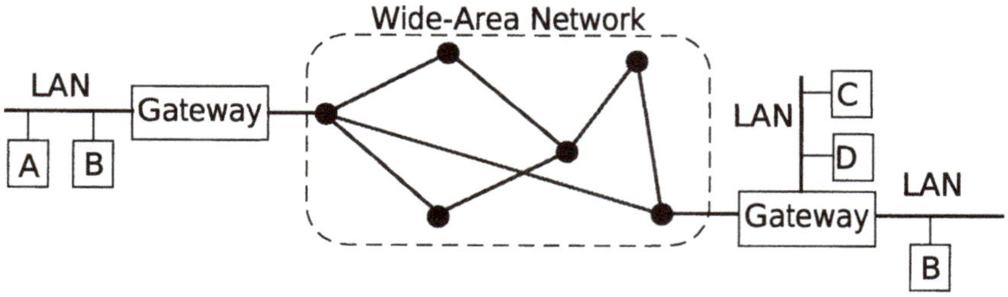

Figure 6.29: The gateway serves as an interface between local area networks and the Internet. The two shown here translate between LAN and WAN protocols; one of these also interfaces between two LANs, presumably because together the two LANs would be geographically too dispersed.

6.36 Ethernet[50]

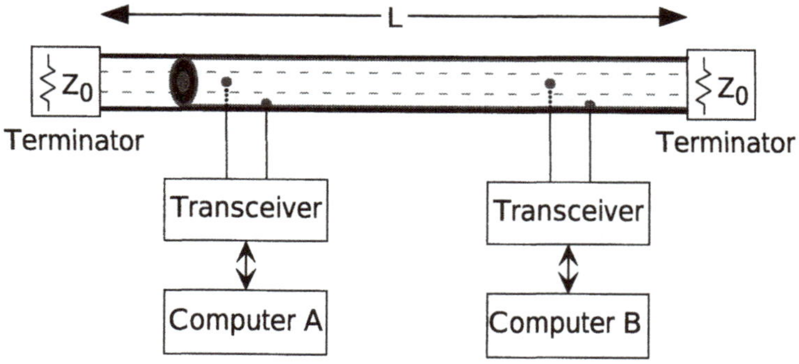

Figure 6.30: The Ethernet architecture consists of a single coaxial cable terminated at either end by a resistor having a value equal to the cable's characteristic impedance. Computers attach to the Ethernet through an interface known as a *transceiver* because it sends as well as receives bit streams represented as analog voltages.

[50] This content is available online at <http://cnx.org/content/m10284/2.12/>.

Ethernet uses as its communication medium a single length of coaxial cable (Figure 6.30). This cable serves as the "ether", through which all digital data travel. Electrically, computers interface to the coaxial cable (Figure 6.30) through a device known as a **transceiver**. This device is capable of monitoring the voltage appearing between the core conductor and the shield as well as applying a voltage to it. Conceptually it consists of two op-amps, one applying a voltage corresponding to a bit stream (transmitting data) and another serving as an amplifier of Ethernet voltage signals (receiving data). The signal set for Ethernet resembles that shown in BPSK Signal Sets, with one signal the negative of the other. Computers are attached in parallel, resulting in the circuit model for Ethernet shown in Figure 6.31.

Exercise 6.37 *(Solution on p. 276.)*

From the viewpoint of a transceiver's sending op-amp, what is the load it sees and what is the transfer function between this output voltage and some other transceiver's receiving circuit? Why should the output resistor R_{out} be large?

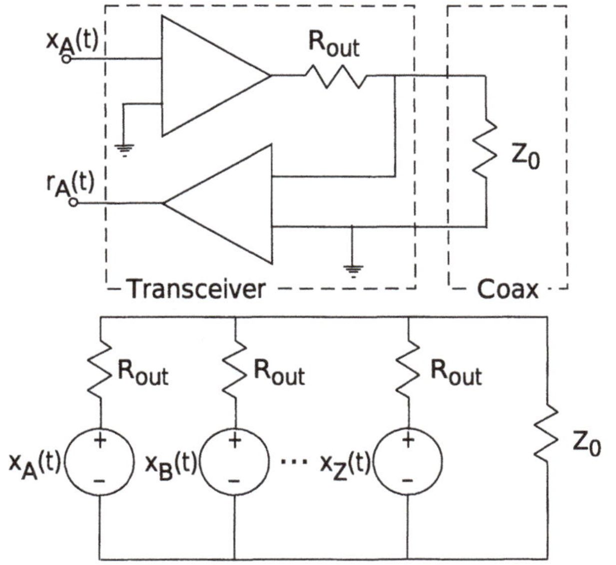

Figure 6.31: The top circuit expresses a simplified circuit model for a transceiver. The output resistance R_{out} must be much larger than Z_0 so that the sum of the various transmitter voltages add to create the Ethernet conductor-to-shield voltage that serves as the received signal $r(t)$ for all transceivers. In this case, the equivalent circuit shown in the bottom circuit applies.

No one computer has more authority than any other to control when and how messages are sent. Without scheduling authority, you might well wonder how one computer sends to another without the (large) interference that the other computers would produce if they transmitted at the same time. The innovation of Ethernet is that computers schedule themselves by a **random-access** method. This method relies on the fact that *all* packets transmitted over the coaxial cable can be received by *all* transceivers, regardless of which computer might actually be the intended recipient. In communications terminology, Ethernet directly supports broadcast. Each computer goes through the following steps to send a packet.

1. The computer senses the voltage across the cable to determine if some other computer is transmitting.

2. If another computer is transmitting, wait until the transmissions finish and go back to the first step. If the cable has no transmissions, begin transmitting the packet.
3. If the receiver portion of the transceiver determines that no other computer is also sending a packet, continue transmitting the packet until completion.
4. On the other hand, if the receiver senses interference from another computer's transmissions, immediately cease transmission, waiting a random amount of time to attempt the transmission again (go to step 1) until only one computer transmits and the others defer. The condition wherein two (or more) computers' transmissions interfere with others is known as a **collision**.

The reason two computers waiting to transmit may not sense the other's transmission immediately arises because of the finite propagation speed of voltage signals through the coaxial cable. The longest time any computer must wait to determine if its transmissions do not encounter interference is $\frac{2L}{c}$, where L is the coaxial cable's length. The maximum-length-specification for Ethernet is 1 km. Assuming a propagation speed of 2/3 the speed of light, this time interval is more than 10 μs. As analyzed in Problem 22 (Problem 6.30), the number of these time intervals required to resolve the collision is, on the average, less than two!

Exercise 6.38 *(Solution on p. 276.)*

Why does the factor of two enter into this equation? (Consider the worst-case situation of two transmitting computers located at the Ethernet's ends.)

Thus, despite not having separate communication paths among the computers to coordinate their transmissions, the Ethernet random access protocol allows computers to communicate without only a slight degradation in efficiency, as measured by the time taken to resolve collisions relative to the time the Ethernet is used to transmit information.

A subtle consideration in Ethernet is the minimum packet size P_{min}. The time required to transmit such packets equals $\frac{P_{min}}{C}$, where C is the Ethernet's capacity in bps. Ethernet now comes in two different types, each with individual specifications, the most distinguishing of which is capacity: 10 Mbps and 100 Mbps. If the minimum transmission time is such that the beginning of the packet has not propagated the full length of the Ethernet before the end-of-transmission, it is possible that two computers will begin transmission at the same time and, by the time their transmissions cease, the other's packet will not have propagated to the other. In this case, computers in-between the two will sense a collision, which renders both computer's transmissions senseless to them, without the two transmitting computers knowing a collision has occurred at all! For Ethernet to succeed, we must have the minimum packet transmission time exceed *twice* the voltage propagation time: $\frac{P_{min}}{C} > \frac{2L}{c}$ or

$$P_{min} > \frac{2LC}{c} \tag{6.63}$$

Thus, for the 10 Mbps Ethernet having a 1 km maximum length specification, the minimum packet size is 200 bits.

Exercise 6.39 *(Solution on p. 277.)*

The 100 Mbps Ethernet was designed more recently than the 10 Mbps alternative. To maintain the same minimum packet size as the earlier, slower version, what should its length specification be? Why should the minimum packet size remain the same?

6.37 Communication Protocols[51]

The complexity of information transmission in a computer network—reliable transmission of bits across a channel, routing, and directing information to the correct destination within the destination computers operating system—demands an overarching concept of how to organize information delivery. No unique set of

[51] This content is available online at <http://cnx.org/content/m0080/2.18/>.

rules satisfies the various constraints communication channels and network organization place on information transmission. For example, random access issues in Ethernet are not present in wide-area networks such as the Internet. A **protocol** is a set of rules that governs how information is delivered. For example, to use the telephone network, the protocol is to pick up the phone, listen for a dial tone, dial a number having a specific number of digits, wait for the phone to ring, and say hello. In radio, the station uses amplitude or frequency modulation with a specific carrier frequency and transmission bandwidth, and you know to turn on the radio and tune in the station. In technical terms, no one protocol or set of protocols can be used for any communication situation. Be that as it may, communication engineers have found that a common thread runs through the *organization* of the various protocols. This grand design of information transmission organization runs through all modern networks today.

What has been defined as a networking standard is a layered, hierarchical protocol organization. As shown in Figure 6.32 (Protocol Picture), protocols are organized by function and level of detail.

Protocol Picture

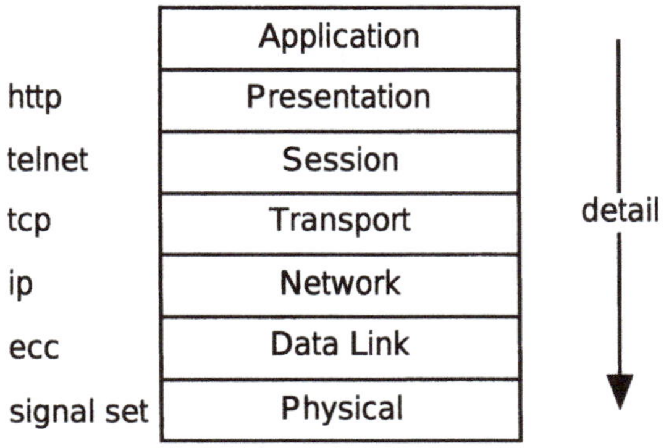

ISO Network Protocol Standard

Figure 6.32: Protocols are organized according to the level of detail required for information transmission. Protocols at the lower levels (shown toward the bottom) concern reliable bit transmission. Higher level protocols concern how bits are organized to represent information, what kind of information is defined by bit sequences, what software needs the information, and how the information is to be interpreted. Bodies such as the IEEE (Institute for Electronics and Electrical Engineers) and the ISO (International Standards Organization) define standards such as this. Despite being a standard, it does not constrain protocol implementation so much that innovation and competitive individuality are ruled out.

Segregation of information transmission, manipulation, and interpretation into these categories directly affects how communication systems are organized, and what role(s) software systems fulfill. Although not thought about in this way in earlier times, this organizational structure governs the way communication engineers think about all communication systems, from radio to the Internet.

Exercise 6.40 ***(Solution on p. 277.)***

How do the various aspects of establishing and maintaining a telephone conversation fit into this layered protocol organization?

We now explicitly state whether we are working in the physical layer (signal set design, for example), the data link layer (source and channel coding), or any other layer. IP abbreviates Internet protocol, and governs gateways (how information is transmitted between networks having different internal organizations). TCP (transmission control protocol) governs how packets are transmitted through a wide-area network such as the Internet. Telnet is a protocol that concerns how a person at one computer logs on to another computer across a network. A moderately high level protocol such as telnet, is not concerned with what data links (wireline or wireless) might have been used by the network or how packets are routed. Rather, it establishes connections between computers and directs each byte (presumed to represent a typed character) to the appropriate operation system component at each end. It is *not* concerned with what the characters mean or what programs the person is typing to. That aspect of information transmission is left to protocols at higher layers.

Recently, an important set of protocols created the World Wide Web. These protocols exist independently of the Internet. The Internet insures that messages are transmitted efficiently and intact; the Internet is not concerned (to date) with what messages contain. HTTP (hypertext transfer protocol) frame what messages contain and what should be done with the data. The extremely rapid development of the Web on top of an essentially stagnant Internet is but one example of the power of organizing how information transmission occurs without overly constraining the details.

6.38 Information Communication Problems[52]

Problem 6.1: Signals on Transmission Lines
A modulated signal needs to be sent over a transmission line having a characteristic impedance of $Z_0 = 50\Omega$. So that the signal does not interfere with signals others may be transmitting, it must be bandpass filtered so that its bandwidth is 1 MHz and centered at 3.5 MHz. The filter's gain should be one in magnitude. An op-amp filter (Figure 6.33) is proposed.

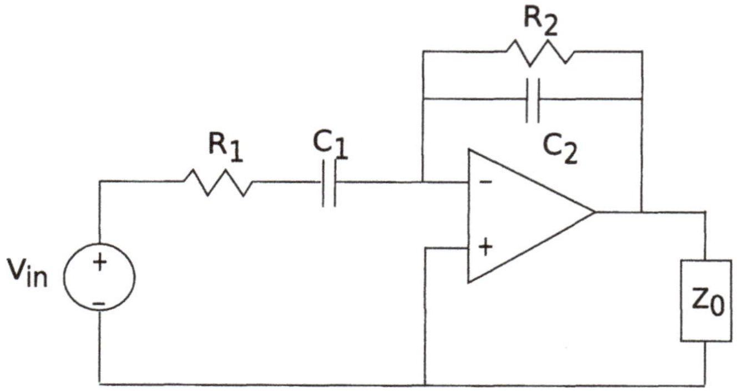

Figure 6.33

a) What is the transfer function between the input voltage and the voltage across the transmission line?
b) Find values for the resistors and capacitors so that design goals are met.

Problem 6.2: Noise in AM Systems
The signal $\hat{s}(t)$ emerging from an AM communication system consists of two parts: the message signal, $s(t)$, and additive noise. The plot (Figure 6.34) shows the message spectrum $S(f)$ and noise power spectrum $P_N(f)$. The noise power spectrum lies completely within the signal's band, and has a constant value there of $\frac{N_0}{2}$.

[52]This content is available online at <http://cnx.org/content/m10352/2.21/>.

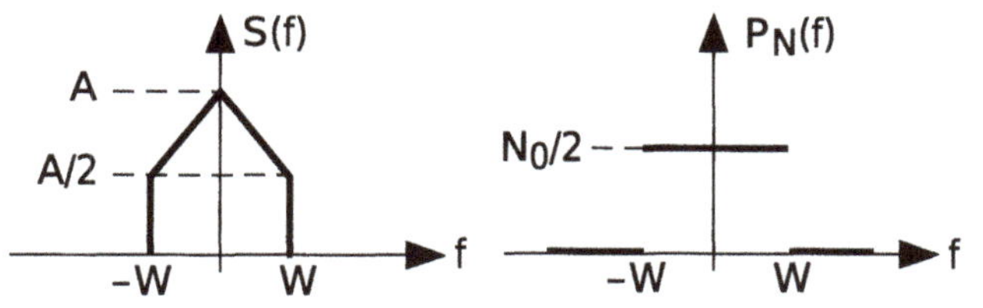

Figure 6.34

a) What is the message signal's power? What is the signal-to-noise ratio?
b) Because the power in the message decreases with frequency, the signal-to-noise ratio is not constant within subbands. What is the signal-to-noise ratio in the upper half of the frequency band?
c) A clever 241 student suggests filtering the message before the transmitter modulates it so that the signal spectrum is *balanced* (constant) across frequency. Realizing that this filtering affects the message signal, the student realizes that the receiver must also compensate for the message to arrive intact. Draw a block diagram of this communication system. How does this system's signal-to-noise ratio compare with that of the usual AM radio?

Problem 6.3: Complementary Filters
Complementary filters usually have "opposite" filtering characteristics (like a lowpass and a highpass) and have transfer functions that add to one. Mathematically, $H_1(f)$ and $H_2(f)$ are complementary if

$$H_1(f) + H_2(f) = 1$$

We can use complementary filters to separate a signal into two parts by passing it through each filter. Each output can then be transmitted separately and the original signal reconstructed at the receiver. Let's assume the mesage is bandlimited to WHz and that $H_1(f) = \frac{a}{a+j2\pi f}$.

a) What circuits would be used to produce the complementary filters?
b) Sketch a block diagram for a communication system (transmitter and receiver) that employs complementary signal transmission to send a message $m(t)$.
c) What is the receiver's signal-to-noise ratio? How does it compare to the standard system that sends the signal by simple amplitude modulation?

Problem 6.4: Phase Modulation
A message signal $m(t)$ **phase modulates** a carrier if the transmitted signal equals

$$x(t) = A\sin(2\pi f_c t + \phi_d m(t))$$

where ϕ_d is known as the phase deviation. In this problem, the phase deviation is small. As with all analog modulation schemes, assume that $|m(t)| < 1$, the message is bandlimited to W Hz, and the carrier frequency f_c is much larger than W.

a) What is the transmission bandwidth?
b) Find a receiver for this modulation scheme.
c) What is the signal-to-noise ratio of the received signal?

HINT: Use the facts that $\cos(x) \approx 1$ and $\sin(x) \approx x$ for small x.

Problem 6.5: Digital Amplitude Modulation
Two ELEC 241 students disagree about a homework problem. The issue concerns the discrete-time signal $s(n)\cos(2\pi f_0 n)$, where the signal $s(n)$ has no special characteristics and the modulation frequency f_0 is known. Sammy says that he can recover $s(n)$ from its amplitude-modulated version by the same approach used in analog communications. Samantha says that approach won't work.

a) What is the spectrum of the modulated signal?
b) Who is correct? Why?
c) The teaching assistant does not want to take sides. He tells them that if $s(n)\cos(2\pi f_0 n)$ and $s(n)\sin(2\pi f_0 n)$ were both available, $s(n)$ can be recovered. What does he have in mind?

Problem 6.6: Anti-Jamming
One way for someone to keep people from receiving an AM transmission is to transmit noise at the same carrier frequency. Thus, if the carrier frequency is f_c so that the transmitted signal is $A_T(1+m(t))\sin(2\pi f_c t)$ the *jammer* would transmit $A_J n(t)(\sin(2\pi f_c t)+\phi)$. The noise $n(t)$ has a constant power density spectrum over the bandwidth of the message $m(t)$. The channel adds white noise of spectral height $\frac{N_0}{2}$.

a) What would be the output of a traditional AM receiver tuned to the carrier frequency f_c?
b) RU Electronics proposes to counteract jamming by using a different modulation scheme. The scheme's transmitted signal has the form $A_T(1+m(t))c(t)$ where $c(t)$ is a periodic carrier signal (period $\frac{1}{f_c}$) having the indicated waveform (Figure 6.35). What is the spectrum of the transmitted signal with the proposed scheme? Assume the message bandwidth W is much less than the fundamental carrier frequency f_c.
c) The jammer, unaware of the change, is transmitting with a carrier frequency of f_c, while the receiver tunes a standard AM receiver to a harmonic of the carrier frequency. What is the signal-to-noise ratio of the receiver tuned to the harmonic having the largest power that does not contain the jammer?

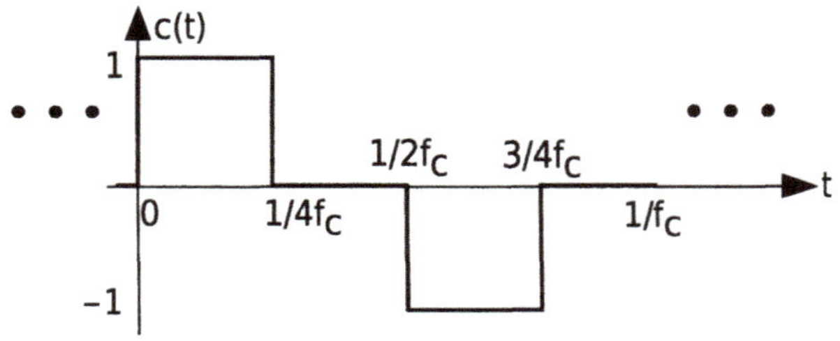

Figure 6.35

Problem 6.7: Secret Comunications
A system for hiding AM transmissions has the transmitter randomly switching between two carrier frequencies f_1 and f_2. "Random switching" means that one carrier frequency is used for some period of time, switches to the other for some other period of time, back to the first, etc. The receiver knows what the carrier frequencies are but not when carrier frequency switches occur. Consequently, the receiver must be designed to receive the transmissions regardless of which carrier frequency is used. Assume the message signal has bandwidth W. The channel adds white noise of spectral height $\frac{N_0}{2}$.

a) How different should the carrier frequencies be so that the message could be received?
b) What receiver would you design?
c) What signal-to-noise ratio for the demodulated signal does your receiver yield?

Problem 6.8: AM Stereo
Stereophonic radio transmits two signals simultaneously that correspond to what comes out of the left and right speakers of the receiving radio. While FM stereo is commonplace, AM stereo is not, but is much simpler to understand and analyze. An amazing aspect of AM stereo is that both signals are transmitted within the same bandwidth as used to transmit just one. Assume the left and right signals are bandlimited to W Hz.

$$x(t) = A(1 + m_l(t))\cos(2\pi f_c t) + A m_r(t)\sin(2\pi f_c t)$$

a) Find the Fourier transform of $x(t)$. What is the transmission bandwidth and how does it compare with that of standard AM?
b) Let us use a coherent demodulator as the receiver, shown in Figure 6.36. Show that this receiver indeed works: It produces the left and right signals separately.
c) Assume the channel adds white noise to the transmitted signal. Find the signal-to-noise ratio of each signal.

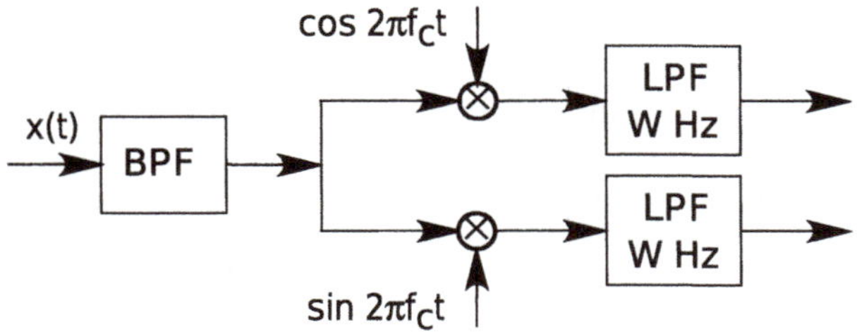

Figure 6.36

Problem 6.9: A Novel Communication System
A clever system designer claims that the depicted transmitter (Figure 6.37) has, despite its complexity, advantages over the usual amplitude modulation system. The message signal $m(t)$ is bandlimited to W Hz, and the carrier frequency $(f_c \gg W)$. The channel attenuates the transmitted signal $x(t)$ and adds white noise of spectral height $\frac{N_0}{2}$.

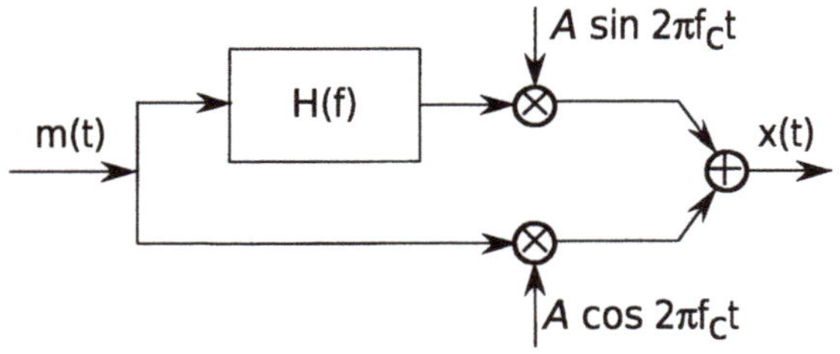

Figure 6.37

The transfer function $H(f)$ is given by $H(f) = \begin{cases} j \text{ if } f < 0 \\ -j \text{ if } f > 0 \end{cases}$

a) Find an expression for the spectrum of $x(t)$. Sketch your answer.
b) Show that the usual coherent receiver demodulates this signal.
c) Find the signal-to-noise ratio that results when this receiver is used.
d) Find a superior receiver (one that yields a better signal-to-noise ratio), and analyze its performance.

Problem 6.10: Multi-Tone Digital Communication
In a so-called multi-tone system, several bits are gathered together and transmitted simultaneously on different carrier frequencies during a T second interval. For example, B bits would be transmitted according to

$$x(t) = A \sum_{k=0}^{B-1} (b_k \sin(2\pi (k+1) f_0 t)) \quad , \quad 0 \le t < T \tag{6.64}$$

Here, f_0 is the frequency offset for each bit and it is harmonically related to the bit interval T. The value of b_k is either -1 or $+1$.

a) Find a receiver for this transmission scheme.
b) An ELEC 241 almuni likes digital systems so much that he decides to produce a discrete-time version. He samples the received signal (sampling interval $T_s = \frac{T}{N}$). How should N be related to B, the number of simultaneously transmitted bits?
c) The alumni wants to find a simple form for the receiver so that his software implementation runs as *efficiently* as possible. How would you recommend he implement the receiver?

Problem 6.11: City Radio Channels
In addition to additive white noise, metropolitan cellular radio channels also contain multipath: the attenuated signal and a delayed, further attenuated signal are received superimposed. As shown in Figure 6.38, multipath occurs because the buildings reflect the signal and the reflected path length between transmitter and receiver is longer than the direct path.

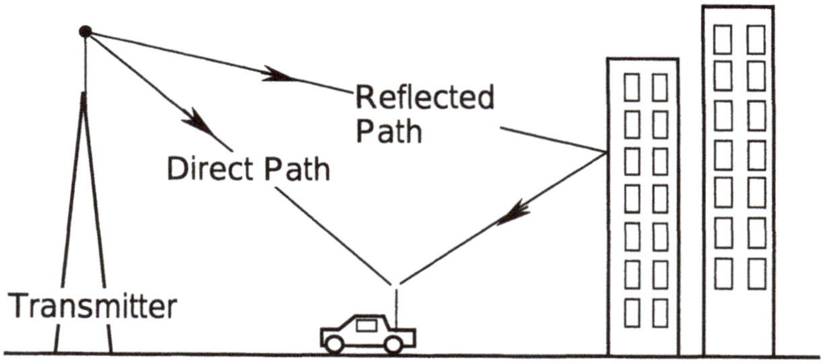

Figure 6.38

a) Assume that the length of the direct path is d meters and the reflected path is 1.5 times as long. What is the model for the channel, including the multipath and the additive noise?

b) Assume d is 1 km. Find and sketch the magnitude of the transfer function for the multipath component of the channel. How would you characterize this transfer function?
c) Would the multipath affect AM radio? If not, why not; if so, how so? Would analog cellular telephone, which operates at much higher carrier frequencies (800 MHz vs. 1 MHz for radio), be affected or not? Analog cellular telephone uses amplitude modulation to transmit voice.
d) How would the usual AM receiver be modified to minimize multipath effects? Express your modified receiver as a block diagram.

Problem 6.12: Downlink Signal Sets
In digital cellular telephone systems, the base station (transmitter) needs to relay different voice signals to several telephones at the same time. Rather than send signals at different frequencies, a clever Rice engineer suggests using a different signal set for each data stream. For example, for two simultaneous data streams, she suggests BPSK signal sets that have the depicted basic signals (Figure 6.39).

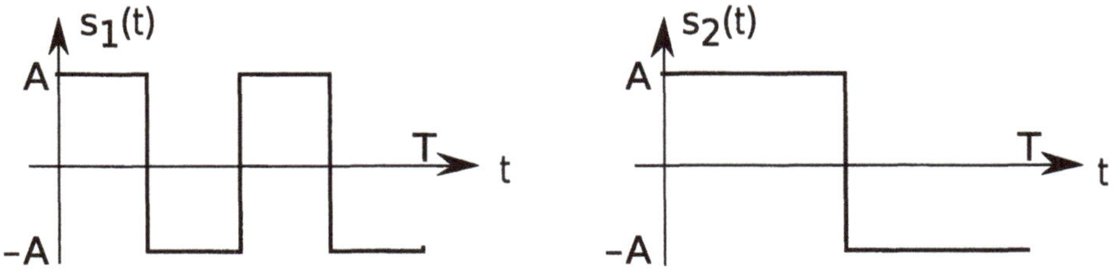

Figure 6.39

Thus, bits are represented in data stream 1 by $s_1(t)$ and $-(s_1(t))$ and in data stream 2 by $s_2(t)$ and $-(s_2(t))$, each of which are modulated by 900 MHz carrier. The transmitter sends the two data streams so that their bit intervals align. Each receiver uses a matched filter for its receiver. The requirement is that each receiver *not* receive the other's bit stream.

a) What is the block diagram describing the proposed system?
b) What is the transmission bandwidth required by the proposed system?
c) Will the proposal work? Does the fact that the two data streams are transmitted in the same bandwidth at the same time mean that each receiver's performance is affected? Can each bit stream be received without interference from the other?

Problem 6.13: Mixed Analog and Digital Transmission
A signal $m(t)$ is transmitted using amplitude modulation in the usual way. The signal has bandwidth W Hz, and the carrier frequency is f_c. In addition to sending this analog signal, the transmitter also wants to send ASCII text in an **auxiliary band** that lies slightly above the analog transmission band. Using an 8-bit representation of the characters and a simple baseband BPSK signal set (the constant signal +1 corresponds to a 0, the constant -1 to a 1), the data signal $d(t)$ representing the text is transmitted as the same time as the analog signal $m(t)$. The transmission signal spectrum is as shown (Figure 6.40), and has a total bandwidth B.

a) Write an expression for the time-domain version of the transmitted signal in terms of $m(t)$ and the digital signal $d(t)$.
b) What is the maximum datarate the scheme can provide in terms of the available bandwidth?
c) Find a receiver that yields both the analog signal and the bit stream.

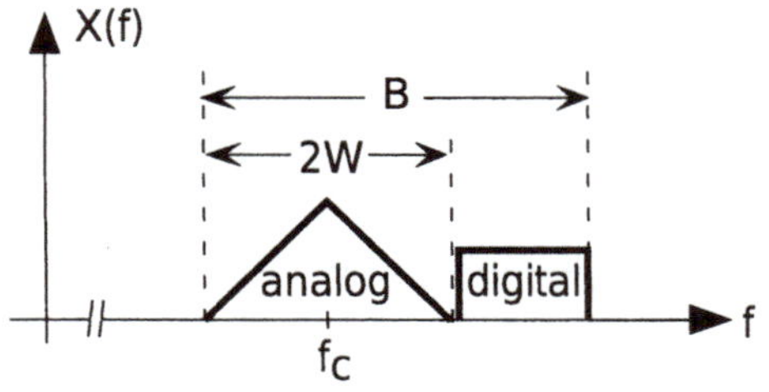

Figure 6.40

Problem 6.14: Digital Stereo
Just as with analog communication, it should be possible to send two signals simultaneously over a digital channel. Assume you have two CD-quality signals (each sampled at 44.1 kHz with 16 bits/sample). One suggested transmission scheme is to use a quadrature BPSK scheme. If $b^{(1)}(n)$ and $b^{(2)}(n)$ each represent a bit stream, the transmitted signal has the form

$$x(t) = A\sum \left(b^{(1)}(n)\sin(2\pi f_c(t-nT))\,p(t-nT) + b^{(2)}(n)\cos(2\pi f_c(t-nT))\,p(t-nT)\right)$$

where $p(t)$ is a unit-amplitude pulse having duration T and $b^{(1)}(n)$, $b^{(2)}(n)$ equal either +1 or -1 according to the bit being transmitted for each signal. The channel adds white noise and attenuates the transmitted signal.

a) What value would you choose for the carrier frequency f_c?
b) What is the transmission bandwidth?
c) What receiver would you design that would yield both bit streams?

Problem 6.15: Digital and Analog Speech Communication
Suppose we transmit speech signals over comparable digital and analog channels. We want to compare the resulting quality of the received signals. Assume the transmitters use the same power, and the channels introduce the same attenuation and additive white noise. Assume the speech signal has a 4 kHz bandwidth and, in the digital case, is sampled at an 8 kHz rate with eight-bit A/D conversion. Assume simple binary source coding and a modulated BPSK transmission scheme.

a) What is the transmission bandwidth of the analog (AM) and digital schemes?
b) Assume the speech signal's amplitude has a magnitude less than one. What is maximum amplitude quantization error introduced by the A/D converter?
c) In the digital case, each bit in quantized speech sample is received in error with probability p_e that depends on signal-to-noise ratio $\frac{E_b}{N_0}$. However, errors in each bit have a different impact on the error in the reconstructed speech sample. Find the mean-squared error between the transmitted and received amplitude.
d) In the digital case, the recovered speech signal can be considered to have two noise sources added to each sample's true value: One is the A/D amplitude quantization noise and the second is due to channel errors. Because these are separate, the total noise power equals the sum of these two. What is the signal-to-noise ratio of the received speech signal as a function of p_e?
e) Compute and plot the received signal's signal-to-noise ratio for the two transmission schemes for a few values of channel signal-to-noise ratios.
f) Compare and evaluate these systems.

Problem 6.16: Source Compression
Consider the following 5-letter source.

Letter	**Probability**
a	0.5
b	0.25
c	0.125
d	0.0625
e	0.0625

a) Find this source's entropy.
b) Show that the simple binary coding is inefficient.
c) Find an unequal-length codebook for this sequence that satisfies the Source Coding Theorem. Does your code achieve the entropy limit?
d) How much more efficient is this code than the simple binary code?

Problem 6.17: Source Compression
Consider the following 5-letter source.

Letter	**Probability**
a	0.4
b	0.2
c	0.15
d	0.15
e	0.1

a) Find this source's entropy.
b) Show that the simple binary coding is inefficient.
c) Find the Huffman code for this source. What is its average code length?

Problem 6.18: Speech Compression
When we sample a signal, such as speech, we quantize the signal's amplitude to a set of integers. For a b-bit converter, signal amplitudes are represented by 2^b integers. Although these integers could be represented by a binary code for digital transmission, we should consider whether a Huffman coding would be more efficient.

a) Load into Matlab the segment of speech contained in `y.mat`. Its sampled values lie in the interval (-1, 1). To simulate a 3-bit converter, we use Matlab's round function to create quantized amplitudes corresponding to the integers `[0 1 2 3 4 5 6 7]`.

   - `y_quant = round(3.5*y + 3.5);`

   Find the relative frequency of occurrence of quantized amplitude values. The following Matlab program computes the number of times each quantized value occurs.

   - `for n=0:7; count(n+1) = sum(y_quant == n); end;`

   Find the entropy of this source.

b) Find the Huffman code for this source. How would you characterize this source code in words?
c) How many fewer bits would be used in transmitting this speech segment with your Huffman code in comparison to simple binary coding?

Problem 6.19: Digital Communication
In a digital cellular system, a signal bandlimited to 5 kHz is sampled with a two-bit A/D converter at its Nyquist frequency. The sample values are found to have the shown relative frequencies.

Sample Value	Probability
0	0.15
1	0.35
2	0.3
3	0.2

We send the bit stream consisting of Huffman-coded samples using one of the two depicted signal sets (Figure 6.41).

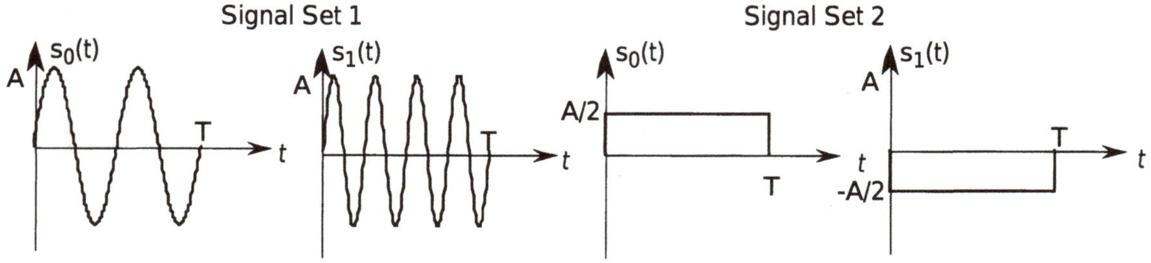

Figure 6.41

a) What is the datarate of the compressed source?
b) Which choice of signal set maximizes the communication system's performance?
c) With no error-correcting coding, what signal-to-noise ratio would be needed for your chosen signal set to guarantee that the bit error probability will not exceed 10^{-3}? If the receiver moves twice as far from the transmitter (relative to the distance at which the 10^{-3} error rate was obtained), how does the performance change?

Problem 6.20: Signal Compression
Letters drawn from a four-symbol alphabet have the indicated probabilities.

Letter	Probability
a	1/3
b	1/3
c	1/4
d	1/12

a) What is the average number of bits necessary to represent this alphabet?
b) Using a simple binary code for this alphabet, a two-bit block of data bits naturally emerges. Find an error correcting code for two-bit data blocks that corrects all single-bit errors.
c) How would you modify your code so that the probability of the letter a being confused with the letter d is minimized? If so, what is your new code; if not, demonstrate that this goal cannot be achieved.

Problem 6.21: Universal Product Code
The Universal Product Code (UPC), often known as a bar code, labels virtually every sold good. An example (Figure 6.42) of a portion of the code is shown.

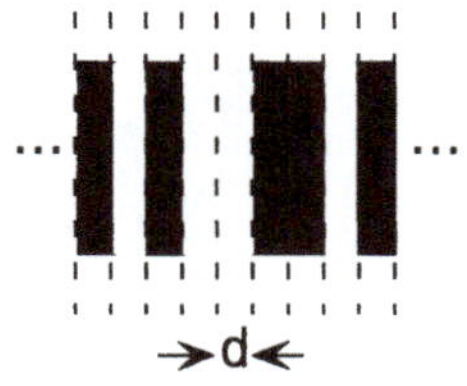

Figure 6.42

Here a sequence of black and white bars, each having width d, presents an 11-digit number (consisting of decimal digits) that uniquely identifies the product. In retail stores, laser scanners read this code, and after accessing a database of prices, enter the price into the cash register.

a) How many bars must be used to represent a single digit?
b) A complication of the laser scanning system is that the bar code must be read either forwards or backwards. Now how many bars are needed to represent each digit?
c) What is the probability that the 11-digit code is read correctly if the probability of reading a single bit incorrectly is p_e?
d) How many error correcting bars would need to be present so that any single bar error occurring in the 11-digit code can be corrected?

Problem 6.22: Error Correcting Codes
A code maps pairs of information bits into codewords of length 5 as follows.

Data	**Codeword**
00	00000
01	01101
10	10111
11	11010

a) What is this code's efficiency?
b) Find the generator matrix G and parity-check matrix H for this code.
c) Give the decoding table for this code. How many patterns of 1, 2, and 3 errors are correctly decoded?
d) What is the block error probability (the probability of any number of errors occurring in the decoded codeword)?

Problem 6.23: Overly Designed Error Correction Codes
An Aggie engineer wants not only to have codewords for his data, but also to hide the information from Rice engineers (no fear of the UT engineers). He decides to represent 3-bit data with 6-bit codewords in which none of the data bits appear explicitly.

$$\begin{aligned} c_1 &= (d_1 \oplus d_2) \\ c_2 &= (d_2 \oplus d_3) \\ c_3 &= (d_1 \oplus d_3) \\ c_4 &= (d_1 \oplus d_2 \oplus d_3) \\ c_5 &= (d_1 \oplus d_2) \\ c_6 &= (d_1 \oplus d_2 \oplus d_3) \end{aligned}$$

a) Find the generator matrix G and parity-check matrix H for this code.
b) Find a 3×6 matrix that recovers the data bits from the codeword.
c) What is the error correcting capability of the code?

Problem 6.24: Error Correction?
It is important to realize that when more transmission errors than can be corrected, error correction algorithms believe that a smaller number of errors have occurred and correct accordingly. For example, consider a (7,4) Hamming Code having the generator matrix

$$G = \begin{pmatrix} 1 & 0 & 0 & 0 \\ 0 & 1 & 0 & 0 \\ 0 & 0 & 1 & 0 \\ 0 & 0 & 0 & 1 \\ 1 & 1 & 1 & 0 \\ 0 & 1 & 1 & 1 \\ 1 & 0 & 1 & 1 \end{pmatrix}$$

This code corrects all single-bit error, but if a double bit error occurs, it corrects using a single-bit error correction approach.

a) How many double-bit errors can occur in a codeword?
b) For each double-bit error pattern, what is the result of channel decoding? Express your result as a binary error sequence for the data bits.

Problem 6.25: Selective Error Correction
We have found that digital transmission errors occur with a probability that remains constant no matter how "important" the bit may be. For example, in transmitting digitized signals, errors occur as frequently for the most significant bit as they do for the least significant bit. Yet, the former errors have a much larger impact on the overall signal-to-noise ratio than the latter. Rather than applying error correction to each sample value, why not concentrate the error correction on the most important bits? Assume that we sample an 8 kHz signal with an 8-bit A/D converter. We use single-bit error correction on the most significant four bits and none on the least significant four. Bits are transmitted using a modulated BPSK signal set over an additive white noise channel.

a) How many error correction bits must be added to provide single-bit error correction on the most significant bits?
b) How large must the signal-to-noise ratio of the received signal be to insure reliable communication?

c) Assume that once error correction is applied, only the least significant 4 bits can be received in error. How much would the output signal-to-noise ratio improve using this error correction scheme?

Problem 6.26: Compact Disk
Errors occur in reading audio compact disks. Very few errors are due to noise in the compact disk player; most occur because of dust and scratches on the disk surface. Because scratches span several bits, a single-bit error is rare; several *consecutive* bits in error are much more common. Assume that scratch and dust-induced errors are four or fewer consecutive bits long. The audio CD standard requires 16-bit, 44.1 kHz analog-to-digital conversion of each channel of the stereo analog signal.

a) How many error-correction bits are required to correct scratch-induced errors for each 16-bit sample?
b) Rather than use a code that can correct several errors in a codeword, a clever 241 engineer proposes *interleaving* consecutive coded samples. As the cartoon (Figure 6.43) shows, the bits representing coded samples are interpersed before they are written on the CD. The CD player de-interleaves the coded data, then performs error-correction. Now, evaluate this proposed scheme with respect to the non-interleaved one.

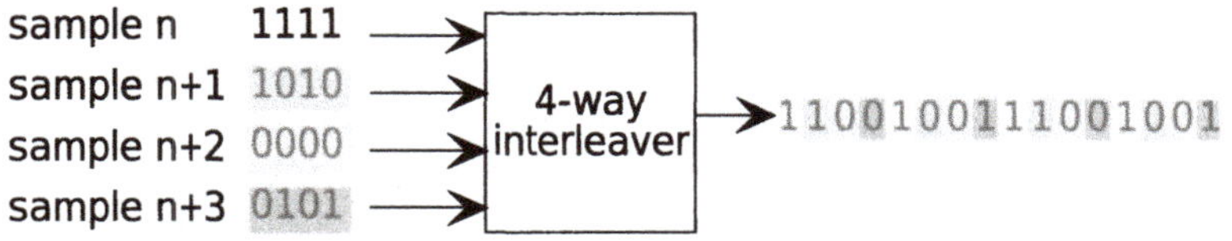

Figure 6.43

Problem 6.27: Communication System Design
RU Communication Systems has been asked to design a communication system that meets the following requirements.

- The baseband message signal has a bandwidth of 10 kHz.
- The RUCS engineers find that the entropy H of the sampled message signal depends on how many bits b are used in the A/D converter (see table below).
- The signal is to be sent through a noisy channel having a bandwidth of 25 kHz channel centered at 2 MHz and a signal-to-noise ration within that band of 10 dB.
- Once received, the message signal must have a signal-to-noise ratio of at least 20 dB.

b	H
3	2.19
4	3.25
5	4.28
6	5.35

Can these specifications be met? Justify your answer.

Problem 6.28: HDTV
As HDTV (high-definition television) was being developed, the FCC restricted this digital system to use in the same bandwidth (6 MHz) as its analog (AM) counterpart. HDTV video is sampled on a 1035×1840 raster at 30 images per second for each of the three colors. The least-acceptable picture received by television sets located at an analog station's broadcast perimeter has a signal-to-noise ratio of about 10 dB.

a) Using signal-to-noise ratio as the criterion, how many bits per sample must be used to guarantee that a high-quality picture, which achieves a signal-to-noise ratio of 20 dB, can be received by any HDTV set within the same broadcast region?
b) Assuming the digital television channel has the same characteristics as an analog one, how much compression must HDTV systems employ?

Problem 6.29: Digital Cellular Telephones
In designing a digital version of a wireless telephone, you must first consider certain fundamentals. First of all, the quality of the received signal, as measured by the signal-to-noise ratio, must be at least as good as that provided by wireline telephones (30 dB) and the message bandwidth must be the same as wireline telephone. The signal-to-noise ratio of the allocated wirelss channel, which has a 5 kHz bandwidth, measured 100 meters from the tower is 70 dB. The desired range for a cell is 1 km. Can a digital cellphone system be designed according to these criteria?

Problem 6.30: Optimial Ethernet Random Access Protocols
Assume a population of N computers want to transmit information on a random access channel. The access algorithm works as follows.

- Before transmitting, flip a coin that has probability p of coming up heads
- If only one of the N computer's coins comes up heads, its transmission occurs successfully, and the others must wait until that transmission is complete and then resume the algorithm.
- If none or more than one head comes up, the N computers will either remain silent (no heads) or a collision will occur (more than one head). This unsuccessful transmission situation will be detected by all computers once the signals have propagated the length of the cable, and the algorithm resumes (return to the beginning).

a) What is the optimal probability to use for flipping the coin? In other words, what should p be to maximize the probability that exactly one computer transmits?
b) What is the probability of one computer transmitting when this optimal value of p is used as the number of computers grows to infinity?
c) Using this optimal probability, what is the average number of coin flips that will be necessary to resolve the access so that one computer successfully transmits?
d) Evaluate this algorithm. Is it realistic? Is it efficient?

Problem 6.31: Repeaters
Because signals attenuate with distance from the transmitter, **repeaters** are frequently employed for both analog and digital communication. For example, let's assume that the transmitter and receiver are D m apart, and a repeater is positioned halfway between them (Figure 6.44). What the repater does is amplify its received signal to exactly cancel the attenuation encountered along the first leg and to re-transmit the signal to the ultimate receiver. However, the signal the repeater receives contains white noise as well as the transmitted signal. The receiver experiences the same amount of white noise as the repeater.

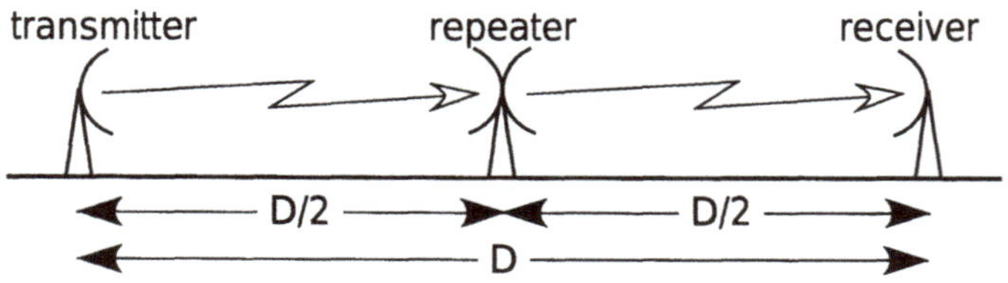

Figure 6.44

a) What is the block diagram for this system?
b) For an amplitude-modulation communication system, what is the signal-to-noise ratio of the demodulated signal at the receiver? Is this better or worse than the signal-to-noise ratio when no repeater is present?
c) For digital communication, we must consider the system's capacity. Is the capacity larger with the repeater system than without it? If so, when; if not, why not?

Solutions to Exercises in Chapter 6

Solution to Exercise 6.1 (p. 214)

In both cases, the answer depends less on geometry than on material properties. For coaxial cable, $c = \frac{1}{\sqrt{\mu_d \epsilon_d}}$. For twisted pair, $c = \frac{1}{\sqrt{\mu\epsilon}}\sqrt{\frac{\operatorname{arccosh}\left(\frac{d}{2r}\right)}{\frac{\delta}{2r}+\operatorname{arccosh}\left(\frac{d}{2r}\right)}}$.

Solution to Exercise 6.2 (p. 214)

You can find these frequencies from the spectrum allocation chart (Section 7.3). Light in the middle of the visible band has a wavelength of about 600 nm, which corresponds to a frequency of $5 \times 10^{14} Hz$. Cable television transmits within the same frequency band as broadcast television (about 200 MHz or $2 \times 10^8 Hz$). Thus, the visible electromagnetic frequencies are over six orders of magnitude higher!

Solution to Exercise 6.3 (p. 215)

As frequency increases, $\left(2\pi f \tilde{C} \gg \tilde{G}\right)$ and $\left(2\pi f \tilde{L} \gg \tilde{R}\right)$. In this high-frequency region,

$$\begin{aligned}\gamma &= j2\pi f\sqrt{\tilde{L}\tilde{C}}\sqrt{\left(1+\frac{\tilde{G}}{j2\pi f\,\tilde{C}}\right)\left(1+\frac{\tilde{R}}{j2\pi f\,\tilde{L}}\right)} \\ &\approx j2\pi f\sqrt{\tilde{L}\tilde{C}}\left(1+\frac{1}{2}\frac{1}{j2\pi f}\left(\frac{\tilde{G}}{\tilde{C}}+\frac{\tilde{R}}{\tilde{L}}\right)\right) \\ &\approx j2\pi f\sqrt{\tilde{L}\tilde{C}}+\frac{1}{2}\left(\tilde{G}\sqrt{\frac{\tilde{L}}{\tilde{C}}}+\tilde{R}\sqrt{\frac{\tilde{C}}{\tilde{L}}}\right)\end{aligned} \quad (6.65)$$

Thus, the attenuation (space) constant equals the real part of this expression, and equals $a(f) = \frac{\tilde{G}Z_0+\frac{\tilde{R}}{Z_0}}{2}$.

Solution to Exercise 6.4 (p. 215)

As shown previously (6.11), voltages and currents in a wireline channel, which is modeled as a transmission line having resistance, capacitance and inductance, decay exponentially with distance. The inverse-square law governs free-space propagation because such propagation is lossless, with the inverse-square law a consequence of the conservation of power. The exponential decay of wireline channels occurs because they have losses and some filtering.

Solution to Exercise 6.5 (p. 217)

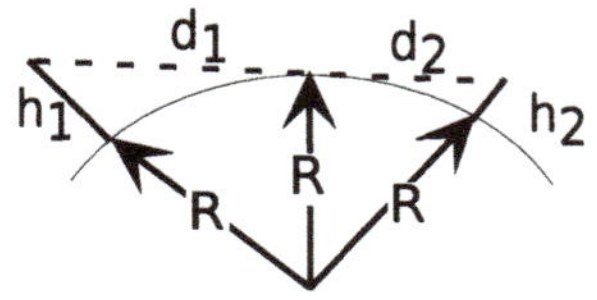

Figure 6.45

Use the Pythagorean Theorem, $(h+R)^2 = R^2 + d^2$, where h is the antenna height, d is the distance from the top of the earth to a tangency point with the earth's surface, and R the earth's radius. The line-of-sight distance between two earth-based antennae equals

$$d_{LOS} = \sqrt{2h_1R+{h_1}^2}+\sqrt{2h_2R+{h_2}^2} \quad (6.66)$$

As the earth's radius is much larger than the antenna height, we have to a good approximation that $d_{LOS} = \sqrt{2h_1R} + \sqrt{2h_2R}$. If one antenna is at ground elevation, say $h_2 = 0$, the other antenna's range is $\sqrt{2h_1R}$.

Solution to Exercise 6.6 (p. 217)

As frequency decreases, wavelength increases and can approach the distance between the earth's surface and the ionosphere. Assuming a distance between the two of 80 km, the relation $\lambda f = c$ gives a corresponding frequency of 3.75 kHz. Such low carrier frequencies would be limited to low bandwidth analog communication and to low datarate digital communications. The US Navy did use such a communication scheme to reach all of its submarines at once.

Solution to Exercise 6.7 (p. 218)

Transmission to the satellite, known as the uplink, encounters inverse-square law power losses. Reflecting off the ionosphere not only encounters the same loss, but twice. Reflection is the same as transmitting exactly what arrives, which means that the total loss is the *product* of the uplink and downlink losses. The geosynchronous orbit lies at an altitude of $35700km$. The ionosphere begins at an altitude of about 50 km. The amplitude loss in the satellite case is proportional to 2.8×10^{-8}; for Marconi, it was proportional to 4.4×10^{-10}. Marconi was *very* lucky.

Solution to Exercise 6.8 (p. 218)

If the interferer's spectrum does not overlap that of our communications channel—the interferer is out-of-band—we need only use a bandpass filter that selects our transmission band and removes other portions of the spectrum.

Solution to Exercise 6.9 (p. 219)

The additive-noise channel is *not* linear because it does not have the zero-input-zero-output property (even though we might transmit nothing, the receiver's input consists of noise).

Solution to Exercise 6.10 (p. 222)

The signal-related portion of the transmitted spectrum is given by $X(f) = \frac{1}{2}M(f - f_c) + \frac{1}{2}M(f + f_c)$. Multiplying at the receiver by the carrier shifts this spectrum to $+f_c$ and to $-f_c$, and scales the result by half.

$$\begin{aligned} \frac{1}{2}X(f - f_c) + \frac{1}{2}X(f + f_c) &= \frac{1}{4}(M(f - 2f_c) + M(f)) + \frac{1}{4}(M(f + 2f_c) + M(f)) \\ &= \frac{1}{4}M(f - 2f_c) + \frac{1}{2}M(f) + \frac{1}{4}M(f + 2f_c) \end{aligned} \quad (6.67)$$

The signal components centered at twice the carrier frequency are removed by the lowpass filter, while the baseband signal $M(f)$ emerges.

Solution to Exercise 6.11 (p. 222)

The key here is that the two spectra $M(f - f_c)$, $M(f + f_c)$ do not overlap because we have assumed that the carrier frequency f_c is much greater than the signal's highest frequency. Consequently, the term $M(f - f_c)M(f + f_c)$ normally obtained in computing the magnitude-squared equals zero.

Solution to Exercise 6.12 (p. 224)

Separation is $2W$. Commercial AM signal bandwidth is $5kHz$. Speech is well contained in this bandwidth, much better than in the telephone!

Solution to Exercise 6.13 (p. 224)

$x(t) = \sum_{n=-\infty}^{\infty} \left(s_{b(n)}(t - nT)\right)$.

Solution to Exercise 6.14 (p. 226)

$k = 4$.

Solution to Exercise 6.15 (p. 226)

$$x(t) = \sum_{n} \left((-1)^{b(n)} A p_T(t - nT) \sin\left(\frac{2\pi kt}{T}\right)\right)$$

Solution to Exercise 6.16 (p. 226)
The harmonic distortion is 10%.

Solution to Exercise 6.17 (p. 226)
Twice the baseband bandwidth because both positive and negative frequencies are shifted to the carrier by the modulation: $3R$.

Solution to Exercise 6.18 (p. 229)
In BPSK, the signals are negatives of each other: $s_1(t) = -(s_0(t))$. Consequently, the output of each multiplier-integrator combination is the negative of the other. Choosing the largest therefore amounts to choosing which one is positive. We only need to calculate one of these. If it is positive, we are done. If it is negative, we choose the other signal.

Solution to Exercise 6.19 (p. 229)
The matched filter outputs are $\pm\left(\frac{A^2T}{2}\right)$ because the sinusoid has less power than a pulse having the same amplitude.

Solution to Exercise 6.20 (p. 231)
The noise-free integrator outputs differ by $\alpha A^2 T$, the factor of two smaller value than in the baseband case arising because the sinusoidal signals have less energy for the same amplitude. Stated in terms of E_b, the difference equals $2\alpha E_b$ just as in the baseband case.

Solution to Exercise 6.21 (p. 232)
The noise-free integrator output difference now equals $\alpha A^2 T = \frac{\alpha E_b}{2}$. The noise power remains the same as in the BPSK case, which from the probability of error equation (6.46) yields $p_e = Q\left(\sqrt{\frac{\alpha^2 E_b}{N_0}}\right)$.

Solution to Exercise 6.22 (p. 234)
Equally likely symbols each have a probability of $\frac{1}{K}$. Thus, $H(A) = -\left(\sum_k \left(\frac{1}{K}\log_2\left(\frac{1}{K}\right)\right)\right) = \log_2 K$. To prove that this is the maximum-entropy probability assignment, we must explicitly take into account that probabilities sum to one. Focus on a particular symbol, say the first. $\Pr[a_0]$ appears *twice* in the entropy formula: the terms $\Pr[a_0]\log_2(\Pr[a_0])$ and $(1-(\Pr[a_0]+\cdots+\Pr[a_{K-2}]))\log_2(1-(\Pr[a_0]+\cdots+\Pr[a_{K-2}]))$. The derivative with respect to this probability (and all the others) must be zero. The derivative equals $\log_2(\Pr[a_0]) - \log_2(1-(\Pr[a_0]+\cdots+\Pr[a_{K-2}]))$, and all other derivatives have the same form (just substitute your letter's index). Thus, each probability must equal the others, and we are done. For the minimum entropy answer, one term is $1\log_2 1 = 0$, and the others are $0\log_2 0$, which we define to be zero also. The minimum value of entropy is zero.

Solution to Exercise 6.23 (p. 237)
The Huffman coding tree for the second set of probabilities is *identical* to that for the first (Figure 6.18 (Huffman Coding Tree)). The average code length is $\frac{1}{2}1+\frac{1}{4}2+\frac{1}{5}3+\frac{1}{20}3 = 1.75$ bits. The entropy calculation is straightforward: $H(A) = -\left(\frac{1}{2}\log\left(\frac{1}{2}\right)+\frac{1}{4}\log\left(\frac{1}{4}\right)+\frac{1}{5}\log\left(\frac{1}{5}\right)+\frac{1}{20}\log\left(\frac{1}{20}\right)\right)$, which equals 1.68 bits.

Solution to Exercise 6.24 (p. 237)
$T = \frac{1}{B(A)R}$.

Solution to Exercise 6.25 (p. 238)
Because no codeword begins with another's codeword, the first codeword encountered in a bit stream must be the right one. Note that we must start at the beginning of the bit stream; jumping into the middle does not guarantee perfect decoding. The end of one codeword and the beginning of another could be a codeword, and we would get lost.

Solution to Exercise 6.26 (p. 238)
Consider the bitstream ...0110111... taken from the bitstream 0|10|110|110|111|.... We would decode the initial part incorrectly, then would synchronize. If we had a fixed-length code (say 00,01,10,11), the situation is *much* worse. Jumping into the middle leads to no synchronization at all!

Solution to Exercise 6.27 (p. 241)

This question is equivalent to $3p_e(1-p_e)+p_e{}^2 \leq 1$ or $2p_e{}^2+(-3)p_e+1 \geq 0$. Because this is an upward-going parabola, we need only check where its roots are. Using the quadratic formula, we find that they are located at $\frac{1}{2}$ and 1. Consequently in the range $0 \leq p_e \leq \frac{1}{2}$ the error rate produced by coding is smaller.

Solution to Exercise 6.28 (p. 242)

With no coding, the average bit-error probability p_e is given by the probability of error equation (6.47): $p_e = Q\left(\sqrt{\frac{2\alpha^2 E_b}{N_0}}\right)$. With a threefold repetition code, the bit-error probability is given by $3p_e'^2(1-p_e')+p_e'^3$, where $p_e' = Q\left(\sqrt{\frac{2\alpha^2 E_b}{3N_0}}\right)$. Plotting this reveals that the increase in bit-error probability out of the channel because of the energy reduction is not compensated by the repetition coding.

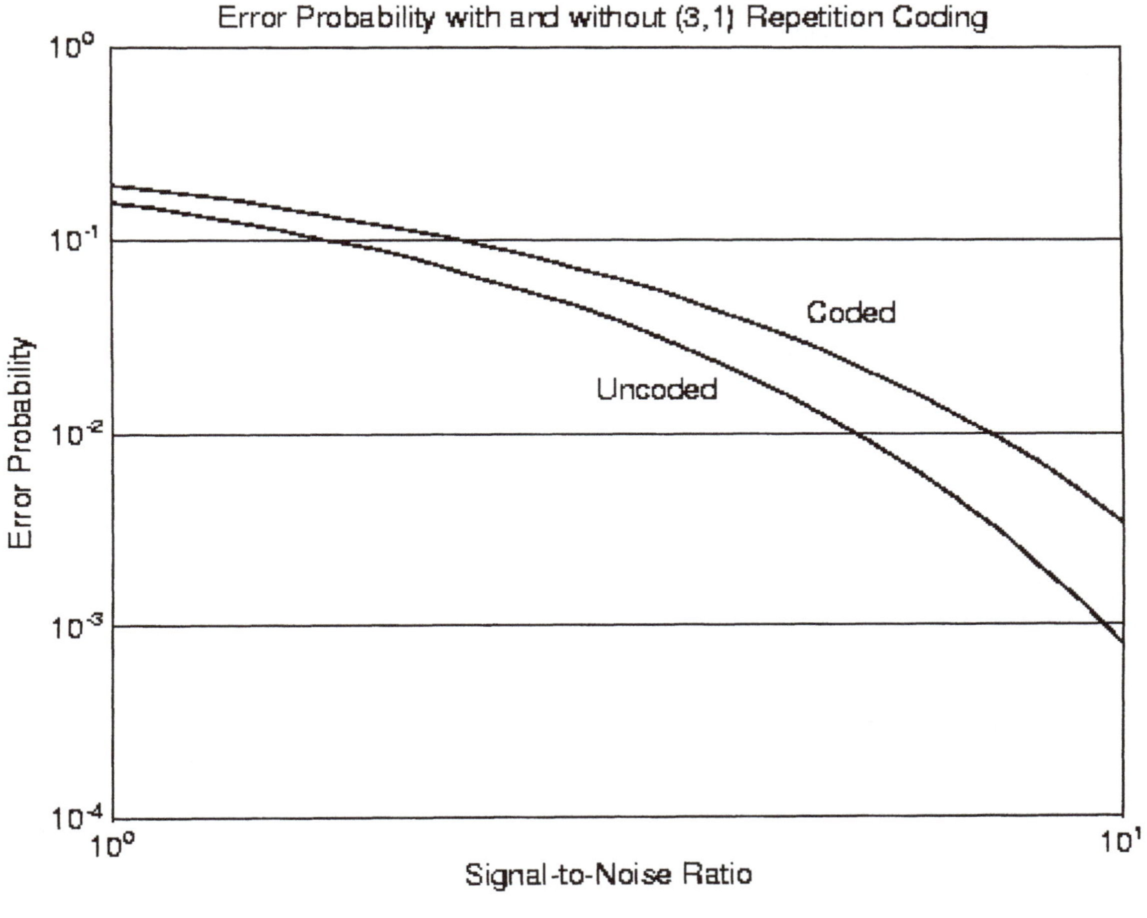

Figure 6.46

Solution to Exercise 6.29 (p. 243)

In binary arithmetic (see Figure 6.23), adding 0 to a binary value results in that binary value while adding 1 results in the opposite binary value.

Solution to Exercise 6.30 (p. 245)

$$d_{min} = 2n + 1$$

Solution to Exercise 6.31 (p. 245)

When we multiply the parity-check matrix times any codeword equal to a column of G, the result consists of the sum of an entry from the lower portion of G and itself that, by the laws of binary arithmetic, is always zero.

Because the code is linear—sum of any two codewords is a codeword—we can generate all codewords as sums of columns of G. Since multiplying by H is also linear, $H\mathbf{c} = 0$.

Solution to Exercise 6.32 (p. 246)

In binary arithmetic see this table[53], adding 0 to a binary value results in that binary value while adding 1 results in the opposite binary value.

Solution to Exercise 6.33 (p. 246)

The probability of a single-bit error in a length-N block is $Np_e(1-p_e)^{N-1}$ and a triple-bit error has probability $\binom{N}{3}p_e{}^3(1-p_e)^{N-3}$. For the first to be greater than the second, we must have

$$p_e < \frac{1}{\sqrt{\frac{(N-1)(N-2)}{6}} + 1}$$

For $N = 7$, $p_e < 0.31$.

Solution to Exercise 6.34 (p. 248)

In a length-N block, N single-bit and $\frac{N(N-1)}{2}$ double-bit errors can occur. The number of non-zero vectors resulting from $H\hat{c}$ must equal or exceed the sum of these two numbers.

$$2^{N-K} - 1 \geq N + \frac{N(N-1)}{2} \text{ or } 2^{N-K} \geq \frac{N^2 + N + 2}{2} \quad (6.68)$$

The first two solutions that attain equality are (5,1) and (90,78) codes. However, *no* perfect code exists other than the single-bit error correcting Hamming code. (Perfect codes satisfy relations like (6.68) with equality.)

Solution to Exercise 6.35 (p. 250)

To convert to bits/second, we divide the capacity stated in bits/transmission by the bit interval duration T.

Solution to Exercise 6.36 (p. 252)

The network entry point is the telephone handset, which connects you to the nearest station. Dialing the telephone number informs the network of who will be the message recipient. The telephone system forms an electrical circuit between your handset and your friend's handset. Your friend receives the message via the same device—the handset—that served as the network entry point.

Solution to Exercise 6.37 (p. 255)

The transmitting op-amp sees a load or $(R_{\text{out}} + Z_0 \parallel R_{\text{out}}/N)$, where N is the number of transceivers other than this one attached to the coaxial cable. The transfer function to some other transceiver's receiver circuit is R_{out} divided by this load.

Solution to Exercise 6.38 (p. 256)

The worst-case situation occurs when one computer begins to transmit just before the other's packet arrives. Transmitters must sense a collision before packet transmission ends. The time taken for one computer's packet to travel the Ethernet's length *and* for the other computer's transmission to arrive equals the round-trip, not one-way, propagation time.

[53] "Error Correction" <http://cnx.org/content/m0095/latest/#table1>

Solution to Exercise 6.39 **(p. 256)**

The cable must be a factor of ten shorter: It cannot exceed 100 m. Different minimum packet sizes means different packet formats, making connecting old and new systems together more complex than need be.

Solution to Exercise 6.40 **(p. 257)**

When you pick up the telephone, you initiate a dialog with your network interface by dialing the number. The network looks up where the destination corresponding to that number is located, and routes the call accordingly. The route remains fixed as long as the call persists. What you say amounts to high-level protocol while establishing the connection and maintaining it corresponds to low-level protocol.

Chapter 7

Appendix

7.1 Decibels[1]

The decibel scale expresses amplitudes and power values **logarithmically**. The definitions for these differ, but are consistent with each other.

$$\text{power}\,(s, \text{in decibels}) = 10\log_{10}\left(\frac{\text{power}\,(s)}{\text{power}\,(s_0)}\right) \tag{7.1}$$

$$\text{amplitude}\,(s, \text{in decibels}) = 20\log_{10}\left(\frac{\text{amplitude}\,(s)}{\text{amplitude}\,(s_0)}\right)$$

Here power (s_0) and amplitude (s_0) represent a **reference** power and amplitude, respectively. Quantifying power or amplitude in decibels essentially means that we are comparing quantities to a standard or that we want to express how they changed. You will hear statements like "The signal went down by 3 dB" and "The filter's gain in the stopband is -60" (Decibels is abbreviated dB.).

Exercise 7.1 *(Solution on p. 283.)*

The prefix "deci" implies a tenth; a decibel is a tenth of a Bel. Who is this measure named for?

The consistency of these two definitions arises because power is proportional to the square of amplitude:

$$\left(\text{power}\,(s) \propto \text{amplitude}^2\,(s)\right) \tag{7.2}$$

Plugging this expression into the definition for decibels, we find that

$$\begin{aligned} 10\log_{10}\left(\frac{\text{power}(s)}{\text{power}(s_0)}\right) &= 10\log_{10}\left(\frac{\text{amplitude}^2(s)}{\text{amplitude}^2(s_0)}\right) \\ &= 20\log_{10}\left(\frac{\text{amplitude}(s)}{\text{amplitude}(s_0)}\right) \end{aligned} \tag{7.3}$$

Because of this consistency, *stating relative change in terms of decibels is unambiguous*. A factor of 10 increase in amplitude corresponds to a 20 dB increase in both amplitude and power!

The accompanying table provides "nice" decibel values. Converting decibel values back and forth is fun, and tests your ability to think of decibel values as sums and/or differences of the well-known values and of ratios as products and/or quotients. This conversion rests on the logarithmic nature of the decibel scale. For example, to find the decibel value for $\sqrt{2}$, we halve the decibel value for 2; 26 dB equals $10 + 10 + 6$ dB that corresponds to a ratio of $10 \times 10 \times 4 = 400$. Decibel quantities add; ratio values multiply.

One reason decibels are used so much is the frequency-domain input-output relation for linear systems: $Y(f) = X(f)H(f)$. Because the transfer function multiplies the input signal's spectrum, to find the output amplitude at a given frequency we simply add the filter's gain in decibels (relative to a reference of one) to the input amplitude at that frequency. This calculation is one reason that we plot transfer function magnitude on a logarithmic vertical scale expressed in decibels.

[1]This content is available online at <http://cnx.org/content/m0082/2.16/>.

Decibel table

Power Ratio	dB
1	0
$\sqrt{2}$	1.5
2	3
$\sqrt{10}$	5
4	6
5	7
8	9
10	10
0.1	−10

Figure 7.1: Common values for the decibel. The decibel values for all but the powers of ten are approximate, but are accurate to a decimal place.

7.2 Permutations and Combinations[2]

7.2.1 Permutations and Combinations

The lottery "game" consists of picking k numbers from a pool of n. For example, you select 6 numbers out of 60. To win, the order in which you pick the numbers doesn't matter; you only have to choose the right set of 6 numbers. The chances of winning equal the number of different length-k sequences that can be chosen. A related, but different, problem is selecting the batting lineup for a baseball team. Now the order matters, and many more choices are possible than when order does not matter.

Answering such questions occurs in many applications beyond games. In digital communications, for example, you might ask how many possible double-bit errors can occur in a codeword. Numbering the bit positions from 1 to N, the answer is the same as the lottery problem with $k = 6$. Solving these kind of problems amounts to understanding **permutations** - the number of ways of choosing things when order matters as in baseball lineups - and **combinations** - the number of ways of choosing things when order does not matter as in lotteries and bit errors.

Calculating permutations is the easiest. If we are to pick k numbers from a pool of n, we have n choices for the first one. For the second choice, we have $n-1$. The number of length-two ordered sequences is therefore be $n(n-1)$. Continuing to choose until we make k choices means the number of permutations is $n(n-1)(n-2)\dots(n-k+1)$. This result can be written in terms of factorials as $\frac{n!}{(n-k)!}$, with $n! = n(n-1)(n-2)\dots 1$. For mathematical convenience, we define $0! = 1$.

When order does not matter, the number of combinations equals the number of permutations divided by the number of orderings. The number of ways a pool of k things can be ordered equals $k!$. Thus, once we choose the nine starters for our baseball game, we have $9! = 362,880$ different lineups! The symbol for the combination of k things drawn from a pool of n is $\binom{n}{k}$ and equals $\frac{n!}{(n-k)!k!}$.

[2]This content is available online at <http://cnx.org/content/m10262/2.12/>.

Exercise 7.2 *(Solution on p. 283.)*

What are the chances of winning the lottery? Assume you pick 6 numbers from the numbers 1-60.

Combinatorials occur in interesting places. For example, Newton derived that the n-th power of a sum obeyed the formula $(x+y)^n = \binom{n}{0}x^n + \binom{n}{1}x^{n-1}y + \binom{n}{2}x^{n-2}y^2 + \cdots + \binom{n}{n}y^n$.

Exercise 7.3 *(Solution on p. 283.)*

What does the sum of binomial coefficients equal? In other words, what is

$$\sum_{k=0}^{n} \binom{n}{k}$$

A related problem is calculating the probability that *any* two bits are in error in a length-n codeword when p is the probability of any bit being in error. The probability of any particular two-bit error sequence is $p^2(1-p)^{n-2}$. The probability of a two-bit error occurring anywhere equals this probability times the number of combinations: $\binom{n}{2}p^2(1-p)^{n-2}$. Note that the probability that zero or one or two, etc. errors occurring must be one; in other words, something must happen to the codeword! That means that we must have $\binom{n}{0}(1-p)^n + \binom{n}{1}p(1-p)^{n-1} + \binom{n}{2}p^2(1-p)^{n-2} + \cdots + \binom{n}{n}p^n = 1$. Can you prove this?

7.3 Frequency Allocations[3]

To prevent radio stations from transmitting signals "on top of each other,"; the United States and other national governments in the 1930s began regulating the carrier frequencies and power outputs stations could use. With increased use of the radio spectrum for both public and private use, this regulation has become increasingly important. This is the so-called **Frequency Allocation Chart**, which shows what kinds of broadcasting can occur in which frequency bands. Detailed radio carrier frequency assignments are much too detailed to present here.

[3]This content is available online at <http://cnx.org/content/m0083/2.10/>.

Frequency Allocation Chart

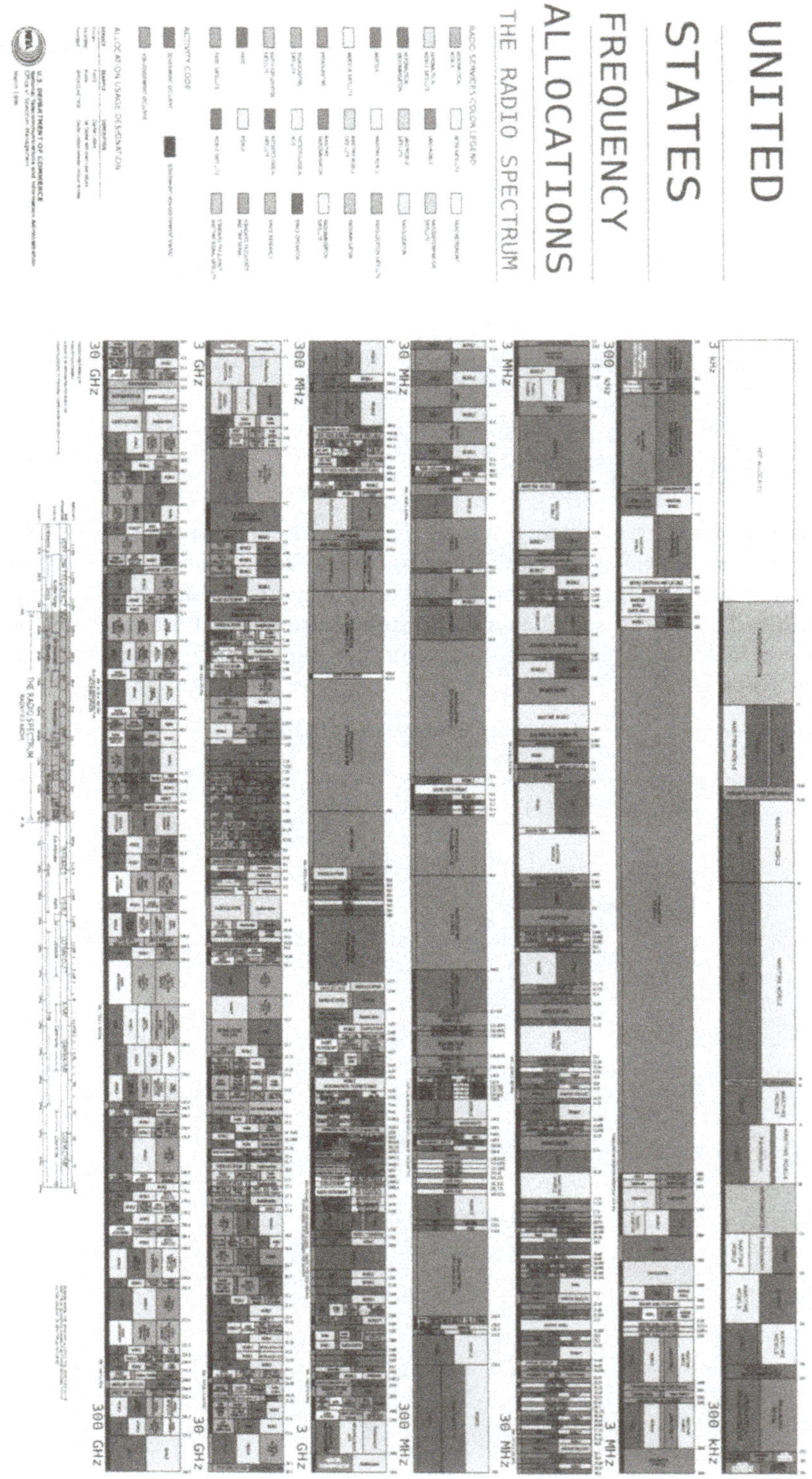

Figure 7.2

Solutions to Exercises in Chapter 7

Solution to Exercise 7.1 (p. 279)
Alexander Graham Bell. He developed it because we seem to perceive physical quantities like loudness and brightness logarithmically. In other words, *percentage*, not absolute differences, matter to us. We use decibels today because common values are small integers. If we used Bels, they would be decimal fractions, which aren't as elegant.

Solution to Exercise 7.2 (p. 280)
$\binom{60}{6} = \frac{60!}{54!6!} = 50,063,860.$

Solution to Exercise 7.3 (p. 281)
Because of Newton's binomial theorem, the sum equals $(1+1)^n = 2^n$.

Index of Keywords and Terms

Keywords are listed by the section with that keyword (page numbers are in parentheses). Keywords do not necessarily appear in the text of the page. They are merely associated with that section. *Ex.* apples, § 1.1 (1) **Terms** are referenced by the page they appear on. *Ex.* apples, 1

Attributions

Module: "Elemental Signals"
By: Don Johnson
URL: http://cnx.org/content/m0004/2.26/
Pages: 17-20
Copyright: Don Johnson
License: http://creativecommons.org/licenses/by/1.0

Module: "Signal Decomposition"
By: Don Johnson
URL: http://cnx.org/content/m0008/2.12/
Pages: 20-21
Copyright: Don Johnson
License: http://creativecommons.org/licenses/by/1.0

Module: "Discrete-Time Signals"
By: Don Johnson
URL: http://cnx.org/content/m0009/2.23/
Pages: 21-23
Copyright: Don Johnson
License: http://creativecommons.org/licenses/by/1.0

Module: "Introduction to Systems"
By: Don Johnson
URL: http://cnx.org/content/m0005/2.18/
Pages: 24-26
Copyright: Don Johnson
License: http://creativecommons.org/licenses/by/1.0

Module: "Simple Systems"
By: Don Johnson
URL: http://cnx.org/content/m0006/2.23/
Pages: 26-29
Copyright: Don Johnson
License: http://creativecommons.org/licenses/by/1.0

Module: "Signals and Systems Problems"
By: Don Johnson
URL: http://cnx.org/content/m10348/2.24/
Pages: 29-33
Copyright: Don Johnson
License: http://creativecommons.org/licenses/by/1.0

Module: "Voltage, Current, and Generic Circuit Elements"
By: Don Johnson
URL: http://cnx.org/content/m0011/2.12/
Pages: 35-36
Copyright: Don Johnson
License: http://creativecommons.org/licenses/by/1.0

Module: "Ideal Circuit Elements"
By: Don Johnson
URL: http://cnx.org/content/m0012/2.19/
Pages: 36-39
Copyright: Don Johnson
License: http://creativecommons.org/licenses/by/1.0

Module: "Ideal and Real-World Circuit Elements"
By: Don Johnson
URL: http://cnx.org/content/m0013/2.9/
Pages: 39-39
Copyright: Don Johnson
License: http://creativecommons.org/licenses/by/1.0

Module: "Electric Circuits and Interconnection Laws"
By: Don Johnson
URL: http://cnx.org/content/m0014/2.26/
Pages: 39-43
Copyright: Don Johnson
License: http://creativecommons.org/licenses/by/1.0

Module: "Power Dissipation in Resistor Circuits"
By: Don Johnson
URL: http://cnx.org/content/m17305/1.5/
Pages: 43-44
Copyright: Don Johnson
License: http://creativecommons.org/licenses/by/2.0/

Module: "Series and Parallel Circuits"
By: Don Johnson
URL: http://cnx.org/content/m10674/2.7/
Pages: 44-49
Copyright: Don Johnson
License: http://creativecommons.org/licenses/by/1.0

Module: "Equivalent Circuits: Resistors and Sources"
By: Don Johnson
URL: http://cnx.org/content/m0020/2.22/
Pages: 49-53
Copyright: Don Johnson
License: http://creativecommons.org/licenses/by/1.0

Module: "Circuits with Capacitors and Inductors"
By: Don Johnson
URL: http://cnx.org/content/m0023/2.11/
Pages: 54-54
Copyright: Don Johnson
License: http://creativecommons.org/licenses/by/1.0

Module: "The Impedance Concept"
By: Don Johnson
URL: http://cnx.org/content/m0024/2.22/
Pages: 54-56
Copyright: Don Johnson
License: http://creativecommons.org/licenses/by/1.0

Module: "Time and Frequency Domains"
By: Don Johnson
URL: http://cnx.org/content/m10708/2.6/
Pages: 56-58
Copyright: Don Johnson
License: http://creativecommons.org/licenses/by/1.0

Module: "Power in the Frequency Domain"
By: Don Johnson
URL: http://cnx.org/content/m17308/1.2/
Pages: 58-59
Copyright: Don Johnson
License: http://creativecommons.org/licenses/by/2.0/

Module: "Equivalent Circuits: Impedances and Sources"
By: Don Johnson
URL: http://cnx.org/content/m0030/2.19/
Pages: 59-60
Copyright: Don Johnson
License: http://creativecommons.org/licenses/by/1.0

Module: "Transfer Functions"
By: Don Johnson
URL: http://cnx.org/content/m0028/2.18/
Pages: 60-64
Copyright: Don Johnson
License: http://creativecommons.org/licenses/by/1.0

Module: "Designing Transfer Functions"
By: Don Johnson
URL: http://cnx.org/content/m0031/2.20/
Pages: 65-66
Copyright: Don Johnson
License: http://creativecommons.org/licenses/by/1.0

Module: "Formal Circuit Methods: Node Method"
By: Don Johnson
URL: http://cnx.org/content/m0032/2.18/
Pages: 66-71
Copyright: Don Johnson
License: http://creativecommons.org/licenses/by/1.0

Module: "Power Conservation in Circuits"
By: Don Johnson
URL: http://cnx.org/content/m17317/1.1/
Pages: 71-73
Copyright: Don Johnson
License: http://creativecommons.org/licenses/by/2.0/

Module: "Electronics"
By: Don Johnson
URL: http://cnx.org/content/m0035/2.8/
Pages: 73-73
Copyright: Don Johnson
License: http://creativecommons.org/licenses/by/1.0

Module: "Dependent Sources"
By: Don Johnson
URL: http://cnx.org/content/m0053/2.13/
Pages: 73-75
Copyright: Don Johnson
License: http://creativecommons.org/licenses/by/1.0

Module: "Operational Amplifiers"
By: Don Johnson
URL: http://cnx.org/content/m0036/2.28/
Pages: 76-81
Copyright: Don Johnson
License: http://creativecommons.org/licenses/by/1.0

Module: "The Diode"
By: Don Johnson
URL: http://cnx.org/content/m0037/2.14/
Pages: 81-84
Copyright: Don Johnson
License: http://creativecommons.org/licenses/by/1.0

Module: "Analog Signal Processing Problems"
By: Don Johnson
URL: http://cnx.org/content/m10349/2.37/
Pages: 84-105
Copyright: Don Johnson
License: http://creativecommons.org/licenses/by/1.0

Module: "Introduction to the Frequency Domain"
By: Don Johnson
URL: http://cnx.org/content/m0038/2.10/
Pages: 109-109
Copyright: Don Johnson
License: http://creativecommons.org/licenses/by/1.0

Module: "Complex Fourier Series"
By: Don Johnson
URL: http://cnx.org/content/m0042/2.24/
Pages: 109-113
Copyright: Don Johnson
License: http://creativecommons.org/licenses/by/1.0

Module: "Classic Fourier Series"
By: Don Johnson
URL: http://cnx.org/content/m0039/2.22/
Pages: 114-116
Copyright: Don Johnson
License: http://creativecommons.org/licenses/by/1.0

Module: "A Signal's Spectrum"
By: Don Johnson
URL: http://cnx.org/content/m0040/2.19/
Pages: 116-117
Copyright: Don Johnson
License: http://creativecommons.org/licenses/by/1.0

Module: "Fourier Series Approximation of Signals"
By: Don Johnson
URL: http://cnx.org/content/m10687/2.8/
Pages: 117-121
Copyright: Don Johnson
License: http://creativecommons.org/licenses/by/1.0

Module: "Encoding Information in the Frequency Domain"
By: Don Johnson
URL: http://cnx.org/content/m0043/2.15/
Pages: 122-124
Copyright: Don Johnson
License: http://creativecommons.org/licenses/by/1.0

Module: "Filtering Periodic Signals"
By: Don Johnson
URL: http://cnx.org/content/m0044/2.9/
Pages: 124-126
Copyright: Don Johnson
License: http://creativecommons.org/licenses/by/1.0

Module: "Derivation of the Fourier Transform"
By: Don Johnson
URL: http://cnx.org/content/m0046/2.19/
Pages: 126-131
Copyright: Don Johnson
License: http://creativecommons.org/licenses/by/1.0

Module: "Linear Time Invariant Systems"
By: Don Johnson
URL: http://cnx.org/content/m0048/2.18/
Pages: 131-133
Copyright: Don Johnson
License: http://creativecommons.org/licenses/by/1.0

Module: "Modeling the Speech Signal"
By: Don Johnson
URL: http://cnx.org/content/m0049/2.25/
Pages: 134-140
Copyright: Don Johnson
License: http://creativecommons.org/licenses/by/1.0

Module: "Frequency Domain Problems"
By: Don Johnson
URL: http://cnx.org/content/m10350/2.32/
Pages: 140-151
Copyright: Don Johnson
License: http://creativecommons.org/licenses/by/1.0

Module: "Introduction to Digital Signal Processing"
By: Don Johnson
URL: http://cnx.org/content/m10781/2.3/
Pages: 155-155
Copyright: Don Johnson
License: http://creativecommons.org/licenses/by/1.0

Module: "Introduction to Computer Organization"
By: Don Johnson
URL: http://cnx.org/content/m10263/2.27/
Pages: 155-159
Copyright: Don Johnson
License: http://creativecommons.org/licenses/by/1.0

Module: "The Sampling Theorem"
By: Don Johnson
URL: http://cnx.org/content/m0050/2.18/
Pages: 159-162
Copyright: Don Johnson
License: http://creativecommons.org/licenses/by/1.0

Module: "Amplitude Quantization"
By: Don Johnson
URL: http://cnx.org/content/m0051/2.21/
Pages: 162-165
Copyright: Don Johnson
License: http://creativecommons.org/licenses/by/1.0

Module: "Discrete-Time Signals and Systems"
By: Don Johnson
URL: http://cnx.org/content/m10342/2.13/
Pages: 165-167
Copyright: Don Johnson
License: http://creativecommons.org/licenses/by/1.0

Module: "Discrete-Time Fourier Transform (DTFT)"
By: Don Johnson
URL: http://cnx.org/content/m10247/2.28/
Pages: 167-172
Copyright: Don Johnson
License: http://creativecommons.org/licenses/by/1.0

Module: "Discrete Fourier Transform (DFT)"
Used here as: "Discrete Fourier Transforms (DFT)"
By: Don Johnson
URL: http://cnx.org/content/m10249/2.26/
Pages: 172-174
Copyright: Don Johnson
License: http://creativecommons.org/licenses/by/1.0

Module: "DFT: Computational Complexity"
By: Don Johnson
URL: http://cnx.org/content/m0503/2.11/
Pages: 174-174
Copyright: Don Johnson
License: http://creativecommons.org/licenses/by/1.0

Module: "Fast Fourier Transform (FFT)"
By: Don Johnson
URL: http://cnx.org/content/m10250/2.15/
Pages: 174-177
Copyright: Don Johnson
License: http://creativecommons.org/licenses/by/1.0

Module: "Spectrograms"
By: Don Johnson
URL: http://cnx.org/content/m0505/2.18/
Pages: 177-180
Copyright: Don Johnson
License: http://creativecommons.org/licenses/by/1.0

Module: "Discrete-Time Systems"
By: Don Johnson
URL: http://cnx.org/content/m0507/2.5/
Pages: 180-180
Copyright: Don Johnson
License: http://creativecommons.org/licenses/by/1.0

Module: "Discrete-Time Systems in the Time-Domain"
By: Don Johnson
URL: http://cnx.org/content/m10251/2.22/
Pages: 181-184
Copyright: Don Johnson
License: http://creativecommons.org/licenses/by/1.0

Module: "Discrete-Time Systems in the Frequency Domain"
By: Don Johnson
URL: http://cnx.org/content/m0510/2.14/
Pages: 185-186
Copyright: Don Johnson
License: http://creativecommons.org/licenses/by/1.0

Module: "Filtering in the Frequency Domain"
By: Don Johnson
URL: http://cnx.org/content/m10257/2.16/
Pages: 186-189
Copyright: Don Johnson
License: http://creativecommons.org/licenses/by/1.0

Module: "Efficiency of Frequency-Domain Filtering"
By: Don Johnson
URL: http://cnx.org/content/m10279/2.14/
Pages: 189-192
Copyright: Don Johnson
License: http://creativecommons.org/licenses/by/1.0

Module: "Discrete-Time Filtering of Analog Signals"
By: Don Johnson
URL: http://cnx.org/content/m0511/2.20/
Pages: 192-193
Copyright: Don Johnson
License: http://creativecommons.org/licenses/by/1.0

Module: "Digital Signal Processing Problems"
By: Don Johnson
URL: http://cnx.org/content/m10351/2.33/
Pages: 193-203
Copyright: Don Johnson
License: http://creativecommons.org/licenses/by/1.0

Module: "Information Communication"
By: Don Johnson
URL: http://cnx.org/content/m0513/2.8/
Pages: 209-209
Copyright: Don Johnson
License: http://creativecommons.org/licenses/by/1.0

Module: "Types of Communication Channels"
By: Don Johnson
URL: http://cnx.org/content/m0099/2.13/
Pages: 210-210
Copyright: Don Johnson
License: http://creativecommons.org/licenses/by/1.0

Module: "Wireline Channels"
By: Don Johnson
URL: http://cnx.org/content/m0100/2.27/
Pages: 210-215
Copyright: Don Johnson
License: http://creativecommons.org/licenses/by/1.0

Module: "Wireless Channels"
By: Don Johnson
URL: http://cnx.org/content/m0101/2.14/
Pages: 215-216
Copyright: Don Johnson
License: http://creativecommons.org/licenses/by/1.0

Module: "Line-of-Sight Transmission"
By: Don Johnson
URL: http://cnx.org/content/m0538/2.13/
Pages: 216-217
Copyright: Don Johnson
License: http://creativecommons.org/licenses/by/1.0

Module: "The Ionosphere and Communications"
By: Don Johnson
URL: http://cnx.org/content/m0539/2.10/
Pages: 217-217
Copyright: Don Johnson
License: http://creativecommons.org/licenses/by/1.0

Module: "Communication with Satellites"
By: Don Johnson
URL: http://cnx.org/content/m0540/2.10/
Pages: 217-218
Copyright: Don Johnson
License: http://creativecommons.org/licenses/by/1.0

Module: "Noise and Interference"
By: Don Johnson
URL: http://cnx.org/content/m0515/2.17/
Pages: 218-219
Copyright: Don Johnson
License: http://creativecommons.org/licenses/by/1.0

Module: "Channel Models"
By: Don Johnson
URL: http://cnx.org/content/m0516/2.10/
Pages: 219-220
Copyright: Don Johnson
License: http://creativecommons.org/licenses/by/1.0

Module: "Baseband Communication"
By: Don Johnson
URL: http://cnx.org/content/m0517/2.17/
Pages: 220-221
Copyright: Don Johnson
License: http://creativecommons.org/licenses/by/1.0

Module: "Modulated Communication"
By: Don Johnson
URL: http://cnx.org/content/m0518/2.24/
Pages: 221-222
Copyright: Don Johnson
License: http://creativecommons.org/licenses/by/1.0

Module: "Signal-to-Noise Ratio of an Amplitude-Modulated Signal"
By: Don Johnson
URL: http://cnx.org/content/m0541/2.16/
Pages: 222-224
Copyright: Don Johnson
License: http://creativecommons.org/licenses/by/1.0

Module: "Digital Communication"
By: Don Johnson
URL: http://cnx.org/content/m0519/2.10/
Pages: 224-224
Copyright: Don Johnson
License: http://creativecommons.org/licenses/by/1.0

Module: "Binary Phase Shift Keying"
By: Don Johnson
URL: http://cnx.org/content/m10280/2.13/
Pages: 224-227
Copyright: Don Johnson
License: http://creativecommons.org/licenses/by/1.0

Module: "Frequency Shift Keying"
By: Don Johnson
URL: http://cnx.org/content/m0545/2.11/
Pages: 227-228
Copyright: Don Johnson
License: http://creativecommons.org/licenses/by/1.0

Module: "Digital Communication Receivers"
By: Don Johnson
URL: http://cnx.org/content/m0520/2.17/
Pages: 228-229
Copyright: Don Johnson
License: http://creativecommons.org/licenses/by/1.0

Module: "Digital Communication in the Presence of Noise"
By: Don Johnson
URL: http://cnx.org/content/m0546/2.12/
Pages: 230-231
Copyright: Don Johnson
License: http://creativecommons.org/licenses/by/1.0

Module: "Digital Communication System Properties"
By: Don Johnson
URL: http://cnx.org/content/m10282/2.9/
Pages: 231-233
Copyright: Don Johnson
License: http://creativecommons.org/licenses/by/1.0

Module: "Digital Channels"
By: Don Johnson
URL: http://cnx.org/content/m0102/2.13/
Pages: 233-233
Copyright: Don Johnson
License: http://creativecommons.org/licenses/by/1.0

Module: "Entropy"
By: Don Johnson
URL: http://cnx.org/content/m0070/2.13/
Pages: 234-234
Copyright: Don Johnson
License: http://creativecommons.org/licenses/by/1.0

Module: "Source Coding Theorem"
By: Don Johnson
URL: http://cnx.org/content/m0091/2.13/
Pages: 235-236
Copyright: Don Johnson
License: http://creativecommons.org/licenses/by/1.0

Module: "Compression and the Huffman Code"
By: Don Johnson
URL: http://cnx.org/content/m0092/2.17/
Pages: 236-237
Copyright: Don Johnson
License: http://creativecommons.org/licenses/by/1.0

Module: "Subtleties of Source Coding"
Used here as: "Subtlies of Coding"
By: Don Johnson
URL: http://cnx.org/content/m0093/2.16/
Pages: 237-238
Copyright: Don Johnson
License: http://creativecommons.org/licenses/by/1.0

Module: "Channel Coding"
By: Don Johnson
URL: http://cnx.org/content/m10782/2.4/
Pages: 238-240
Copyright: Don Johnson
License: http://creativecommons.org/licenses/by/1.0

Module: "Repetition Codes"
By: Don Johnson
URL: http://cnx.org/content/m0071/2.21/
Pages: 240-241
Copyright: Don Johnson
License: http://creativecommons.org/licenses/by/1.0

Module: "Block Channel Coding"
By: Don Johnson
URL: http://cnx.org/content/m0094/2.14/
Pages: 241-242
Copyright: Don Johnson
License: http://creativecommons.org/licenses/by/1.0

Module: "Error-Correcting Codes: Hamming Distance"
By: Don Johnson
URL: http://cnx.org/content/m10283/2.28/
Pages: 242-245
Copyright: Don Johnson
License: http://creativecommons.org/licenses/by/1.0

Module: "Error-Correcting Codes: Channel Decoding"
By: Don Johnson
URL: http://cnx.org/content/m0072/2.20/
Pages: 245-246
Copyright: Don Johnson
License: http://creativecommons.org/licenses/by/1.0

Module: "Error-Correcting Codes: Hamming Codes"
By: Don Johnson
URL: http://cnx.org/content/m0097/2.24/
Pages: 246-248
Copyright: Don Johnson
License: http://creativecommons.org/licenses/by/1.0

Module: "Noisy Channel Coding Theorem"
By: Don Johnson
URL: http://cnx.org/content/m0073/2.11/
Pages: 248-249
Copyright: Don Johnson
License: http://creativecommons.org/licenses/by/1.0

Module: "Capacity of a Channel"
By: Don Johnson
URL: http://cnx.org/content/m0098/2.13/
Pages: 249-250
Copyright: Don Johnson
License: http://creativecommons.org/licenses/by/1.0

Module: "Comparison of Analog and Digital Communication"
By: Don Johnson
URL: http://cnx.org/content/m0074/2.11/
Pages: 250-251
Copyright: Don Johnson
License: http://creativecommons.org/licenses/by/1.0

Module: "Communication Networks"
By: Don Johnson
URL: http://cnx.org/content/m0075/2.10/
Pages: 251-252
Copyright: Don Johnson
License: http://creativecommons.org/licenses/by/1.0

Module: "Message Routing"
By: Don Johnson
URL: http://cnx.org/content/m0076/2.8/
Pages: 252-253
Copyright: Don Johnson
License: http://creativecommons.org/licenses/by/1.0

Module: "Network architectures and interconnection"
By: Don Johnson
URL: http://cnx.org/content/m0077/2.9/
Pages: 253-254
Copyright: Don Johnson
License: http://creativecommons.org/licenses/by/1.0

Module: "Ethernet"
By: Don Johnson
URL: http://cnx.org/content/m10284/2.12/
Pages: 254-256
Copyright: Don Johnson
License: http://creativecommons.org/licenses/by/1.0

Module: "Communication Protocols"
By: Don Johnson
URL: http://cnx.org/content/m0080/2.18/
Pages: 256-258
Copyright: Don Johnson
License: http://creativecommons.org/licenses/by/1.0

Module: "Information Communication Problems"
By: Don Johnson
URL: http://cnx.org/content/m10352/2.21/
Pages: 258-271
Copyright: Don Johnson
License: http://creativecommons.org/licenses/by/1.0

Module: "Decibels"
By: Don Johnson
URL: http://cnx.org/content/m0082/2.16/
Pages: 279-279
Copyright: Don Johnson
License: http://creativecommons.org/licenses/by/1.0

Module: "Permutations and Combinations"
By: Don Johnson
URL: http://cnx.org/content/m10262/2.12/
Pages: 280-281
Copyright: Don Johnson
License: http://creativecommons.org/licenses/by/1.0

Module: "Frequency Allocations"
By: Don Johnson
URL: http://cnx.org/content/m0083/2.10/
Pages: 281-283
Copyright: Don Johnson
License: http://creativecommons.org/licenses/by/1.0

Fundamentals of Electrical Engineering I

The course focuses on the creation, manipulation, transmission, and reception of information by electronic means. Elementary signal theory; time- and frequency-domain analysis; Sampling Theorem. Digital information theory; digital transmission of analog signals; error-correcting codes.

About Connexions

Since 1999, Connexions has been pioneering a global system where anyone can create course materials and make them fully accessible and easily reusable free of charge. We are a Web-based authoring, teaching and learning environment open to anyone interested in education, including students, teachers, professors and lifelong learners. We connect ideas and facilitate educational communities.

Connexions's modular, interactive courses are in use worldwide by universities, community colleges, K-12 schools, distance learners, and lifelong learners. Connexions materials are in many languages, including English, Spanish, Chinese, Japanese, Italian, Vietnamese, French, Portuguese, and Thai. Connexions is part of an exciting new information distribution system that allows for **Print on Demand Books**. Connexions has partnered with innovative on-demand publisher QOOP to accelerate the delivery of printed course materials and textbooks into classrooms worldwide at lower prices than traditional academic publishers.